Satellite Communications for the Nonspecialist

Satellite Communications for the Nonspecialist

Mark R. Chartrand

SPIE PRESS
A Publication of SPIE—The International Society for Optical Engineering
Bellingham, Washington USA

Library of Congress Cataloging-in-Publication Data

Chartrand, Mark R.
Satellite communications for the nonspecialist / Mark R. Chartrand.
p. cm. — (SPIE Press monograph ; PM128)
Includes bibliographical references and index.
ISBN 0-8194-5185-1
1. Artificial satellites in telecommunication—Popular works. I. Title. II. Series.

TK5104.C47 2003
621.382'5--dc22 2003064648

Published by

SPIE—The International Society for Optical Engineering
P.O. Box 10
Bellingham, Washington 98227-0010 USA
Phone: (1) 360.676.3290
Fax: (1) 360.647.1445
Email: spie@spie.org
Web: www.spie.org

Printed in the United States o

About the cover: The image on the bottom of the front cover shows an "antenna farm" (photograph courtesy of SES Astra). The satellite in the top right is an example of a "Big LEO" satellite that is part of the Iridium constellation (photograph courtesy of Iridium). Between the title and the antenna farm is a glimpse of a regional beam footprint over Europe (photograph courtesy of SES Astra).

Contents

Preface

The primary purpose of this book, and the introductory seminars out of which it grew, is to explain *to nontechnical people* the concepts, terminology, buzzwords, and jargon of the commercial satellite communications industry, and to show the interconnections between its various technologies and components. This book attempts to survey the entire commercial satellite telecommunications industry, and to that end discusses everything from decibels; to principles of radio propagation; launch vehicles; link budgets; business, legal, and regulatory issues; and services provided by satellites. The content and presentation have been refined through seminars given for two decades by the author to thousands of businesspeople and hundreds of organizations in public and on-site sessions all over the world. The people who take the seminar, and at whom this book is aimed, come primarily from such occupations as banking and investment, public relations, marketing, writing, insurance, law, and corporate management. These are people who are involved with or interested in the satellite communications industry, but have not had specialized training. In other words, these are the nonspecialists.

Part 1 of this book sets out the historical, business, and regulatory background of the commercial satellite industry. Some examples show how the technology is used to provide specific satellite applications and build markets. Part 2 covers the technical background necessary for understanding the concepts and applications of the entire field of telecommunications. Part 3 continues by explaining the space segment part of the business. Part 4 covers the ground side. Part 5 puts it all together, linking satellites and Earth. Finally, Part 6 contains an overview of the major satellite systems and applications, and a brief discussion of some of the issues and trends affecting the industry.

The chapters of the book contain very little mathematics, just enough to convey the concepts necessary for understanding rather complicated technical concepts. Most of the equations and formulas have been banished to Appendix E, which is provided for those with more mathematical preparation who are interested in the details. For nontechnical readers, the formulas are largely unnecessary for basic understanding of concepts, providing official definitions rather than as something the typical reader is expected to master in detail. The major exceptions to this are a few definitions of terms and the technical climax of the text in Chapter 19 dealing with the all-important topic of link budgets.

Complex technical concepts are explicated using everyday analogies that allow a nonspecialist to extrapolate from familiar situations to unfamiliar technical applications.

Another objective of the book is to help make palatable the thick, often murky and indigestible alphabet soup of acronyms and abbreviations that are a part of any technical endeavor. Engineers are fine and useful people, but they often forget that not everyone has had the training they have had, and thus they often speak in their own language. (A particular favorite is TDMA, which engineers in the field know to mean "Time-Division Multiple Access" but which many nonengineers consider refers to the "Too Damn Many Acronyms" encountered in the field!) An even more appropriate example is the self-referential TLA, which stands for "Three-Letter Abbreviation!" The glossary contains a long list of TLAs, FLAs, and so on.

This book is designed to help nonspecialists understand and converse with the technical personnel in the satellite industry, to be able to read and understand the relevant trade journals and exhibits at trade shows, and to help them explain the concepts to their colleagues and customers. (As I jokingly tell my students, by the end of the course they should be able to converse with technical personnel and construct entire sentences containing only jargon and acronyms, and no real English whatsoever!)

Sometimes attendees at the seminars remark at the start that they are interested in a particular aspect of the industry, and not interested in some other topic. By the end of the seminar, they realize that a synoptic view is important. For instance, even if you personally never plan to have to buy a launch vehicle, knowing about launchers is important because what a satellite system operator has to pay for launch partially determines what it has to charge for satellite services.

A very common misconception among nontechnical people who are trying to understand technical issues is that there are generic answers to many technical questions or generic technical solutions to desired services. This desire for simplicity is understandable, but such simplicity is seldom possible. There are always trade-offs, balancing such issues as cost, quality, and speed. For instance, I have had seminar attendees and consulting clients ask such questions as "Which is better, satellites or optical fibers?" or "Which is better, FDMA or TDMA?" or "Which is better, low orbits or high orbits?" The only generic answer to all these questions is "No," because until you know what the detailed requirements of the system are, you cannot specify the technical details. This is another reason that seeing the interconnections of all the parts of the industry is important. You always need to start with the users' requirements and work to find the appropriate technology. This book sets out many of the trade-offs that need to be considered when planning a specific satellite telecommunication service.

A secondary audience is technical people who are entering the commercial satellite communications business for the first time from other fields such as computers, broadcasting, or military satcomms, and for those in the satellite industry who need a refresher to show the applications, relevance, and interconnections of the various parts of the industry. Often technical people work in one small specialized area of the industry, and while they know their "trees" well, they have little knowledge of the "forest" of the industry around them. Thus the book includes an overview of the business, technical, and regulatory issues of satellite communications, which often have their own specialized jargon. I have also had many technical people take the seminar in order to learn how to communicate concepts and what they are doing to their less-technical colleagues and customers. More-technical readers will find Appendix E gives them a bit more mathematical depth, but it is not in any way designed as a substitute for more technical engineering references.

As in any introduction, some explanations cannot be as complete or as comprehensive as in more detailed (and highly mathematical) technical references, and some details an engineer might want included have been left out in the interests of simplicity. Nevertheless, it is hoped the examples and analogies will prove useful and instructive, and references are given in the appendices to further information.

Since not every example nor company and its products and services can be included, it is hoped those chosen for illustrative purposes are representative and helpful. The mention of any particular company, product, service, or organization is not intended to be a recommendation of that entity, just as the omission of any particular company, product, or service is not meant as a slight nor negative recommendation, but merely as an indication of availability of appropriate material, or of the reality of constrained space. An extensive appendix points to other, more detailed sources of information. Readers should also search the vast resources of the World Wide Web; some useful addresses are given in the Appendix, with the caveat that websites seem to come and go irregularly.

Further, although this text is intended to be up-to-date, change in telecommunication is rapid, continual, and global. Changes in technology, laws, regulations, politics, corporate structures, and users' needs occur frequently. Sudden events, from launch of a new satellite, to the failure of a satellite, to the purchase of one company by another, can cause instantaneous alterations in the industry. Only periodicals can (attempt to) keep up with this, and you are urged to refer to the magazines and newsletters in Appendix D for the latest information.

Mark R. Chartrand

July 18, 2003

Acknowledgments

The production of this book has been the result of almost two decades of well-attended public seminars and on-site seminars to individual firms. I am greatly indebted to the comments and questions from all of the thousands of past attendees who have helped me perfect the order and depth of the explanations, and the usefulness of the analogies. The public seminars are now arranged under the auspices of the Applied Technology Institute.

Particular appreciation is directed to the people who helped get these seminars started, the redoubtable Ellen Hamm Stuhlmann and Mark Kimmel (now both gone on to bigger and better endeavors). Many valuable discussions and ideas have come over the many years from editors and executives, including the redoubtable Scott Chase, Cynthia Boeke, Paul Dykewicz, and David Bross.

The author is also especially grateful for many years of professional and collegial discussions with Richard T. Cassidy, president of the consulting firm Blue Mountain Group and one of the engineers who pioneered the first satellite-distributed radio network. Industry doyen Dr. Joseph N. Pelton supplied many helpful comments. Tim Lamkins at SPIE Press is to be thanked for initiating the publication of this book at the suggestion of one of my erstwhile students, Gunar Fedosejevs. Thanks, too, to F. William Chickering for his helpful comments. Thanks should go also to my partner and my close friends, for putting up with the constraints finishing this book put on my schedule.

In order to illustrate the topics covered herein, I have approached many firms and individuals for assistance. Of particular help were the following, in no particular order: Richard Daniels, AMSAT; Joanne Welsh and Washington Wedderburn at Analytical Graphics; Michelle Lyle and Orly König-Lopez at International Launch Services; Susan Gordon and Max Saffell at Intelsat; Fran Slimmer at Lockheed Martin; Giles Khong at Thales Electrondevices; and Wende Cover at Verestar.

I hope you find this book useful. Thank you.

Part 1

Telecommunications and Satellites

Chapter 1

Introduction and Some Historical Background

"How I Lost a Billion Dollars in My Spare Time" is the partial title of (now Sir) Arthur C. Clarke's 1965 essay on how and why he did not (and probably could not) patent the idea of a geostationary communications satellite, which he detailed and publicized in 1945. For one thing, he expected it to be at least 50 years in the future! The marvel is that Sir Arthur (Fig. 1.1) survived to see his concept fulfilled by approximately 250 geostationary commercial communications satellites ringing the globe.

Most people consider Clarke to be the father of the communications satellite. He, however, considers himself the godfather, and considers the fathers to be the two scientists who more fully developed the technical concepts, Dr. John R. Pierce (Fig. 1.2) and Dr. Harold Rosen. Nonetheless, the unique orbit around our planet

Figure 1.1 Sir Arthur C. Clarke (Photograph courtesy of the Arthur C. Clarke Foundation.)

Figure 1.2 John R. Pierce (Photograph courtesy of Lucent.)

where satellites seem to be stationary as seen from the surface is universally honored as the Clarke orbit.

As Clarke himself says, the concept is simple and capable of being understood through orbital physics. Although Newton could have come up with the idea, he doesn't seem to have done so. In the nineteenth century, a foresighted writer described a (brick!) satellite that communicated to Earth by having people on the satellite jump up and down (they didn't know much about the vacuum of space then). The brilliant and pioneering German, Hermann Oberth, wrote of communicating with manned satellites by mirrors and lights in 1923 when radio was still in its infancy. Other writers, including a little-known Austrian army officer named Hermann Potocnik (who also wrote science fiction under the pen name of Hermann Noordung) had proposed a manned space station in his 1928 book *The Problem of Spaceflight*, placing it in a geostationary orbit to facilitate radiocommunications with Earth. Finally, in 1942, engineer-writer George O. Smith proposed a radio relay satellite in Venus' orbit to permit communication between that planet and Earth when they were on opposite sides of the sun.

Clarke's contribution is his description of the technical characteristics of a geostationary communications satellite (Fig. 1.3). First published in 1945, his articles brought together his interests and technical knowledge of radio and space flight (not to mention science fiction, of which genre he is a master) with his ability to write. In fact, Clarke's article is still somewhat prescient: he not only suggested radio links with Earth, but also suggested what we call intersatellite links today. While some government satellites make use of intersatellite links in geostationary orbits, commercial satellites in geosynchronous orbit are just beginning to use them because they make the satellite system much more expensive. Some low-orbit satellites do use such links.

The major advantages of satellites in geostationary orbit, as we will detail more fully in later chapters, is twofold: coverage and simplicity. Each Clarke-orbit satellite sees about 44% of the total surface of the earth, which aggregates a huge

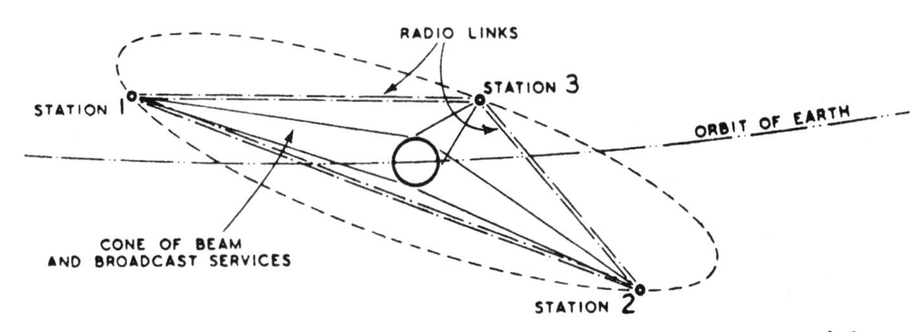

Fig. 3. Three satellite stations would ensure complete coverage of the globe.

Figure 1.3 Arthur C. Clarke's diagram from his seminal article in *Wireless World* magazine, October 1945, showing the concept of geostationary radio relay stations. (Photograph courtesy of Electronics World, Cambridge, U.K.)

market. Since the satellites appear stationary in the sky as seen from an earthstation, the stations can be simpler, and thus less expensive, as they can be pointed once at a satellite and not have to track it across the sky.

Rocketry and electronics progressed much faster than even futurists like Clarke foresaw. Twelve years after Clarke's article, on October 4,1957, the Soviet Union launched Sputnik-1 as the first artificial satellite. The first active communications satellite came only 14 months later. Note that by "active" we mean a satellite that has a transmitter aboard to relay a radio signal to Earth, not just a passive reflector like the Echo balloon satellite that came a couple of years later.

Project SCORE, which stood for "signal communications by orbital relay equipment," was a wire recorder in the nosecone of an Atlas missile, launched December 18, 1958, to an orbit just a few hundred miles up. It was battery-powered and broadcast holiday greetings recorded by President Dwight Eisenhower to the world. The mission lasted only a few weeks before the batteries ran out and the missile plunged back to Earth. Simple and crude, it was the first active commsat (as opposed to Comsat, which is a trademarked company name).

The next step in satellite communications was the Echo balloon, launched August 12, 1960. This 30-m-diameter aluminum-covered sphere reflected radio waves back to Earth. (It also reflected light well and was one of the visually brightest satellites ever to appear in our skies.) It stayed in orbit until 1968.

The first commercial satellite went into orbit on July 10, 1962, when AT&T launched Telstar-1 (Fig. 1.4). At that time, rockets still did not have the power to reach the geosynchronous orbit, and Telstar orbited only several hundred miles above

Figure 1.4 Telstar-1, the first commercial (but non-geostationary) communications satellite, built by AT&T. (Photograph courtesy of Lucent Technologies, Bell Labs.)

Earth. Because of its low altitude, like Echo and Project SCORE before it, it could link sites on Earth only a few thousand miles apart. (It's still up there, for another 100 centuries at least.) It required a huge tracking antenna to communicate with the satellite.

Finally, on February 14, 1963, the Syncom-1 ("synchronous communications") reached the geosynchronous altitude of roughly 36,000 km (23,000 miles). Although the satellite went around Earth in one day—and was therefore geosynchronous—its orbit was not over the equator so it was therefore not truly geostationary. It was followed later that year by Syncom-2, and the next year by Syncom-3, which finally achieved an almost perfectly circular orbit over the equator, making it the first geostationary commsat. Syncom-3 could carry only one television signal, and was used to relay the opening ceremonies of the 1964 Tokyo Olympics to North America.

On April 6, 1965, the Communications Satellite Corporation, later named Comsat, launched "Early Bird" (Fig. 1.5). At 8:40:25 EST, Early Bird was injected into the Clarke orbit, with a final small rocket firing to make it the first truly commercial geostationary communications satellite. Later renamed Intelsat-1, the 39-kg satellite could carry 240 telephone calls or 1 television signal. That may not sound like a lot of telephone circuits now, but at the time it was almost 10 times the number of telephone circuits available on the trans-Atlantic analog cables. Commercial users paid $4,200 per month ($23,000 in today's dollars) for a two-way leased telephone circuit on "Early Bird," and paid $2,400 for a half-hour television transmission.

The year 1965 also saw the launch of the U.S. Department of Defense's first military communications satellite and the first Molniya satellite from the Soviet Union.

Figure 1.5 Intelsat-1, nicknamed "Early Bird." This was the first commercial geostationary communications satellite, launched in 1965. It could carry 240 telephone channels, or one television picture. (Photograph courtesy of Intelsat Global Service Corp.)

In 1972, Canada launched its first Anik satellite, the world's first commercial domestic communications satellite.

In those decades, only nations owned launch vehicles and launch pads, and only nations or huge companies or organizations could afford satellites. Because satellites of the era were low-powered, earthstations were huge and expensive, costing tens to hundreds of millions of dollars each.

The early predictions of the future of satellite communication repeated in ironic detail the predictions of the future of telephony made about a century earlier: According to one story, at a conference of mayors of U.S. cities held in Washington, D.C., part of the program was a demonstration of the new technical device, the telephone, by Alexander Graham Bell himself. Some of the mayors were interviewed for their reactions to the invention by a local newspaper. Some saw it as only a kind of parlor trick; more thoughtful ones predicted it would be useful as an interoffice intercom system; and one foresightful mayor proclaimed it a marvel and that he could foresee the time in which every city would have one.

One!

So a century later, the expectation was that most countries might have one earthstation, while countries bordering on two oceans might have two. International organizations were founded to provide satellite communications, such as Intelsat, Inmarsat, Intersputnik, and Eutelsat.

Today there are around 250–300 communications satellites in the Clarke orbit plus 100–200 or so in lower orbits. In addition, there are dozens of military communications satellites, military and civilian meteorological satellites, and "dead" satellites dating from the days before we consigned retiring satellites to more distant "graveyard orbits." In general, they are too far away and too small to be seen except with a telescope. (One exception to this is the so-called "Iridium flares" that are seen when the low-orbit Iridium satellites reflect sunlight back to Earth.) The Clarke orbit is getting crowded. Proof of this shows up well in a remarkable photograph taken at the National Optical Astronomy Observatories (Fig. 1.6).

As communications satellites developed over the past decades, other technologies advanced as well. There is no separating communication from computation, so advances in computation contributed to both satellite and terrestrial technologies. Powered by engineering progress in launch vehicles, the push of Moore's "Law" (which states that processing power and memory capacity double about every 18 months), and the consumer pull for more and more capabilities, satellite communication has become an indispensable part of global telecommunications in only four decades. It has become a huge business of around $70 billion in 2000, and even at that large figure, remains only a small part of the telecommunications industry.

Today there are uncounted millions of earthstations, ranging from tiny antennas for receiving digital radio in automobiles to 30-m-diameter gateway systems. Many of them little resemble the common notion of a satellite dish. Today, one can buy a GPS receiver to pick up satellite signals for less than $100, a satellite-delivered television for a couple hundred dollars, and a two-way data terminal for about a couple thousand dollars. International organizations have almost all become private

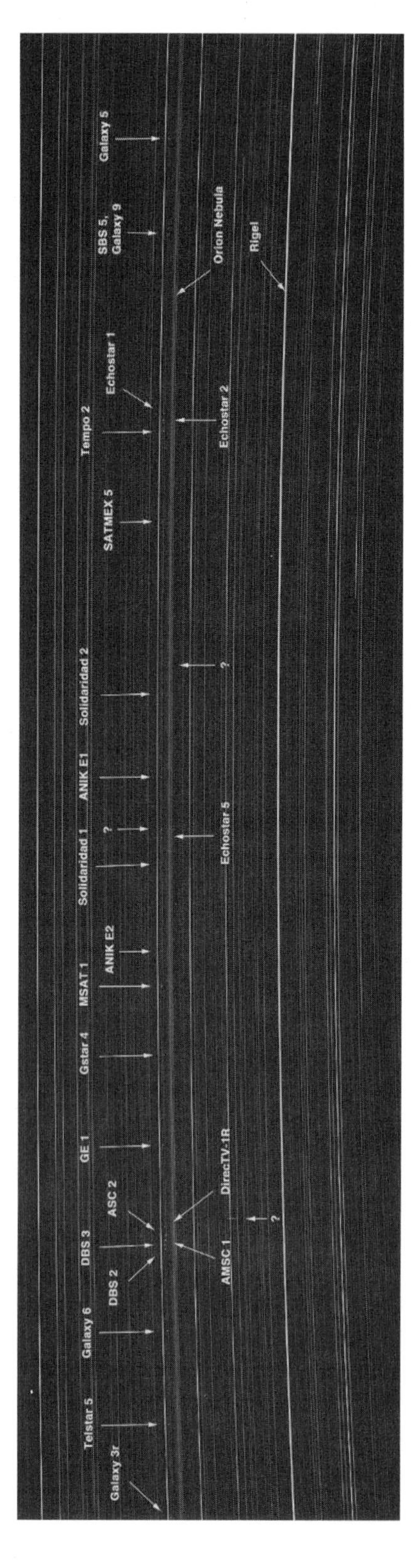

Figure 1.6 A remarkable time-lapse photograph along a portion of the celestial equator. Stars make trails, but the geostationary satellites show up as points of light. (Photograph courtesy of W. Livingston/NSO/AURA/NSF.)

companies competing in a busy market. About 250 operational geostationary satellites are in space, hundreds more are in lower orbits, and that number is increasing.

During these same decades, optical fiber technology has made huge strides and is now capable of carrying staggering amounts of information at low cost. Over the years, fiber has overtaken satellites (and even most other terrestrial technologies) for point-to-point communications. In the 1980s, fiber promoters boasted that they would crush the satellite industry. In fact, this has happened in some areas. To understand why, consider that a single transoceanic optical fiber has a greater telecommunications capacity than the total capacity of all communications satellites now operational! In the early years of this decade, user costs for optical fiber communications were decreasing by 50% per year.

Most of the heavily telecommunicating regions of the world are fibered together, and satellites are left to specialize in the huge (but more diffuse) markets of medium- and thin-density routes, preeminently in mobile communications. Satellite-delivered television battles cable-delivered television in heavily wired locations. The next big application for satellites is worldwide delivery of the broadband, multimedia applications pioneered across the Internet. (In later chapters we will explore the applications where satellites are competitive with other technologies.)

But satellite communication—and telecommunications in general—is not just about technology. Over the years, the legal, regulatory, and political systems have struggled to keep up with the technology. As usual, laws lag technology. The global reach of satellites has caused major disruptions. The 1990s liberation of Eastern Europe owes much to the global reach of satellite-delivered television. That decade saw the changeover from economies based on goods to those based on information and services.

There are also the very human issues of people, personalities, opinions, vested interests, politics, religion, and culture. (As I tell my clients, "If you think dealing with technology is confusing and difficult, try dealing with people!")

There are many examples of advances in technologies being thwarted or delayed by preexisting legal or cultural institutions (and, to be fair, sometimes this has been a good thing). AT&T's Bell Laboratories developed the first commercial communications satellite, but antimonopoly laws prevented this company from going further with the technology. This allowed for the later development of other, smaller companies to pioneer the field. As another example, the early mind-set that only nations and their monopoly telecom carriers could afford satellite communications resulted in laws that gave satellite monopolies to certain firms; it took years of legal and political battles for entrepreneurial companies, pioneered by Rene Anselmo and his company PanAmSat, to break open these restrictions. In other countries, we have seen examples of a dictatorial religious or political authority decreeing that communications (usually television) from other parts of the world is incompatible with that country's particular mythology, demanding—and obtaining—the destruction of all satellite stations. Even today, some nations completely ban reception of satellite-delivered television, simply because some program offends the ruling political machine.

The nontechnical aspects of telecommunications by satellite demand some consideration as the arena within which the satcomms game is played out. Thus, the next two chapters explore the legal and regulatory regimes within which this industry functions, and the global telecommunications marketplace of which satcomms is but a small part.

Chapter 2

The Legal and Regulatory Environment of Telecommunications

When the telegraph was first used in the United States and other nations, the systems were completely independent of one another and evolved somewhat differently. All of these systems carried pulses of electricity by wire to send letters of the alphabet to distant places. However, how the pulses were encoded to represent the alphabet(s), the voltages and currents carried in the wires, and the cost to send a message differed from system to system. When these systems begun to be interconnected, the issues of technical compatibility and of accounting arose, prompting lawmakers and international organizations to set standards and procedures. On the technical side, such issues as commonality of the letter coding and voltages allowed independent national systems to be connected to one another. On the business side, companies and nations were forced to decide how to account and charge for a telegram that originated in one country for delivery to another, perhaps after transiting a third country. The legal and regulatory problems are a complex, messy mixture of engineering and sociopolitical philosophy that is changing with time. In this attempt to give an overview of these aspects of satellite telecommunications, you may note that this is one of the longest chapters in the book, and yet only skims the surface.

After a century of slow changes in telecommunications, the last two decades of the twentieth century saw increasingly fast changes. This trend is continuing and poses a huge challenge for bureaucratic regulatory agencies that are accustomed to a more leisurely deliberative pace.

In the 150 years or so since the first signals were sent electrically, telecommunications—like other technologies—has acquired a large body of laws, regulations, treaties, standards, and customs that control the business to a large degree. Although the laws of physics allow a wide variety of types of communication, the laws of humans are much more restrictive. With the globalization of telecommunications brought on by the interconnection of the world's telephone systems and expanded by the advent of satellites in the 1960s, legal and regulatory structures have grown and changed continually. As demand has soared, so has the necessity for regulation and for standards of interconnection and operation.

Telecommunications and global trade have significantly diluted the concept of totally independent, sovereign nations. The two simple facts that you cannot stop a radio wave at a national border, and that a single geostationary satellite can see 44% of the entire surface of the planet, mean that even nations wishing to cut themselves off from others find it impossible to do so completely. For better or worse, the United States is often the leader in trends and policy, not always with the acquiescence of other nations. This is due to the fact that much of the telecommunications technology developed in the United States, and that the U.S. has such a huge economic base. European countries, especially after the formation of the European Union, have also taken a major role in standards and regulation. Since the U.S. telecommunications industry has always been one of regulated but free markets, becoming even more market-oriented in the last quarter of a century, many countries have felt impelled to take their lead from the U.S. to liberalize their markets. This is a process still ongoing. Conversely, if U.S.-based multinational corporations wish to have global markets, they must work within the constraints of the other 200+ nations of the world—and U.S. national policies and laws must accommodate them.

2.1 Telecommunications issues

Telecommunication regulation basically deals with three major kinds of issues:

1. Technical issues, such as codes, voltages, addresses, and shapes of connectors; these go under the rubrics of "standards" and "protocols";
2. Resource issues, such as the limited spectrum, orbital space, and interference; these issues fall under such rubrics as "allocation" and "coordination"; and
3. Money issues, such as availability of capital, risk management, insurance, and who gets paid how much and in what manner; these issues are those of business strategy and "tariffs."

In addition, there are many less quantifiable issues that can vary greatly from country to country, and these often conflict with the desire for more efficient and inexpensive communication:

- National security issues;
- Trade issues;
- Individual privacy and security issues;
- Intellectual property issues; and
- Social, political, and cultural issues.

It is no exaggeration to say that dealing with legal and regulatory issues among more than 200 nations is very often much more difficult, time-consuming, and money-consuming than dealing with technical details. Furthermore, the regulations within a country are frequently different from those controlling communication country-to-country. Thus, many telecommunications services that are technically possible are prohibited by rules and regulations.

2.1.1 Resource allocation

Some things in nature are limited. Mark Twain reportedly advised, "Buy land: they ain't making any more of it!" because there are only 14,824,392,000 hectares of land on the planet (ignoring small changes caused by geologic events).

In satellite telecommunications, the two limited—and thus limiting—physical resources are the electromagnetic spectrum and the space around Earth. While the spectrum is technically infinite—extending from infinitely long wavelengths to infinitely small ones—we will see in Chapter 5 that only some waves are useful for telecommunication. Furthermore, many wavelengths are already in use, limiting the rise of new services. Also, as we will explore in Chapter 10, there is only one geostationary orbit, and it, of course, is of fixed length (about 264,000 km around). Since satellites placed there must be spaced apart to avoid physical or electronic collisions, there is only so much useable space in space. Even though there are technically an infinite number of lower orbits, satellites must avoid one another. We will discuss limitations on orbits and frequencies in much more detail later.

There are other items in telecommunication that may be in short supply, and therefore can be considered limited resources. A simple example is telephone numbers. In the U.S., consider the rapid growth in the number of telephone area codes. Internationally, until recently the maximum number of digits that telecommunications switches could handle while placing a call was 12. That has now been increased to 15, but many older switches still cannot handle more than 12. While this was a "resource" that became limited because of earlier decisions, it is nonetheless a real one that causes problems as more and more people communicate. (While there are fewer than 10 billion people in the world, and thus you would simplistically think that 10 digits would be enough to assign to everyone, this is not so. Since much of our telecommunications is geographically oriented, and many people have more than one number, 10 digits is very inadequate.) Another pressing current similar problem is identification numbers, called IP addresses, for computers connected to the Internet.

There are some situations when resources are not so limited, generally in situations in which one user has no possibility of interfering with another. For example, every cable television distribution system in the United States uses the same standard channel frequencies to send video to homes. They can reuse these frequencies because the waves travel only through conducted transmission media such as coaxial cables, optical fiber cables, waveguides, etc., and cannot get out to interfere with other systems. Consequently, such systems need less regulation, and what national regulation they are subject to is largely to promote standardization.

2.1.2 Money allocation

In terms of practical business issues, money is a resource used to found and expand telecommunications systems. At times, money is easily available, and at other times it is hard to come by for some projects. The disappointing problems with mobile satellite systems such as Iridium, GlobalStar, and OrbComm soured the investment community on investing in other satellite ventures, and failures of launch vehicles raise insurance rates and thus limit the risks that can be underwritten.

A *tariff* is a specification of a telecommunications service and the charge for it. Some telecommunications services are tariffed, while some are not. Different countries tariff differently. Not all telecommunications carriers have published tariffs. In the U.S. and some other nations, common carriers (companies that must serve everyone) must publish their tariffs, which can be challenged by would-be customers, and even by competitors. Other, private (noncommon) carriers, may offer services under individual contracts with users, and are subject to less regulation.

In some nations, telecommunications is still the task of a government agency or of a government-designated monopoly. In such cases, the telecommunications service supplier is also the telecommunications regulator (and sometimes is also the broadcaster, the post office, and other services provider). In nations with a more free-market telecommunications economy, the government usually regulates the monopoly or near-monopoly carrier(s) in one of two ways: either controlling and capping their rates, or controlling their rate-of-return. The former control is more objective, but provides no incentive for the provider to be efficient. The latter method provides this incentive, but is more subject to "creative accounting" practices subject to debate.

Internationally, the situation is more complicated, because not only can telecommunications traffic—say a telephone call—originate in one nation and terminate in another nation, the traffic may pass through one or more other nations. Each of these nations will have different regulatory structures, rules, and rates. A country through which communications passes will want a cut of the revenues. Free-market nations must work with monopoly nations such that each nation's policies are adhered to.

The governing regulations for international telecommunications accounting is called the International Settlements Policy (and sometimes the Uniform Settlements Policy). It arose in the 1930s out of U.S. anti-trust policies, to prevent carriers in the U.S. from stifling competition.

Tariffs (and other policies) may be set for other than purely economic and business reasons. They may be set to promote sociopolitical agendas. For example, for many years, the bulk of international telecommunications was carried on satellites via the Intelsat system. Intelsat was established specifically with the intent of promoting the growth of international telecommunications, and to that end, its charter specified that its tariffs will be uniform, based on the average cost of providing services. This policy in effect charged users along heavy routes (such as New York to London) more than it actually cost to provide a service to subsidize thin routes (mostly to, from, and between developing nations).

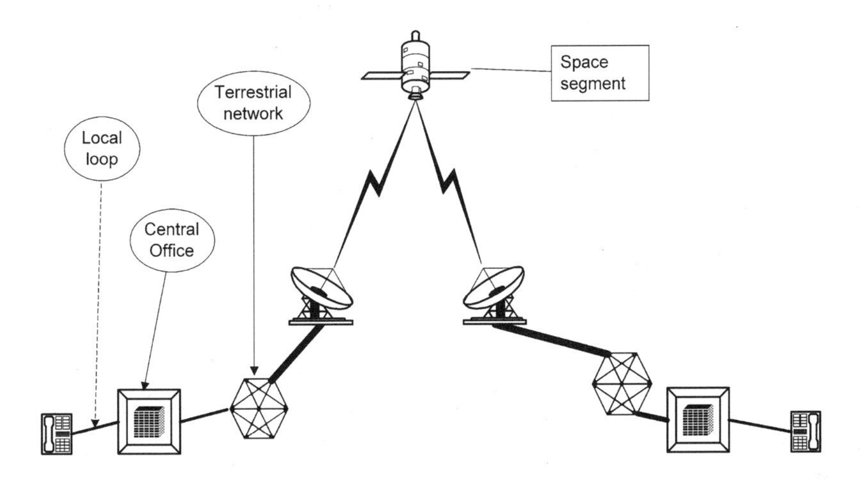

Figure 2.1 Even a "simple" telephone call via satellite is carried by many disparate parts of a telecommunications system. Your telephone is connected by a pair of twisted wires to a local central office, which connect through (perhaps several) other offices and the Public Switched Telephone Network (PSTN) on its way to a gateway earthstation. This station links to a satellite, and another gateway station in the country you are calling connects your call to that country's PSTN, and finally to the called telephone.

One must keep in mind that many telecommunications tasks go through many steps involving different technologies, companies, and nations (see Fig. 2.1). For example, a satellite telephone call from country A to country Z will have (at least) these parts:

- The calling telephone is connected via a local loop to a local central office in country A.
- The central office is connected by a long-distance carrier to the international gateway earthstation in country A.
- There is a half-circuit from the gateway earthstation in country A to the satellite.
- The call is carried through the satellite (called the "space segment").
- There is a half-circuit to the gateway earthstation in country Z.
- The call is passed through a long-distance company in country Z to a central office near the called number.
- The local loop is connected to the called telephone.

Each of these steps may involve a different company and thus may be subject to a different tariff or charge. If you are the calling party, the bill you receive for the call will be composed of your telephone company's charges for local service, long distance charges, uplink and downlink charges, space segment charges (usually with a mark-up added), and termination charges from country Z passed along to you.

For this reason, various companies may decide to provide services only to some parts of the telecommunications chain. Some may decide to become common carriers, while others become carriers' carriers, or providers of value-added services such as format conversion, while others undertake to provide end-to-end connectivity for clients. Each type of service may be regulated differently.

2.1.3 The World Trade Organization

The WTO is an international negotiating body that deals with international trade, both goods and services. As almost all satellite operators have an international—even global—scope of activities, WTO policies are extremely important to the industry. One such policy document, the General Agreement on Trade in Services (GATS), concerns the satellite industry. Negotiations begun in the 1990s and continuing today deal with how nations will (or will not) allow satellite services based in other countries to operate in their own. Within this operational framework, nations make commitments to liberalize the restrictions on foreign operators, the level of competition they will allow, and what restrictions they will place on foreign operators as compared to domestic ones.

These policies potentially affect huge markets, not only for "traditional" telecommunications services, but also for more recent ones such as Internet access and direct-to-consumer broadcasting. For more details, see the WTO website at www.wto.org.

2.1.4 Other users

Keep in mind that commercial satellite telecommunications companies are not the only users of electromagnetic and orbital resources. This greatly complicates coordination and allocation. There are nongeostationary satellites, military and government telecommunications services, and even non-communications uses of the electromagnetic spectrum, such as radar and microwave ovens.

For example, besides the civilian communications satellites in the geostationary orbit, there are military communications satellites, military surveillance satellites, military weather satellites, civilian weather satellites, and other satellites launched by national space agencies such as NASA, ESA, and JAXA, plus dozens to hundreds of now-dead satellites launched since the 1960s. Military and civilian agencies also use low-orbit and higher-orbit spacecraft with which they must communicate using radio, with potential interference to civilian communications users. Space-based navigation systems—GPS from the United States and GLONASS from Russia—are important systems with interference issues with both space-based communications and terrestrial services.

There are billions of terrestrial users of the spectrum, many of them having begun operation well before the era of satellites. These frequently cause interference to satellite systems, and satellite systems interfere with them. For example, it is often difficult to place an earthstation in a large city because of competing microwave tower users. Sometimes new services must be given portions of the spectrum in use by another older service which must then be moved to another band. This is sometimes called spectrum "re-farming." Often the new service is required to pay the older one to relocate, since the costs can run into the billions.

Because of the growing complexity of allocating precious natural resources, international, national, and regional regulatory bodies have been formed to try to bring order and fairness to the process.

2.2 Telecommunication statistics

Obviously, for telecommunications system operators to provide services, they must know the size and distribution of the market and keep up with its changes. The very disparate levels of economic development, and consequent markets, of the world's more than 200 nations makes for a complicated situation. Those of us living and working in the highly-developed nations, with our cellphones, Internet access, wireless PDAs, and other paraphernalia, have a very skewed view of the global situation.

Just about every telecommunication statistic, from the number of telephone lines per person, to cellphone usage, to number of websites, is highest in North America and Western Europe. Other areas, such as Eastern Europe, Latin America and Asia are catching up. At the bottom of almost every list of usage is the huge continent of Africa.

Despite how much you may use telecommunications, the ITU estimates that about two-thirds of the total population of our planet has never made a telephone call! More than half, it is estimated, live more than a hour's travel from the nearest telephone. Major cities in North America and Western Europe may have dozens of radio and television stations. Much of Africa has none at all.

As we will explore later, satellites offer one of the best ways to upgrade the telecommunications offerings in a country because of their huge coverage areas. Broadcast services from satellites, such as various DTH services and the systems of WorldSpace, are bringing information and entertainment to nations and regions with no terrestrial communications at all.

The disparate status of the various countries again raises the question of how to fairly allocate scarce resources among them.

2.2.1 What is fair?

To start on a philosophical note, just what is "fair" is the subject of a centuries-old continuing debate. To take a simple example: just how should the finite number of orbital locations along the geostationary orbit be apportioned? Should each of the 200+ nations in the world receive the same number? Should Andorra (population 55,000) get one and neighboring France (population 60 million) also get one?

Andorra is unlikely to establish its own satellite system anytime soon. On the other hand, Luxembourg (population 400,000) has an active satellite program, while neighboring Belgium (population 10 million) has none.

For years, the smaller developing nations have worried that the developed nations will use up all of the resources. The thrust of their argument is that all of the frequencies and orbits will be taken by the time they are ready to take advantage of them. The developed nations parry with the claim that technology is improving so that more telecommunications will be able to be done with the same resources by the time the developing nations are ready. The developing nations typically riposte that they don't believe it.

Most satellite services today, with one major exception, are on a first-filed, first-served basis (the details of this process are explained at the end of this chapter). The process is both technical and political, and there have been cases in which proposed systems with technical merit have been canceled because the companies proposing them did not have the financial resources to pursue the years of negotiation. Unfortunately, some companies have filed for systems unlikely to be built just to tie up orbital slots and keep potential competitors out. These are called "paper satellites."

The one major exception to this first-filed-first-served system is the service broadcasting television entertainment directly to households. It is described as being *preallocated*. The official designation for this is BSS-TV, for broadcast satellite service television, but it is more popularly known as DBS, for direct broadcast satellites, or DTH, for direct to home.

True BSS-TV service was rigidly preallocated in 1977 for regions 1 and 3, and in 1983 for region 2. These preallocations reserve a certain fixed number of orbital slots and even specific channels to each country then existing, regardless of the nation's need, desire, or capability of using the resources. A nation wishing to establish a domestic DBS service within these guidelines may do so without further consultation or notification to anyone. The reserved orbital slots and channels were done mostly on the basis of equality, at least within Europe: every Western European nation got five channels, regardless of population or size.

Of course, many complications have arisen, which again brings up the concept of what fair allocation is. For instance, in 1977, both East Germany and West Germany received channels. Now that there is one Germany, their allocations are combined. Is that fair? On the other side, in 1977, Yugoslavia was given a reservation for the one nation it was then, but now it has split into several nations. Who gets the resources? And if all of the slots and channels are allocated, how do new nations get anything? Because of both technological and geopolitical changes, these issues are being reconsidered by various countries and by the ITU.

This raises a related issue: while BSS-TV is a very tightly defined service specifying orbital slots, channel frequencies, interference criteria, etc., the same BSS-like kinds of services can be provided by other satellites at other orbital slots using other frequencies. And these, of course, are established on the priority system like most satellite services.

So what if, say, Andorra, wished to sell off its BSS-TV channels to, say, Spain? This is illegal under ITU rules, and is called "trafficking in slots." But that is not to say that such has not happened. In one example, a tiny nation applied for more than two dozen orbital slots, and after other nations complained, was actually allocated fewer slots. The nation then leased some of the slots to commercial firms that employed used satellites moved to these slots. (Note that most of this scheme fell apart when disagreements among the companies and individuals resulted in a spate of lawsuits and countersuits.)

Other nations have applied for authorization to have exclusive use of frequencies or orbital locations greatly in excess of their needs. This is called "warehousing" of spectrum or slots, and is prohibited and discouraged by international regulations. However, this is not to say that it has not happened.

2.3 Standards and protocols

Standards are the agreements that allow useful connections to be made. Another related term is *protocol*. They are concerned with the use of scarce resources, interoperability of systems, and performance. These can be simple agreements, dealing with the highest "level" of communications, such as "We will speak Japanese" or "We will use the WordPerfect document format to exchange files." Or, they can be very complex, specifying how many bits make up a computer "word," how many words make up a "frame," which bits represent the identity of the sender and receiver for a digital transmission, the shape of the connectors, the voltages and currents to be used, how to compress video signals, and many other details.

Standards are important to facilitate the exchange of information, to enable mobility, and to expand markets. The lack of a single global (analog) television standard means that every earthstation dealing with international exchange of video must have equipment to convert from one television system to another. Terrestrially, the fact that Europe has adopted a single standard for cellular telephony, while the U.S. has several incompatible systems, means that Europeans can roam all over that continent using the same cellphone, while in North America you have pockets of availability and unavailability.

Standards can evolve from a company, a nation, or from international agreement. If a company pioneers a technology, it often sets the standards that may or may not become globally accepted later. When other companies or nations catch up or surpass the pioneer, coordinating new technologies with old standards can make for very contentious and expensive debates. (For example, in the 1950s, the United States FCC decided that the newer color television system had to be compatible with the older black-and-white system; this compatibility greatly reduced the quality of the color system.)

Standards can be promulgated by law and regulation. Examples include the (more than two dozen) television standards in use in various countries, the frequency bands and orbital slots assigned to various direct-to-home television broadcast satellites, and the various cellular telephony systems in different countries.

Standards can also arise out of voluntary consensus. Examples include the use of the code called ASCII in all personal computers. Standards may also come from a pioneering or dominant organization in an industry. Examples include the standard-type earthstations defined by Intelsat or Inmarsat.

Standards often begin in one nation. Only when the systems of nations must be connected do they become international. Often different nations evolve different standards, making connections difficult and/or expensive. For example, intercontinental exchange of telegraph signals has been going on for more than a century, as the first transatlantic telegraphy cable was laid in the mid-nineteenth century; as a consequence, there is a single worldwide standard for telegraphy. However, national telephone systems separated by oceans evolved separately and differently in Europe and North America, and were not connected until the advent of transatlantic telephone cables in the 1950s, by which time each region's standards were well entrenched. Thus, telephone calls between these continents must go through standards-converting equipment, making the connection more complicated and expensive. In a similar way, television systems developed differently in Europe and in North America, and so exchange of television among nations usually requires format conversion.

2.4 The International Telecommunication Union

To coordinate the multitude of telegraph systems which had arisen, twenty European nations founded the International Telegraph Union in 1865. It is the world's oldest technical coordination organization. Other countries soon joined in. As submarine telegraph cables interconnected the continents, the organization helped unify the standards for telegraph traffic. It pretty much stuck to only telegraphy for a while: although the telephone was invented only a decade afterward, the organization had little to do with the new technology until the turn of the twentieth century, and did not set up a standards committee, called the Consultative Committee for International Telegraphy and Telephony (CCITT), until the 1920s, and a regulatory structure in 1932. In that same year, a separately-formed body, the International Radiotelegraph Union, merged with the telegraph union and the overall body became the International Telecommunication Union (ITU). The radiotelegraph functions became a committee called the Consultative Committee for International Radio (CCIR). The CCITT is now called ITU-T, and the CCIR is called ITU-R.

The standards promulgated by these committees go by such designations as X.25 (pronounced "X-dot-25," which deals with packet switching), X.400 (which deals with electronic mail), V.90 (a standard for 56 kbps modems), and the like.

In the area of telecommunication by radio waves, the ITU members have produced a legal regime codified in the Radio Regulations (RR). These regulations allocate different frequencies of the radio spectrum to different uses. The ITU also maintains a comprehensive list, called the Master International Frequency Register of all uses of radio frequencies, and of orbital slots in the Clarke orbit. Terrestrial, geostationary satellites, and nongeostationary satellites fall under different regulatory regimes. The regulations set out standards and procedures to minimize interference among various users and provide for coordinating the entry of new users.

A note for purists and copyreaders: the word "telecommunication" in the English version of the ITU's name is singular: no "s" on the end. The official French version of the name does have the "s." Much information about ITU standards and activities (some only for members or by subscription) is available online at www.itu.org.

While the pre-1932 ITU dealt largely with standards and accounting, the addition of radio greatly complicated things, forcing it to deal with issues of interference. The problem has grown exponentially since. When communications satellites became possible, it seemed a logical extension to let the ITU allocate these limited resources as well. (On a historical note, since telecommunications evolved from private companies in North America, while it was developed and provided by government monopolies in Europe, it was not until after World War II that the ITU became a specialized agency of the United Nations and North American nations agreed to be bound by ITU regulations.)

Today the ITU is the dominant international telecommunications regulatory body. It has treaty status, meaning that the provisions of the ITU's Constitution and Convention are binding on ITU member countries. It should be emphasized, however, that it is a policy organization only, and has no enforcement powers. It decides policies on the basis of one-nation, one-vote, and any nation willing to adhere to its rules may join. Standards and procedures are only (strong) recommendations to member nations. In fact, any member may object to adhering to any specific regulation by filing what are called "exceptions" to the rules. The supreme governing body of the ITU is the Plenipotentiary Conference, and meets every four years. (Insiders refer to it as the "plenipot.") Figure 2.2 shows the structure of the ITU.

It should be emphasized again that the ITU has no enforcement powers of its own. As long as procedures are followed, it does not have the authority to approve or reject applications for satellites that may be dubious. It simply accepts them and enters them into its registry.

2.4.1 ITU regions

The ITU considers the world divided into three large geographic regions, referred to as "ITU Regions," or "WRC Regions" (for World Radio Conference). Regulations and standards may be set differently in different regions, set the same in two regions, or even set globally. As the world becomes more interconnected, the need for global allocations and regulations is increasing, particularly for the new satellite-delivered mobile services. This is complicated and slowed down by legacy systems. For example, some of the frequencies used for global mobile satellite communications are allocated in some regions, but not in several specific countries because those nations or regions have been using those frequencies for other purposes. Time must be allowed for these incumbent users to move to other bands.

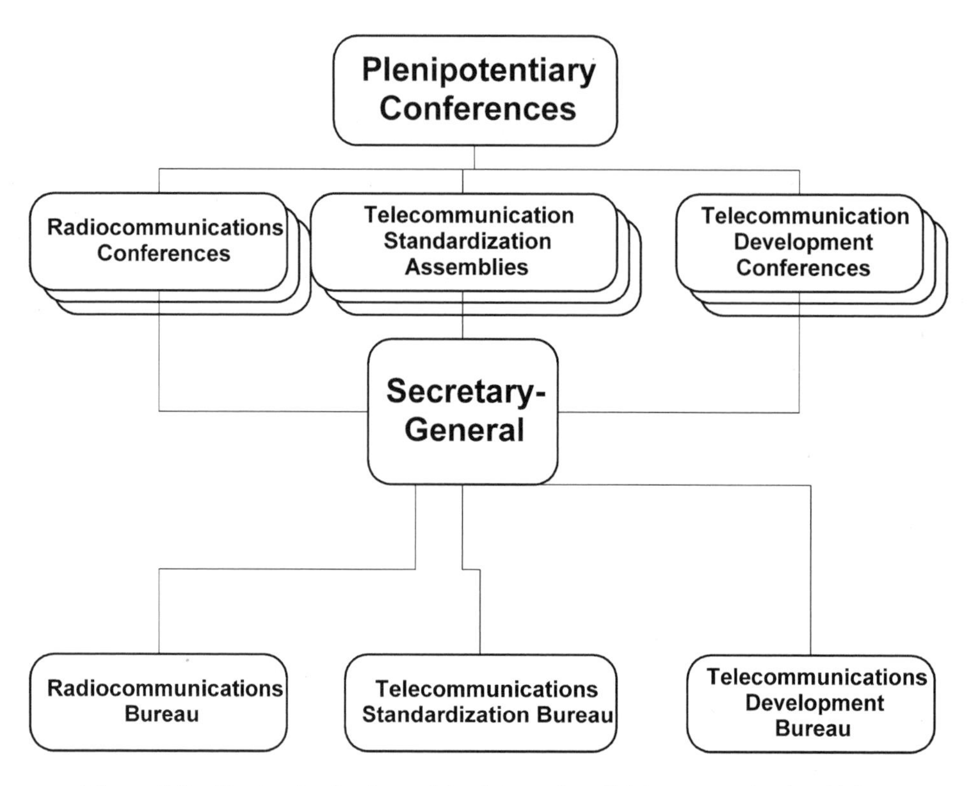

Figure 2.2 The basic structure of the International Telecommunication Union.

Here are the approximate descriptions for the three regions. The definitive regulations can be found either in the ITU Radio Regulations or in Part 47 of the United States Code of Federal Regulations. (Specifically, *47CFR* Chapter 1, §2.104.) See the accompanying map shown in Fig. 2.3.

- ITU region 1 consists of all of Africa, Europe, the Middle East, and most of what was the Soviet Union.
- ITU region 2 includes the Americas, Greenland (but not Iceland), the Aleutians, and the ocean regions adjacent to North and South America.
- ITU region 3 encompasses everything else, including most of Asia, the East Indies, India, Australia and New Zealand, Japan, Korea, and much of the South Pacific.

Standards are debated and recommended by various ITU bodies. The ITU has recently opened up its procedures to input from telecommunications firms and other nongovernmental organizations, although only the nation members may vote.

ITU-T is the telecommunications standards organization, dealing mostly with communications traveling through wires and optical fibers. A world telecommunications standardization conference meets every four years, with small relevant bodies meeting more frequently as needed. Such standards as V.90 for 56-kbps modems, and X.25 for packet switching came out of ITU-T.

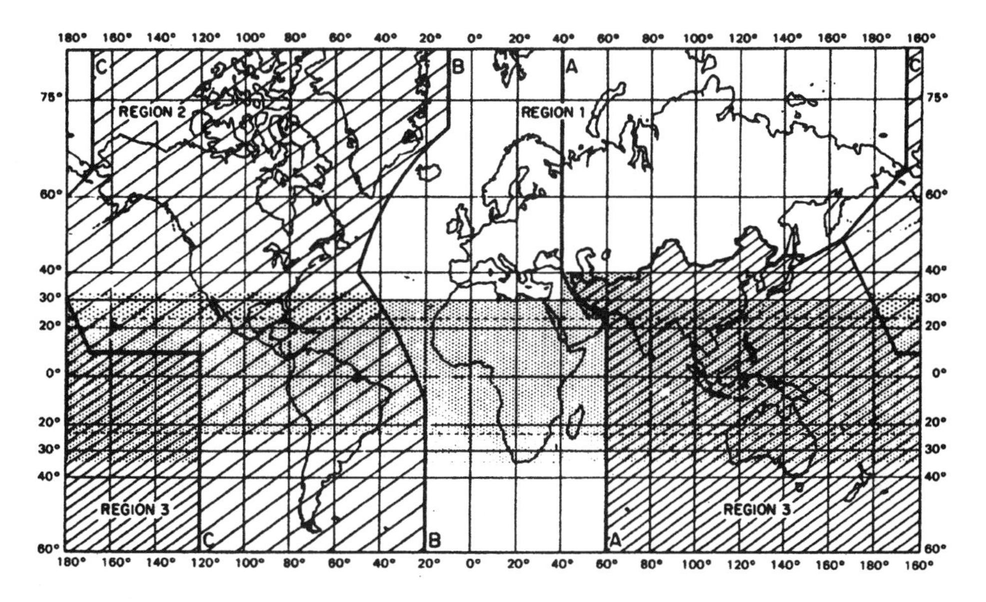

Figure 2.3 The three regions defined by the International Telecommunication Union. The shaded area straddling the equator is known as the "tropical zone." (Source: FCC.)

In the area of radio that concerns the satellite industry, there is the ITU-R committee. Part of ITU-R is the Radio Regulations Board. It interprets the Radio Regulations, sets policies for registering frequency uses, and maintains the Master International Frequency Register and the Table of Frequency Allocations. It holds the biennial World Radiocommunications Conference (WRC) at which radio regulations are debated and adopted. Radio regulations have more stature than standards recommendations, due to the potential for interference. Once adopted in documents called "final acts," the regulations have treaty status like the ITU's founding documents, meaning that they are binding on member nations; but again, a nation has the right to take exception to a regulation provided it does not cause interference. Another newer body, the ITU-D, encourages telecommunications development in the developing nations.

That is not to say that all nations always adhere strictly to the rules. Within the past decade, we have seen several instances where the "comity of nations" has deteriorated into disputes that involved not just words, but the implementation of services that conflict with others.

For example, a while ago, two nations disputed the assignment of an orbital slot serving part of Asia, and also disputed whether the satellite systems planned would interfere with one another. Failing to resolve the issue, both nations launched satellites to orbital locations too close to one another, resulting in both systems being impaired in their operations by mutual radio interference. With hundreds of millions of dollars worth of satellites and earthstations going to waste, they then sat down to debate further, and finally one satellite was moved away from its original orbital slot.

In another case, one satellite operator had filed to launch a satellite, but did not do so within the time limit required by the rules of filing, so another operator launched a satellite to the requested orbital slot. The first organization maintained that it should have that slot, since it had briefly tested a satellite at that location well within the deadline, but the rival claimed that brief testing is not "entering into operation" of the filed-for service and thus the first organization was in default.

One big problem that has increasingly plagued the ITU and would-be satellite operators is the "paper satellite." These are filings for future systems that are made either on far-out speculation or as a ploy to preempt other systems from filing. Since ITU rules give priority to the first systems filed, subsequent systems must first prove that they will not interfere with the first system, even though all parties know that it may never be launched. To partially solve this problem, the ITU has recently shortened the time a filing remains in effect from nine years to six, and is demanding more detailed information and progress reports on the progress of filings. Some people have proposed multimillion-dollar application fees as well as tightening and enforcing the due-diligence requirements to reduce the number of specious filings. The FCC in the United States has changed its licensing procedures to consider applications on a first-come, first-served schedule, and to require a substantial bond from applicants to try to constrain frivolous applications (however, these bonds are opposed by some in the industry).

2.5 Other standards and regulatory organizations

There are a number of other international and regional standards-setting and regulatory organizations whose members are usually the standards-setting bodies and regulators in individual nations. The most important of these is the ISO, the International Standards Organization, which covers areas of telecommunications and other areas far removed from it. In particular, its Technical Committee 97 sets standards for data processing and data communications. Its best-known standard in this field is the Open Systems Interconnection model, known commonly as OSI, which defines seven "layers" of data handling during transmission.

Headquartered in the United States, the Institute of Electrical and Electronic Engineers (IEEE) has been instrumental in setting standards, many of which have gone on to become global standards.

Another relevant body is the International Electrotechnical Commission, which deals with electrical and electronic issues not covered by the ISO. It works closely with the ITU. Technical Committee 46 handles various telecommunications transmission technologies.

Within Europe, the Conference of Post and Telecommunication Administrations, known as CEPT, is a coordinating body established in 1959 to unify the various services and technologies provided by the PTTs of Europe. Much of its work has become part of the European Union as the nations of Europe merge or unify many of their technical and business practices. Some of its work has been superceded by the European Telecommunications Standards Institute, ETSI, which is becoming the dominant European standards entity. ETSI is a member of ITU-T to help ensure global compatibility of standards.

Of rising importance is the Internet Engineering Task Force, IETF, which establishes operational standards for the Internet, such as the Transmission Control Protocol/Internet Protocol, TCP/IP. One of the continuing problems is the coordination of activities and standards of the IETF and the ITU. In both organizations, the satellite issues sometimes take a backseat to those of other parts of the telecommunications industry.

2.5.1 National regulations and standards

Each nation has some standards-setting organization(s) and telecommunications regulator(s). Sometimes these are a part of the PTT and sometimes they are an independent body. Most nations have committees that represent the country at the ITU, ISO, and other international bodies.

National telecommunications regulators often take recommendations from regional and international bodies and incorporate them into the corpus of their national laws and regulations.

In the U.S., there are two such regulatory administrations, the Federal Communications Commission (FCC), which regulates nongovernmental uses, and the National Telecommunications and Information Agency (NTIA), which regulates government users. The two organizations coordinate their policies to a greater or lesser extent. The telecommunication regulations for commercial users in the United States are set out in Part 47 of the *Code of Federal Regulations*, referred to as *47CFR*, available in print and online.

Other nations typically have a single telecommunications regulator. The administration often goes under the sobriquet of PTT, for Post, Telegraph, and Telephone administration, harking back to the (bad old) days when one government bureau was responsible for all of these functions. As the trend toward more commercial, market-based regimes has grown, an increasing number of nations have separated their regulatory bodies from their operating entities (such as broadcasters and postal services), so that the old PTT term is less and less applicable except in the more backward nations. Many have their regulations available online as well.

2.5.2 Some regulatory jargon

Telecommunications laws and regulations use everyday words in very specific ways that you must understand. At times the distinctions between terms are subtle. Many of the terms and definitions come from the International Telecommunication Union.

Allocation of spectrum is the process and result of the ITU setting aside certain frequency bands for particular types of telecommunications services. The ITU's Radio Regulations and the FCC define it as follows:

> *Allocation (of a frequency band). Entry in the Table of Frequency Allocations of a given frequency band for the purpose of its use by one or more terrestrial or space radiocommunications services or the radio astronomy service under specified conditions. This term shall also be applied to the band concerned. [Code of Federal Regulations, Part 47, Section 2.1, (47 CFR 2.1)]*

For example, the frequency band from 88 to 108 MHz is allocated for the use of broadcasting radio stations, specifically, those called FM stations. For an example in the satellite industry, the frequency band of 3.7 to 4.2 GHz is allocated for the radio link from satellites to earthstations (technically called the space-to-earth link) in what is termed the C-band of frequencies (to be detailed in Chapter 5).

Allotment is the process by which the ITU or other bodies divide up a spectrum allocation among nations to avoid interference. This is necessary only when there exists the possibility of interference across national borders. The formal ITU and FCC definition is

> *Allotment (of a radio frequency or radio frequency channel). Entry of a designated frequency channel in an agreed plan, adopted by a competent conference, for use by one or more administrations for a terrestrial or space radiocommunication service in one or more identified countries or geographical area and under specified conditions. [47 CFR 2.1]*

For example, each Western European nation was allotted five video channels from satellites located at one or more specified orbital longitudes for use for direct-to-home television broadcasting.

Assignment is the particular authorization by a national telecommunications regulatory body to a specific user. Again, from the ITU and FCC:

> *Assignment (of a radio frequency or radio frequency channel). Authorization given by an administration for a radio station to use a radio frequency or radio frequency channel under specified conditions. [47 CFR 2.1]*

For example, within the VHF television broadcasting band allotment to the U.S., the band of frequencies is divided into smaller bandwidths called television channels, which are then assigned to companies to operate in different parts of the country. For a satellite example, in accordance with international allocations, the FCC in the U.S. has assigned the right to use the downlink frequencies of 12.2 to 12.7 GHz to Echostar Communications Corporation to provide a BSS-TV service from a satellite at the orbital location of 241° longitude.

As is typical of bureaucratese, definitions both depend on and beget other definitions. In the previous one, an *administration* is "Any governmental department or service responsible for discharging the obligations [of a country] undertaken in the Convention of the International Telecommunication Union and the Regulations."

Some important specific terminology arises with respect to frequency assignments. If more than one service has been authorized to use a part of the spectrum, each service may be classified in one of three ways. *Primary services* have priority in choosing which frequencies to use within the allocated band. *Permitted services* have much the same rights, but come second in choices of frequencies. *Secondary services* may operate within the same band, but "[s]hall not cause harmful interference to stations of primary or permitted services to which frequencies are already assigned or to which frequencies may be assigned at a later date."

Coordination is the process by which multiple users of a frequency band analyze their usage and potential mutual interference, and revise their systems to minimize or eliminate it.

Due diligence is a process and a proof to an administration or other regulatory body that a company that has filed to provide a telecommunications service is actually making progress toward providing the service, and is not just warehousing spectrum or orbital resources. For example, if a U.S. company seeks and is authorized to build a satellite service, it has one year to prove to the FCC that it has contracts for the construction and launch of the satellite(s).

Landing rights, as used in telecommunication, is the legal authority for a company to connect telecommunications traffic to a country from outside that country. It applies to such things as transoceanic cables, and for our purposes, the authority for a satellite system operator to provide uplinks and downlinks from satellite(s) into a nation. Such landing rights must typically be negotiated by a satellite system individually with each nation within which it wishes to do business.

The legal and financial details of telecommunications consume thousands of volumes of legalese and accountingese, and are much too much to go into in this text. For a good quick overview, consult the book by Kennedy and Pastor listed in the bibliography of Appendix C.

2.6 Satellite services and applications

Telecommunications for both terrestrial applications and for satellite-delivered uses are grouped into what are called *services*. This term actually has two related meanings in the industry: first is the official legal definition of a type of satellite service, which may be characterized by the nature of the earthstations using the service, or by the nature of the traffic; second is the commercial meaning of the word, referring to specific applications offered to users.

Applications are tasks such as a telephone call, video broadcast, network contribution feed, or Internet access. Many technologies, satellites included, can provide these services to users. In many cases a variety of technologies is used, satellites sometimes being only one of them. Signals often get to and from users and earthstations by terrestrial technologies such as coaxial cable and fiber optic lines.

In this section we will explain the official satellite service designations established by the ITU and generally followed by national telecommunication laws and regulatory authorities. Then, within each legal service definition, we will give some examples of specific commercial telecommunications applications.

§2.106 444 47 CFR Ch. I (10–1–99 Edition)

International table			United States table		FCC use designators	
Region 1—allocation MHz	Region 2—allocation MHz	Region 3—allocation MHz	Government Allocation MHz	Non-Government Allocation MHz	Rule part(s)	Special-use frequencies
(1)	(2)	(3)	(4)	(5)	(6)	(7)
2160–2165 FIXED MOBILE S5.388 S5.392A	2160–2165 FIXED MOBILE MOBILE-SATELLITE (space-to-Earth) S5.388 S5.389C S5.389D S5.389E	2160–2165 FIXED MOBILE S5.388	2160–2165	2160–2165 FIXED MOBILE NG23 NG153	DOMESTIC PUBLIC FIXED (21) FIXED MICROWAVE (101) PUBLIC MOBILE (22)	EMERGING TECH-NOL-OGIES
2165–2170 FIXED MOBILE S5.388 S5.392A	2165–2170 FIXED MOBILE MOBILE-SATELLITE (space-to-Earth) S5.388 S5.389C S5.389D S5.389E	2165–2170 FIXED MOBILE S5.388	2165–2170	2165–2170 MOBILE-SATELLITE (space-to-Earth) NG23	FIXED MICROWAVE (101) PUBLIC MOBILE (22) SATELLITE COMMU-NICATIONS (25)	
2170–2200 FIXED MOBILE MOBILE-SATELLITE (space-to-Earth) S5.388 S5.389A S5.389F S5.392A	2170–2200 FIXED MOBILE MOBILE-SATELLITE (space-to-Earth) S5.388 S5.389A	2170–2200 FIXED MOBILE MOBILE-SATELLITE (space-to-Earth) S5.388 S5.389A	2170–2200	2170–2200 MOBILE-SATELLITE (space-to-Earth) NG23	FIXED MICROWAVE (101) PUBLIC MOBILE (22) SATELLITE COMMU-NICATIONS (25)	
2200–2290	FIXED SPACE RESEARCH (space-to-Earth) (space-to-space) SPACE OPERATION (space-to-Earth) (space-to-space) EARTH EXPLORATION-SATELLITE (space-to-Earth) (space-to-space) MOBILE 747A 750A		2200–2290 FIXED MOBILE SPACE RESEARCH (space-to-Earth) (space-to-space) US303 G101	2200–2290 US303		

Figure 2.4 One page of the dozens of pages outlining the radio spectrum allotments for terrestrial and satellite communications, from the *Code of Federal Regulations,* §2.106. These tables are followed by hundreds of footnotes giving additional details.

This is not to say, however, that all real-world applications are rigidly confined to specific service categories. For instance from the satellite area, while there is a tightly defined television broadcasting service, virtually identical services can and are being provided by operators using frequencies outside the designated satellite broadcast band. Similarly, applications intended to serve mobile users, such as paging, can be and are being done using satellites originally intended to serve stationary users.

Information on what telecommunications services may be provided in which frequency bands can be obtained from ITU publications, and in the United States in Part 47 of the *Code of Federal Regulations*, §2.106. Fig. 2.4 shows an example of just one of the 103 pages of frequency tables, which are followed by 54 pages of footnotes elaborating them!

2.6.1 Satellite services

The satellite-delivered telecommunications applications fall into 12 service categories, with some subcategories, designated by the ITU. Of these, only five are of major interest to commercial end users:

- Fixed Satellite Service (FSS)
- Broadcast Satellite Service (BSS)

- Mobile Satellite Service (MSS)
- Radiodetermination Satellite Service (RDSS)
- Radionavigation Satellite Service (RNSS).

One service is primarily of importance to the operators of (some) satellite systems:

- Inter-Satellite Service (ISS).

Six designated services you rarely hear about are of interest primarily to specialized users, such as scientific researchers, space exploration agencies, meteorological agencies, satellite operators, and amateur radio operators:

- Amateur Satellite Service
- Earth-exploration Satellite Service
- Meteorological Satellite Service
- Space Operation Service
- Space Research Service
- Standard Frequency and Time Signal Satellite Service.

We will give a brief description of only one of the services that are of little commercial importance first, and then concentrate on those of primary interest to providers and users of satellite communication. It should be kept in mind, however, that while these services have few commercial users, they may compete with commercially important services for the scarce resources of orbits and spectrum.

With a couple of exceptions, these noncommercial satellite services are almost exclusively used by governmental and intergovernmental entities. One such exception is the service used by amateur radio operators. The other major exception is that of satellite control and operations by satellite system operators. In the definitions, you will see the term "space station." In radio regulatory parlance, this does not mean manned orbital outposts such as the Mir or the International Space Station, but only a radio transmitter in space.

2.6.1.1 Amsat

Many people are unaware that the "ham" radio community has built, launched, and operated more than two dozen small communications satellites in an allocation known as the Amateur Satellite Service. The ITU's Radio Regulations define it as "A radiocommunications service using space stations on earth satellites for the same purposes as those of the amateur service." Designed and built using private funds from hams in many nations, they are launched as space-available payloads aboard the same launch vehicles that put other payloads into space. The first one was OSCAR 1 launched December 12, 1961. (OSCAR stands for "orbiting satellite carrying amateur radio.") The first amateur calls relayed from space went through the 30-pound OSCAR 3 launched on March 9, 1965 (before Early Bird!).

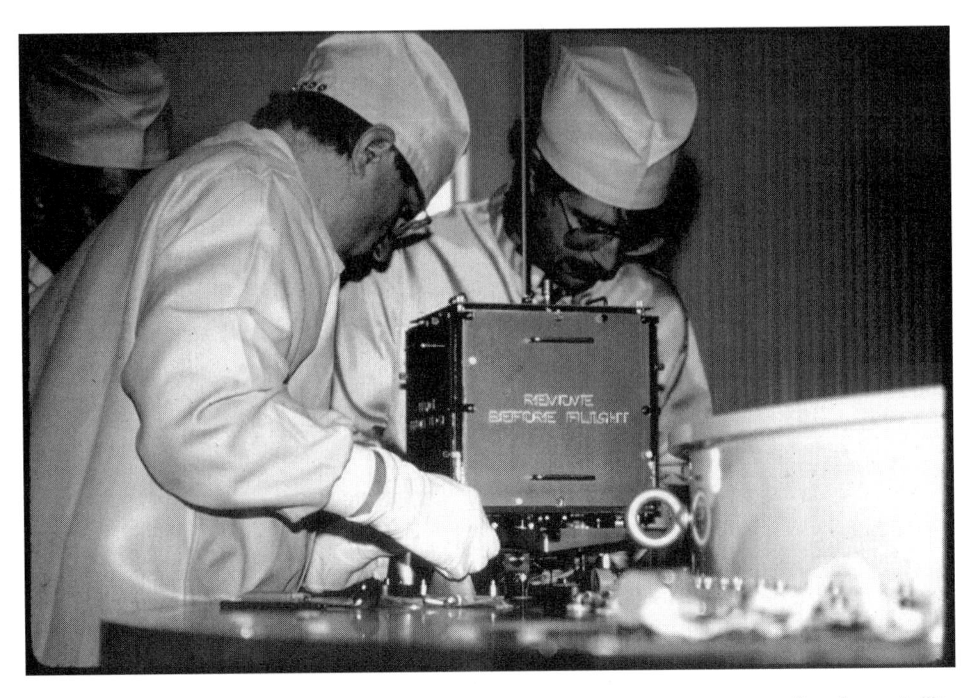

Figure 2.5 Amateur satellite (AMSAT) volunteers mounting an amateur radio microsatellite on a launcher, one of four simultaneously launched on an Ariane in 1990. AMSAT satellites are designed and built by "ham" radio operators. This series provided store-and-forward digital messaging capability of radio amateur operators around the world. (Photograph courtesy of Richard L. Daniels/AMSAT.)

These satellites operate mostly in the HF and VHF frequency band. They provide a variety of types of communication between amateurs, including Morse code, voice, packet data, and even some video. These satellites are controlled by the hams as well. They are typically less than a few hundred pounds in weight, and are all in low orbits. A not-for-profit company, Amsat, the Radio Amateur Corporation, was formed in 1969 to foster these efforts.

Figure 2.5 shows a recent amateur communications satellite. For more information, consult www.amsat.org.

2.6.2 The major commercial satellite services

Various satellite system operators will refer to an FSS satellite or earthstation, or a BSS orbital slot, or an MSS standard terminal, or an RDSS service. These designate the kinds of services for which a number of frequency bands were allotted. We have already noted, however, that BSS-like services are being provided by FSS satellites, and FSS-like services are being provided by MSS systems. While the legal definitions of these services are fixed, the actual user applications provided are broad and fluid. In the following section, we will describe the formal definitions of the five major commercial services. The following chapter will give some examples of what actual user applications they are being used for.

2.6.2.1 Fixed Satellite Service (FSS)

This is the oldest and most used of all of the satellite services. It is intended for communication through a satellite between earthstations that are fixed, or which are within a specified area. Examples include point-to-point communications, corporate networks, very small aperture terminal (VSAT) networks, and data distribution networks. "Fixed" does not mean that the earthstations are not moveable, just that they are not usually moving when in use. Thus, such temporary applications as transportable uplinks from events, portable downlinks to teleconferencing sites, and satellite newsgathering (SNG) are all examples of FSS services.

The FSS frequency allocations also specify the direction of signal travel, assigning one frequency band for uplinks (or "earth-to-space" as the formal definitions call it), and another complementary band for downlinks ("space-to-earth"). Uplink and downlink frequencies are different in order to avoid interfering with one another, but typically have the same total bandwidth so the same amount of information can be transmitted. FSS frequency allocations have much larger bandwidths, usually 500 MHz, than MSS, RNSS, and RDSS applications because FSS services are intended for such things as television or highly multiplexed telephony or data signals. The major FSS frequency allocations are in the C-, Ku-, and Ka-bands (see Chapter 5 for definitions of these bands).

2.6.2.2 Broadcast Satellite Service (BSS)

The Broadcast Satellite Service is actually three services specifically designed to provide audio and video entertainment, possibly along with ancillary services, directly to end users: BSS-TV, BSS-HDTV, and BSS-Sound. BSS-TV is by far the most highly specified of all of the satellite services, and is preallocated in most countries to specify orbital slots, frequencies, and channels. BSS-TV also specifies many details of operations, such as interference ratios, minimum powers to be sent to users, and other issues.

During the 1990s, BSS-TV became the greatest money-maker in the commercial satellite industry, gradually overtaking and then greatly surpassing the previously largest income producer, telecomms trunking and other "traditional" relay services.

BSS is often called Direct Broadcast Satellite or Direct Broadcast Service, DBS, an abbreviation which has no official international legal meaning. Another similar term is DTH, for Direct-to-Home. An older but still existing version is called TVRO, Television Receive Only. DBS and DTH can mean any service that sends video to consumers, including true BSS, into small (typically 1 m or smaller) dishes. TVRO implies use of a larger, much more expensive "backyard dish" often several meters in diameter, and mostly in C-band.

BSS-TV is the service of providing conventional television entertainment directly from orbit to a small antenna at a user's home. "Conventional" means television of at least the quality provided by today's terrestrial broadcast systems, although they are not required to use one of these three broadcast standards. Some DTH services use MAC or DVB television standards. If services other than television are provided, they are incidental to the television. BSS-TV has frequency allocations only in the Ku-band, both for the feeds from the providers to the satellites and for the downlinks to the consumers.

BSS-HDTV is intended for the coming so-called "high-definition television," although several of the current BSS providers intend to provide HDTV signals on their existing frequencies. These BSS-HDTV allocations are little-used thus far.

BSS-Sound is designed to provide high-quality audio signals to both fixed and mobile consumers. For a while this service was unofficially called DAB for Digital Audio Broadcast, but that term and abbreviation have now been taken over by terrestrial providers of digital terrestrially-broadcast audio services. The newer and more common term for BSS-Sound is *SDARS,* for *Satellite Digital Audio Radio Service.*

2.6.2.3 Mobile Satellite Service (MSS)

Satellite-delivered services to users on the go are divided into three major categories, depending on where the user is: MMSS, Maritime Mobile Satellite Service; AMSS, Aeronautical Mobile Satellite Service; and LMSS, Land Mobile Satellite Service. The major supplier of MSS services is Inmarsat.

It is worth emphasizing again that all of these mobile services are primarily intended to carry telephone calls, while sometimes carrying low-speed digital services such as facsimile, telex, and pure low-bitrate data. Video and high-speed data are not a part of these services, but highly-compressed video can be sent to provide satellite newsgathering functions.

The *MMSS, Maritime Mobile Satellite Service* part of the business got its start in 1979 with the formation of the International Maritime Satellite Organization (Inmarsat) to provide telephony, very-low-speed data, telex, and facsimile to ships at sea. (In April, 1999, Inmarsat converted from its treaty-based structure to become a commercial entity, leaving behind a small treaty-based organization to provide safety-at-sea functions.)

In the early 1980s, terminals cost around a quarter of a million dollars per ship, could provide only one call to a ship at a time, and a call cost many dollars a minute. This was worth it for very large ships, which may cost $100,000 a day to operate, if it saved them time in transit. As with all technology, the price has declined steeply since then, and today ship terminals and other terrestrial terminals using Inmarsat services can be had for a few thousand dollars. There are now more users on land than at sea.

The inauguration of Inmarsat services in the early 1980s introduced a big change in the way that users had communicated via satellite: whereas before the end user had always had to send traffic through gateway stations, by necessity the individual end-user ships had to link directly to the space segment of the Inmarsat system. However, the terrestrial ends of the telephone calls do go through Inmarsat's land earthstations.

The *Aeronautical Mobile Satellite Service, AMSS,* has several parts. These are the general AMSS, an Aeronautical Mobile-Satellite Route Service (for aircraft along national and international civil air routes), and the Aeronautical Mobile-Satellite Off-Route Service (for aircraft not on established national and international civil air routes).

Mobile satellite services now also include the data-only so-called Little LEO (Low Earth Orbit) systems, such as OrbComm (officially called Nonvoice, Nongeostationary Mobile Satellite Service), and the Big LEO telephony-and-data systems such as Iridium and GlobalStar.

Mobile satellite systems operate at lower frequencies than most other satellite services. This is partially because a user could be anywhere, under any kind of weather conditions, and lower frequencies are not much obstructed by the atmosphere. Little LEO systems use VHF and UHF bands of frequency, while Big LEOs use L-band and S-band. As we will see in Chapter 18, atmospheric problems are worse at higher frequencies. Further, antennas are less directional at lower frequencies, reducing the requirement for precise pointing at the satellite. Thus, the mobile services use L-band and S-band links to the ships and C-band or Ku-band links to land gateway stations. Mobile satellite services operate in the VHF and UHF bands for the Little LEO systems, and in the L-band and S-band for telephony systems.

2.6.2.4 Radiodetermination Satellite System (RDSS)

RDSS is designed to provide the ability to users to determine and, if desired, report their location to someone else, such as a fleet dispatcher. In some cases, the mobile unit uses conventional terrestrial radionavigation services, such as LORAN, to find its position, or it can use the navigation link that is part of RDSS. The unit then relays that information via the RDSS satellite service back to the office. The links from the mobile user to and from the satellite operate in the L-band for the uplink and S-band for the navigation downlink, while the link from the satellite to the control facility uses C-band frequencies.

The RDSS service was intended to provide a global ability to locate mobile users and report their positions. Since its inauguration, however, two other world-spanning systems of navigation have become available, the Global Positioning System supplied by the Navstar satellites of the United States Department of Defense, and the GLONASS service from Russia. Both were originally intended primarily for use by those nations' military services, but both have become available to civilian users, and have become very popular because of their accuracy and ease of use. Because of their common incorporation into mobile terminals, while they are not strictly a part of the civilian commercial satellite industry, they deserve special mention in Chapter 3.

2.6.2.5 Radionavigation Satellite Service (RNSS)

These services are designed specifically for navigation, in contrast to radiolocation and position reporting. These services are little-used at present.

2.6.2.6 Inter-Satellite Service (ISS)

This designation defines radiocommunication between satellites. As such, it is not a service for end users, but one way for satellite operators to route traffic. It is authorized in the Ka-band.

Until the late 1990s, intersatellite links (ISL; also called "satellite-to-satellite links," SSL) had not been a part of the civilian communications satellite industry. This was because it was not much needed, and because it was very expensive, especially for ISLs between geostationary satellites. The military uses 60-GHz ISLs, and the Space Shuttle uses ISLs though the TDRS satellites to earthstations to stay in touch with controllers during times when it is out of line-of-sight communications with NASA's earthstations.

The advent of commercial ISL came with the beginning of service of the Iridium "Big LEO" mobile telephony service.

2.7 Steps to licensing a satellite system

So you have a great idea for a satellite-delivered communications service? You think you can invent or acquire the technology, hire the management and technical expertise, secure the financing, manage the risks, and market your service to an expectant world? Your problems are just beginning!

Before you can build and launch that system, you must plan for how it will consume orbital and spectrum resources, and how it may potentially interfere with the systems of others that are already operational or are farther along in the planning and implementation. This is the huge, complicated, and expensive topic of "coordination."

If a company wishes to establish a new satellite service, using one or more satellites in geostationary orbit, it is embarking on a three-to-five year process of garnering the financial and technical resources and regulatory approvals. Leaving aside the business aspects, here are the steps usually encountered from conception to "throwing the switch" on a satellite service:

- **Stage 0: Conceptual design**
  The satellite system operator determines the technical requirements and parameters of the system, such as the desired orbital location(s) of the satellite(s), frequencies, polarizations, timing, beam coverage, and other operational parameters. Since the typical lifetime of GSO satellites is 9–15 years, the would-be operator must try to make estimates of the amount of traffic the satellite(s) will carry 9–15 years into the future. The operator must also plan for the number and characteristics of the earthstations intended to link with the satellite(s). This can take from months to years.
- **Stage 1: Planning the filing**
  The would-be system operator selects a country that it wants to be the "host" for the system and approaches the telecommunications regulatory authorities in that nation to work with it in preparing a document for the ITU called the "advance publication," also know as an AP4. The host country does not have to be the country in which the company is incorporated. The typical time for this is six months.
- **Stage 2: Filing the AP4**
  The host country files the AP with the ITU, beginning the period of time that the filing is on record with the ITU. A period of nine years is allowed for the system to "come into use"; if it is not up and running by then, the process must be started all over again. The AP4 contains a general description of the system, such as dates, orbital details, frequencies, power levels, and other items. The ITU publishes this information, and any owner of a prior system who thinks that the applicant's satellite might interfere with it has four months to notify the applicant and the ITU of potential problems. The typical time for this is six months.
- **Stage 3: Filing the AP3 (supplying the details)**
  The satellite operator, through the host country, no earlier than six months after the AP4, files a "Request for Coordination" or AP3 with the ITU, giving more details of the system, such as the transponder frequency plan, terminal equipment specifications, and the other systems and nations that it thinks will be affected by the new system. Usually at the same time one asks the ITU for a "notification" to enter the system in its Master Register. The date of the system's entry onto the Master Register establishes its priority against later-filed systems. The typical time for this is six months.
- **Stage 4: Coordination**
  The system operator and host country enter into a lengthy series of bilateral meetings with possibly affected systems and countries to determine the effects each will have on the other(s). This usually takes place in the country whose system has priority of filing date. The delegation consists of officials of the telecommunications administration of each country (sometimes several countries) and technical representatives of the system seeking coordination. It can happen that agreement is not reached on all issues. At the end of the process, the delegation compiles and signs a "summary record" that details all of the changes that

have been agreed upon for each system. (This also serves an opportunity for the applicant to legally modify his system application with the telecommunications administration without having to file it all over again.) It should be noted that the result of these negotiations may alter or even reduce the service offerings of the applicant's or the prior system. While no one likes this, negotiators of good will and pragmatism recognize that this is often necessary for the continued "comity of nations." The typical time for this is 12–36 months.

- **Stage 5: Notification**
  The system operator and host country revise the filing to take into account any modifications in the system dictated by the coordination process, and the system is refiled as a notification. The typical time for this is 3–6 months.
- **Stage 6: Review and entry into registry**
  The ITU staff reviews the revised filing for conformity to ITU rules and enters the applied-for frequencies and orbital slots into the ITU master register of frequencies. The typical time for this is 2–4 months.

Thus, there has been a balancing of technical possibility against political reality. After all of this, the satellite operator begins building the system and contracting for satellite construction, launch, insurance, and earthstation construction if necessary.

This chapter can only hint at the complex, multilayered, highly political, contentious, and continually changing nature of the legal and regulatory environment within which civilian satellite telecommunications is only a part. For more details and depth, there is a plethora of books, telecommunications attorneys, and consultants available.

Chapter 3

Satellite Telecommunications: Users, Applications, and Markets

Telecommunication means communication at a distance electrically or electronically. As such, the technology is only about a century and a half old, its birth marked by the invention of the telegraph by Samuel F. B. Morse. By encoding letters of the alphabet into pulses of electricity, Morse code could, at first, transmit a few words per minute over distances of a few miles. Today, we can transmit images across the Solar System.

(By the way, for some historical trivia, the term "telegraph" was first used not for an electrical system of communications, but for a mechanical system of shutters and semaphore arms on towers, invented by Frenchman Claude Chappe in 1791. The system was used in that country until 1852.)

Since Morse's time, both our demand for communication and our technology for providing communication have grown exponentially. There has never been a long-term excess of transmission capacity, for as capacity has grown and costs have continually come down, so too have new users found things that they want to communicate. People communicate to fill the channels allotted, just as they drive to fill the freeway lanes built, and, as Parkinson's famous law states, "work expands to fill the time allotted for its completion."

That is not to say that systems cannot be overbuilt, providing capacity ahead of market demand. In the mid-to-late 1980s, many satellites had excess capacity, but newer services, such as VSATs, came into being and eventually used up the available capacity, so that some years in the 1990s saw acute shortages and concomitant price rises in certain frequency bands and geographic regions. The first half of the first decade of the 21st century also saw an excess of satellite transponder capacity, made more acute by the huge excess of terrestrial optical fiber carrying capacity. But we can be rather confident that as the overcapacity leads to reductions in the cost of communication, new and expanded uses will reduce the oversupply, and, if history is any guide, produce a shortage.

We have seen 150 years of this growth, most especially rapid in the last few decades since the advent of communications satellites and of one of their major competitors, optical fibers. The development of high-speed computers not only parallels, but made possible, the telecommunications of today. Not only has the total amount of communications grown to a level Morse could never have imagined, the kinds of things people want to communicate has grown as well. First it was words, then in an effort to get more telegraph channels over one pair of wires, voice transmission over wires became possible with the invention of the telephone by Alexander Graham Bell, and simultaneously by Elisha Gray. (Another bit of historical trivia: Bell's patent application for the telephone was filed only a few hours before Gray filed his patent caveat, on 14 February 1876.) Then came radio, capable of sending signals through empty space by "wireless." At first, again, it was only words via radiotelegraphy, then voices and music. Later came facsimiles of pages and photographs; then, moving pictures, black and white at first, then in color. Today the telecommunication market is developing the ability to transmit and receive almost any kind of signal, to anyone, anywhere, anytime.

One of the big problems facing some of the older telecommunications companies is accommodating new demands with old technology. Originally telephone companies handled only voice calls. Now people want to send facsimile, music, pictures, and high-speed data over these old networks. The recent explosion in applications over the Internet is severely challenging the old telephony paradigm of circuit-switched technology with newer packet switching.

One of the things complicating the business of telecommunications is the broad range of criteria that different users may apply to evaluate a technology. Universal, one-technology-fits-all solutions are impossible to invent. While one user may think cost is the most important determinant for making a call, another may place reliability or security above all else. The communications needs of one business may require only intermittent broadcasts from headquarters to stores. Another company may need full-time two-way connectivity to every one of its offices. One user may need only a very few bits of information, infrequently, and another may require high-speed permanent connections.

These (often conflicting) demands apply to all of telecommunications, including satellites. No technology, method, protocol, or company is best for all applications. Specifically, there are times when satellites are a good solution, and others in which other technologies are better. Understanding the range of demands made by people who want to communicate will help make appropriate choices. The important thing is to start with the users' requirements, and work toward finding the technologies that will meet them—not the other way around, shoving a technology down a user's throat when there may be better ones available.

There are few generic solutions to telecommunications tasks.

3.1 Carrying capacity

There are huge differences in demands that the different types of signals place on telecommunications systems. Just using examples of signals originated and received by humans (as opposed to computer-to-computer data), textual information (such as e-mail) is least demanding; audio signals contain about 1000 times as much information as text and are therefore about 1000 times more demanding;

video, with 1000 times more information than an audio signal, is the most demanding of all. Thus, a television signal has about 1,000,000 times as much data as an e-mail, and obviously a telecommunications system designed for the former will not be able to transmit the latter.

Later we will quantify signal quantities in terms of bandwidth and bitrate, but for now, a couple of human-scale analogies will provide the sense of scale: If the capacity required to send one page of electronic mail is represented by the single sheet of paper that it could typically be printed on, then a telephone call is represented by a stack of paper about as thick as a big-city telephone directory, and a normal television broadcast would then be a stack of paper 318 feet high; high-definition television is at least four times higher (see Fig. 3.1). Alternatively, if the flow of electronic signals is represented by the flow of water and one could transmit an e-mail through a quarter-inch-diameter soda straw, a telephone call would require an 8-inch-diameter pipe, while a video signal would demand a 21-foot-diameter storm sewer conduit to carry it.

When you want to plan a telecommunications task, you would expect to have to prepare a budget for the project, adding up all of the items that contribute, subtracting all the items that detract, and seeing if what you have left over is sufficient to get the job done. This "bottom line" will vary from application to application. And this is just talking about money.

Similarly, there is a more technical budget that must be calculated, called the link budget. It is a summation of all of the technical parameters that enable communication, minus those items that inhibit communication, resulting in another kind of bottom line often called the "carrier-to-noise ratio," usually denoted by the mathematical notation C/N. Just as the monetary budget calculation tells you if the task

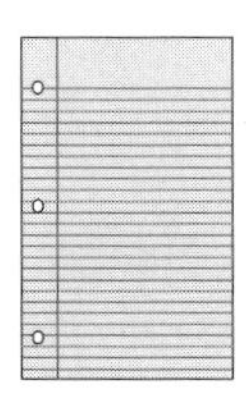

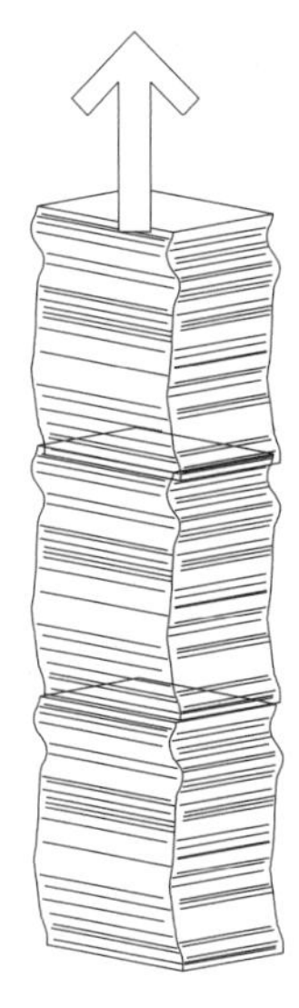

Figure 3.1 The kinds of signals we wish to send span a huge range of capacity. There is about a million-fold difference between the simplest communications, such as an e-mail or telemetry signal, and high-definition television.

is economically sound, the link budget "bottom line" calculation, which we will detail in Chapter 19, will tell you if the task you wish to perform is technically possible.

It is the job of the telecommunication engineer to configure systems to meet this wide range of demands for the various criteria that may be important to the variegated population of users. This is why there are so many possible ways of telecommunicating, of which satellites are only one way—and not always the best.

3.2 The place of satellites in telecommunications

Satellites are only one niche technology within the enormous field of telecommunications. They may or may not satisfy all of the important criteria demanded by a user for a particular telecommunications task. The enormous variety of such tasks requires that we spend some time looking at the kinds of criteria users may apply.

3.2.1 User criteria

If you look at any instance of a telecommunications task, you will find that it can be characterized by selecting one (sometimes more than one) description from under each heading of this list of some of the possible properties and criteria such as those shown in Fig. 3.2.

Cost may be the most important criterion, but perhaps not. And cost is not just the cost of equipment or service, but perhaps the cost of energy (electricity) to run the system, or the cost of time delays in accessing the system (think of the money lost due to the inability to make a stock transaction on time, or interest lost due to a delay in depositing funds).

For some types of systems, convenience is important, which means such things as location and mobility of equipment, and possible wait for capacity on a system. This has given rise to the mobile communications industry of cellphones and satellite phones, fly-away terminals, and wireless handheld information devices like the Visor and other PDAs.

For others, reliability is paramount, and a closely related criterion is system security. Military communication systems are a particularly good example of these.

Some systems are designed to be versatile, carrying a variety of kinds of signals of various bandwidths and speeds. Others are intended to do one thing optimally, but are not versatile. We will see examples of both kinds of satellite communications systems later. Many real-world situations have a variety of requirements. For instance, your company may need to communicate low-demand fax and telephone calls, as well as somewhat more demanding computer-to-computer data, as well as much more demanding video from corporate headquarters to field offices.

As you can see, there are many combinations of possible requirements. If we look at the three generic kinds of transmissions: audio, video, and ("pure") data in

- **Nature of Service**
  - Telephony
  - Program audio
  - Low-speed data
  - High-speed data
  - Multimedia
- **Bandwidth**
  - Narrowband
  - Medium bandwidth
  - Broadband
- Timeliness
  - Store-and-forward
  - Near realtime
  - Realtime
  - Dedicated circuit
  - Demand-assigned
  - Connection set-up time
  - Sensitivity to interruption
- Transmission Mode
  - Wires
  - Coaxial cable
  - Fiber cable
  - Waveguide
  - Radiated
- **Type of Signal**
  - Analog
  - Digital
  - Mixed
- ***Directionality***
  - *One-way (simplex)*
  - *Two-way half-duplex*
  - *Two-way full duplex*
  - *Multicast*
  - *Broadcast*
- Network geometry
  - Point-to-point
  - Point-to-multipoint (star)
  - Multipoint-to-multipoint
  - Ring
  - Tree
  - Bus
  - Circuit-switched
  - Packet-switched
- **Geographic coverage**
  - Local
  - Metropolitan
  - Regional
  - National
  - Global-regional
  - Global
- COST PARAMETERS
  - TOTAL COST MOST IMPORTANT
  - SENDER'S COST MOST IMPORTANT
  - RECEIVER'S COST MOST IMPORTANT
  - COST IS MINOR CONCERN

Figure 3.2 Different kinds of telecommunications signals and different users have widely varying criteria for their applications. This range of characteristics and demands makes it challenging to design telecommunications systems, and impossible to design one that is ideal for all applications.

use (realizing that there are exceptions and special cases), we see that they make different demands on telecommunications systems.

For instance, in terms of circuit continuity, most audio and video communications intended for use by an end user take place as continuous, noninterruptible sessions. They must use a system that provides for a continuous connection for the duration of the call or broadcast, or at least give the user the impression of such a connection. On the other hand, pure data, which is just a stream of numbers, can be started, paused, stopped, and continued at will as long as the data bits finally get to the destination in time, accurately, and in the correct order. In contrast, consider downloading music from a website for storage on a portable player. The compressed music, often in what is called MP3 format, is sent from the website to your computer using a packet-switched network (the Internet) at any datarate, and if the network gets congested, the data flow can start and stop until the music clip is finished. Continuous data is not required because it is going to be stored in your player. Then the whole music selection is played in a continuous way at the proper datarate when you want to listen to it. Another example is the distribution of motion pictures to cinemas. Some is now done digitally via satellite, taking many hours intermittently over several days to send one film, but avoiding the cost of film printing and shipping.

But using the criterion of circuit availability, most audio and video communications are done by humans, so if it takes a few seconds after you dial for the called telephone to ring, or for your television picture to appear a few seconds after you turn it on, this is usually acceptable. However, data communications usually take place between computers operating at millions to billions of operations a second, so quicker connections are needed. To a computer, a second is an eternity.

From a user's point of view as well as from that of the carrier, the nature of the communication is usually paramount. Because of the great disparity between the amount of information in a moving picture and in audio or textual communications, both users and providers often divide the telecommunications world into two kinds of traffic: video and everything else. That "everything else" is also sometimes called "voice and data" or something similar. Another rubric under which you will see switched traffic bundled is the term "telecomms traffic."

Of course, the boundaries are not rigid. There is digital compressed video for a teleconference, for instance, which is much the same in properties as a medium-speed datastream with the additional requirement that it needs to be uninterrupted. Highly multiplexed telephone calls have much the same traffic characteristics as high-speed data. A new rubric, "multimedia," has recently arisen as a catch-all term for a mix of types and speeds of traffic, often synchronized multipart signals, particularly when you don't know just exactly what the traffic is in detail.

If you think of any particular telecommunication application, you will see that you can take one choice (sometimes more) from each of the foregoing characteristics. This produces the wide variety of telecommunications systems that have been in use, are now used, and those planned for future requirements. We will see in later chapters that satellite systems, in particular, which have been designed to do one kind of task at maximal efficiency, and can do nothing else; and we will see other systems that have been designed as "jack of all trades, master of none" to be as versatile as possible. What you design and use depends on your communication needs. Most real-world tasks are complicated by the fact that most networks have a variety of demands, mixing low- and high-speed requirements, real-time and non–real-time demands, etc.

3.3 Global markets for various applications

3.3.1 Television

No one really knows just how many television sets there are in the world, but guesstimates place the number at one and a half billion. That is about one television set per four people, averaged globally, but of course many homes have several sets, while undeveloped countries have very few.

Nonetheless, there is no doubt about the commercial market for television programming. In fact, the statistics show that most people would rather do without telephone than without television. This is why satellite-delivered television broadcast services have proven popular everywhere that they can be supplied at a reasonable price. It is a huge consumer of the total transponder capacity on orbit. Not only are transponders used for the broadcast to the end users, but the television providers use more space segment for backhaul and contribution feeds. In calendar year 2000, the world's teleports got 59% of their revenues from relaying video signals.

Television is probably the most demanding kind of signal, both because of its large bandwidth and because of the wide variety of television systems used to send video to users. It is also the most confusing, because of the wide range of video standards. Since "conventional" broadcast television (the older but still common analog systems) arose independently in several parts of the world, the systems became largely incompatible.

Like motion pictures, it works by projecting many images per second onto a screen to give the illusion of motion. There are two major kinds of analog television signals, based on two technical parameters: the number of lines that make up one screen of images, and the number of images that are projected each second. Roughly speaking, North America, Japan, and a few other nations use a system called NTSC (for National Television Standards Committee, its authors) that breaks each screen image into 525 horizontal scan lines and refreshes the screen 30 times a second. In video shorthand it is referred to as a 525/30 system. (Actually the screen is made up of two interlaced scans, each taking 1/60 of a second, so it is also sometimes dubbed a 525/60 system.) In NTSC, the video part of the signal is amplitude-modulated (AM) and the soundtrack is a frequency-modulated (FM) signal alongside the much wider video signal. Because of the susceptibility of AM signals to noise and distortion, critics quip that NTSC should stand for "never twice the same color."

The rest of the world uses a video standard which places 625 lines on the screen but refreshes the screen 25 times a second; called 625/25, or 625/50. But of this standard there are two major variants. One, used mainly in France and Eastern Europe, is called SECAM, sequential color with memory; the other, used by the rest of Europe, is called PAL, phase alternation by line. (North American engineers sometimes retort that SECAM really means "system essentially contrary to the American method.")

You might think that having three major analog video standards was bad enough, as it means that when broadcasters want to exchange programming, each must have the ability to deal with all three standards and convert from one to another. This means more equipment and expense. But the situation is actually worse, as over the years since the systems were devised some nations have slightly altered the parameters. Most commonly altered have been the bandwidth and placement of the soundtrack carrier, so there are really more than a dozen video "standards" that programmers must deal with.

In addition to these three conventional systems, there is another set of video standards called MAC: multiplexed analog component. There are several variants of this; the ones in use are B-MAC, C-MAC, D-MAC, and D2-MAC, with a newer high-resolution HD-MAC. These systems are used for some DBS services, terrestrial cable systems, and some video teleconferences. They work by multiplexing the analog parts of a television picture together and adding a stream of digital data that can carry audio tracks (perhaps several) and ancillary information. MAC systems are incompatible with the conventional systems and need their own type of equipment.

In addition to all of this, the trend now is to digital television. For satellite services, two major standards have emerged. One, called DSS, is used only by the DirecTV systems; the other is called DVB-S, digital video broadcast by satellite, and is used by several other DBS systems.

Figure 3.3 shows a typical home dish for direct reception of television.

3.3.2 Audio broadcast

Direct-to-consumer audio services, analogous to broadcast radio, is a newer offering than direct-to-home video. The market is still unproven but growing—several SDARS systems are operating in various parts of the world. In the U.S., Sirius and XM Radio are beginning to build customer bases, mostly with in-car receivers. More globally, the use of the Worldspace satellites AfriStar, AsiaStar, and the anticipated AmeriStar, has been slowed by the (relatively) high cost of the boom-box-sized mobile radios. Some other individual nations also have radio systems with various amounts of user acceptance.

Figure 3.3 The most popular satellite communications application is direct-to-home video entertainment, received in small dishes on homes and apartment buildings, and now even on vehicles. (Photograph courtesy of SES-Astra.)

3.3.3 Telephony

There are an estimated one billion telephone sets in the world, again most highly concentrated in developed countries. Global telephone traffic continues to increase each year, and the number of mobile users of terrestrial cellular networks continues to increase quickly. Telephone users make over 100 billion minutes of calls a year, increasing at a rate of 8–16% per year. In developed countries, telephone charges are mostly declining. Some terrestrial telephone traffic is now being routed over the Internet using what is called *voice-over IP* (VoIP) techniques, and telephone networks are increasingly carrying digital traffic (faxes, low-speed data such as Internet connections, highly-compressed video). VoIP carries about 10% of all international voice traffic, according to researchers' estimates. Thus, telephone usage statistics do not always measure actual telephone conversations.

One statistic used is *teledensity*, the number of telephones per 100 people. Telephone usage statistics show the huge disparity in telephone usage between the developed and undeveloped countries. The ITU estimates that perhaps as many as two-thirds of the world's population has never made a telephone call and lives more than a hour from the nearest telephone. The following Table 3.1 shows examples of this disparity. Note that some regions and nations have more mobile lines than fixed lines, usually because fixed lines are more regulated or because it is easier and faster to obtain a mobile line than a fixed line.

Table 3.1 Population and teledensity for selected regions and countries. (Data from the ITU for the year 2000 or 2001.)

Region	Population (millions)	Fixed Lines per 100 People	Mobile Lines per 100 People
World	6100	17.21	15.57
Africa	799	2.62	2.95
Americas	842	35.21	26.35
Arab countries	290	7.52	5.57
Asia-Pacific	3525	10.96	9.52
Europe-CIS	800	40.62	43.77
USA	286	66.45	44.42
Haiti	9	0.97	1.11
Eritrea	4	0.84	0

Between and within developed countries, the vast majority of telephone traffic has been captured by the fiber optics industry. But communications to and from lesser-developed countries often relies, sometimes exclusively, on satellite links. A few countries have no connections to the global PSTN other than through their gateway earthstations. Thus, the demand for telephony over satellite continues to grow, but at a slower pace than before. The total number of satellite-carried telephone calls, using such systems as Inmarsat, Iridium, Globalstar, and Thuraya is small compared to the number or number of minutes of terrestrially carried calls.

3.3.4 Data

Data, also called digital, is the fastest growing segment of telecommunications traffic. In the first half of this decade, data traffic surpassed voice traffic, and telephony is now less than 10% of total data traffic. Total global data transmission is estimated at over one and a half trillion minutes a year, and is growing.

But "data" is not in itself a service: it is a way of sending the information contained in an application. To a telecommunications transmission technology, a signal may be a stream of bits, but to a user it is a telephone call, a webpage, a music program, or a television program. Thus, data may be carried over terrestrial or satellite-based telephone networks, over public or private terrestrial data networks, or over satellites.

Demands for data transmission are increasing at double-digit rates.

3.4 Satellite services

Satellite communications capacity exists to provide users with services useful to them. End users, however, do not care what technology is used to provide them with those services, as long as quality and financial desiderata are met. To the end user, telecommunications should be a "black box" that works without worrying about the details.

Markets thus depend on users, and thus on the terminal hardware that the end users have available to them. Among the approximately 6 billion people on Earth, only a small proportion use most of the total telecommunications capacity. User hardware is unevenly distributed around the world, with North America and western Europe the heaviest users, and Africa in general the lightest.

The rest of this chapter will give some examples of real-world telecommunications applications and services provided by satellites. New applications are continually arising, some old ones waning.

3.4.1 Broadcast video applications

Satellites have become very important in all aspects of broadcast video. Beginning in the 1980s, when many television networks abandoned terrestrial coaxial cables for satellite distribution of programming, and as international exchange of video grew tremendously, satellites were well-placed to grab a large share of the market, geographically and financially. The wide coverage and wideband nature of the signal were perfect for satellites.

Distribution is the sending of video signals to end users or intermediary distributors. Satellite distribution of broadcast-quality video includes television sent directly to customers' homes and offices, distribution from television network programming to network affiliate stations, and the distribution of all of the various cable channels to the cable headends for final distribution by terrestrial networks.

Contribution is the process television networks use to receive all of the many video feeds that make up a television show, particularly, a news broadcast. News and events do not always happen near a "point-of-presence" of a terrestrial network, so transportable uplinks and satellite newsgathering trucks have revolutionized the business. Demands for satellite transponder time peak every afternoon as every news program requires video feeds to use on that day's programs, so contributions to a particular news program may be live from reporters all over the world.

Backhaul is the television industry jargon for intranetwork communications. This may be the exchange of audio or video footage for incorporation in later programs. It may also involve taking the video feed from the originating site of a program and sending it to a network's central studio where commercials and other items are inserted before the finished program is distributed to end viewers.

3.4.2 Audio applications

DARS, Digital Audio Radio Service, also called Satellite Digital Audio Radio Service, SDARS, sends CD-quality music and talk shows, usually by subscription, directly to end users primarily in vehicles, but also in homes and offices (Fig. 3.4). The market is still developing for this application, particularly in the United States where there is an abundance of terrestrially available radio stations. In parts of the world where consumer programming is missing or limited, such as Africa, this kind of service may be the only one available. In particular, the company WorldSpace is trying to provide audio information and entertainment services to such underserved regions through satellites called AfriStar, AsiaStar, and CaribStar. The antennas to receive the satellites are small domes or disks or rods a few inches in size.

Station Relay is a service often used by radio stations or programming providers to send their programs to distant terrestrial distribution points. For example, in a very large country with limited or expensive terrestrial networks, a radio station in one city may uplink its programming to a satellite for downlinking in distant small towns and cities. Or, a major national network such as National Public Radio may send its feeds to affiliate stations. Typical terminals are VSAT-sized, one to three meters in diameter for the dish.

Figure 3.4 A receiver for another satellite-delivered consumer application, satellite digital audio radio, providing about 100 channels of CD-quality radio. Intended primarily for vehicles, some units can be used as part of indoor sound systems as well. (Photograph courtesy of XM Satellite Radio.)

Background music, the musical "wallpaper" used in stores, malls and elevators, is largely distributed by satellite, often offering many channels attuned to the variety of environments that receive it. These services were originally distributed by terrestrial means, but satellites allow a much more economical distribution because of the broadcast nature of a satellite signal. Small dishes receive the satellite feeds.

Retail audio or *audiovisual information* is the distribution of company-specific information to its affiliate stores. This can include updating each store's inventory and price files, providing in-store background music interspersed with company-specific commercials, and even displaying audiovisual commercials through kiosks around a store or at the checkout lines. Again, a small dish with appropriate receiver is all that is needed.

3.4.3 Telephony applications

Telecommunications via satellite began by carrying telephone calls across oceans because the transoceanic coaxial cables were of low capacity and mediocre quality. From the 1960s to the 1980s, satellites had the lion's share of the transoceanic telephony market. This slowly changed with the advent of fiber optic cables, at least on the heavily used routes. Now satellites only carry a small percentage of such calls except on routes where no fibers have yet been laid. Some underdeveloped countries, particularly in Africa, still have no international telephony connections other than satellites.

Trunking is the carriage of many telephone calls combined (multiplexed) as a single broadband signal. Satellites are used for such applications primarily in developing nations and regions. Terminals range from a few meters in diameter to several tens of meters, depending on the amount of traffic carried and the power of the satellites involved.

Mobile telephony aims to provide the ability to an individual user to make a telephone call from anywhere in the world without direct access to a terrestrial telephone network. Such services began with Inmarsat providing a single direct-dialed call to ships, using expensive terminals (around a quarter-million dollars per ship) with high per-minute charges. The call from or to a ship connected through a satellite to a large earthstation, which was then connected to the telephone network. Technology improvements have greatly decreased the cost of both the terminals (now in the low thousands of dollars) and the per-minute tariffs. Improvements have also allowed such terminals to carry not just voice calls, but facsimiles, low-speed data, and even very highly compressed video. Terminals range in size from laptop-computer-sized for some Inmarsat services to large handheld telephone units with sausage-sized antennas for the mobile telephone systems like Iridium and GlobalStar. (See Figs. 3.5 and 3.6.)

Figure 3.5 Mobile satellite telephony, also capable of providing low-speed data services. The two-way satellite "dish" is the flat object seen on the ground just beyond the table. (Photograph courtesy of Thrane & Thrane.)

3.4.4 Motion picture distribution to cinemas

One of the newer and still nascent applications, starting in the early "naughts" of this century, is that of sending major motion picture releases directly to cinemas as digital datastreams. More and more pictures are being shot digitally, and it is both expensive and degrading in quality to make film prints to be shipped to each cinema showing a motion picture. A print of a typical 35-mm film costs thousands of dollars to make, and hundreds to ship. When a major release may take place initially at more than 5000 cinemas, the costs rapidly add up. Of course, after the big opening, many fewer prints are needed, so the prints are then discarded.

The alternative that some studios are examining is the distribution of the digital copy of a motion picture to each movie house via satellite downlink into a small dish. Moreover, this can be done at off-hours when transponder cost is low, saving money. Some of the first trials of this technology required days to send entire picture to the cinemas for storage on their hard disks, but as digital data can be easily started and stopped, this caused no problem. The major problem is that cinemas wishing to use this method must install downlink earthstations and data storage equipment, not to mention that they must install high-resolution digital projectors

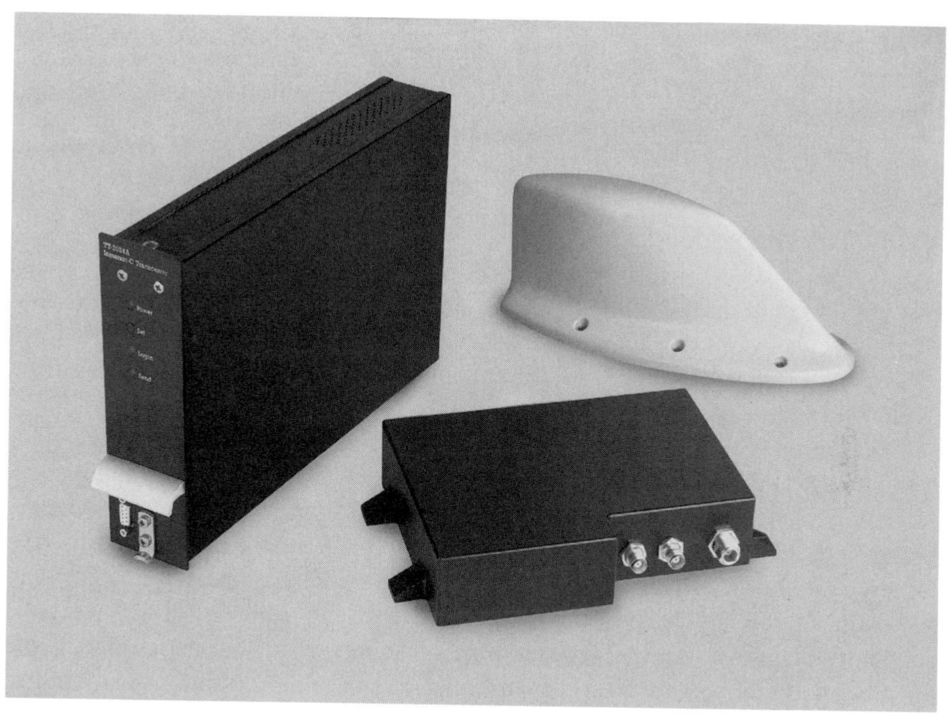

Figure 3.6 Satellite-delivered telephony is entering the cockpits of even small aircraft. The white fin-shaped object in the photograph is the external antenna to be mounted to the airplane's upper fuselage. The black boxes are the in-plane equipment. (Photograph courtesy of Thrane & Thrane.)

at even higher expense. More cinemas may be expected to go to this technology as those costs decline. Eventually the term "film" will be a misnomer, as motion pictures will be shot, transmitted, and projected digitally, moving nothing larger than electrons.

3.4.5 In-flight entertainment and information

Airplane travel is often tedious and boring, and airlines are increasingly investigating ways to distinguish themselves from their competitors. In-flight films and other entertainment and information are possible solutions. Also, business travelers want continual connectivity with their offices, clients, and information sources, and flight crews want communications with their company dispatchers.

Thus, some nascent services offer DBS-like programming to each seat of an airplane, allowing the passenger to choose an audio or video channel. Other services allow business passengers to plug their portable computers into a network connected to the Internet. In both of these cases, the wide footprint of satellite signals makes them the best choice for delivering the signal to the airplane.

The airborne terminals needed range from simple fixed-strip antennas hidden under aerodynamically styled radio-transparent covers on the upper fuselage of the aircraft to small directional dishes which are gyro-controlled and gimbaled under a dome on the fuselage, providing a higher-capacity signal.

3.4.6 Telepresence

Modern telecommunications, whether terrestrial or satellite, allows people to extend their presence to remote locations for a variety of purposes. This is sometimes called *telepresence*, or some similar term, and other, more specific terms are used to denote specific applications, such as working or teaching.

Education is a growing industry, and so-called "distance learning" or tele-education is growing as a way of cutting down on travel expenses in time and money. Many companies inform, train, and encourage their employees via company programs. The conferences may be full two-way video (usually compressed to save money and bandwidth), one-way broadcast (compressed) video with two-way audio links to all sites, or just audio, depending on the need. If the connection is simply between two sites that are near fiber optic networks, fiber optics may be the most economical means. However, if connectivity is a problem, or if there are many sites, satellites are an excellent choice for the service.

Teleconferencing is a similar application to telelearning in that it allows people to meet electronically, exchange images and documents, and avoid the disadvantages of physical travel.

Telehealth or telemedicine allow interaction among distant healthcare providers, facilities, and patients. This includes not only conferencing, but remote monitoring of a patient's status or even remote manipulation of equipment such as surgical operating tools.

Employers are increasingly finding that employees need not always be physically present in their offices to get work done. Many workers now work from home (at least part of the time) in an application called telework. This can benefit the company in the reduction of needed office space, benefit society by the reduction of rush-hour congestion, and benefit the employee with a more flexible lifestyle. Some companies farm out routine work on such tasks as reconciling used tickets with orders or staffing help desks to other nations around the globe. For example, in the specific case of help desk staffing, a company located in Europe can provide twenty-four hour service by employing call centers in countries on the other side of the globe. A related application, telecommunity, can unite distant participants in common tasks, such as clubs, discussion groups, or game-playing. Satellite industry seer Dr. Joseph N. Pelton predicts that the number of people in the United States alone taking advantage of the possibilities of telecommunity will rise from 12 million to more than 50 million in the next decade.

A similar application is the "video press release" in which a company holds an informative session for such outsiders as reporters or stock analysts. Datastreams can go along with the video and audio to send graphical illustrations, facsimiles, or other information. In some cases, two-way data allows input from the audience back to the presenter, as in opinion polling or testing. Most such services use VSAT-sized dishes.

3.4.7 Data distribution and exchange

Many industries have a need to distribute data. Many others want to interconnect their local area networks into wide-area networks. These data may be measurements, knowledge of events, reference information, or any other information. Many such applications are one-way and are of low demand in terms of bandwidth or bitrate, yet are essential to particular users. Satellites can be competitive when around 50 sites or more need to be interconnected, compared to running terrestrial T-1 rate circuits to all of them.

Stock market ticker news is of interest to investors and brokerages around the world. The datarate is low, but the information is vital. To send this information to each potential user through wires or fibers would be expensive, so many brokerages receive this data via satellite. In one particular application, the ticker data is sent to a satellite transponder as a low-datarate signal appended as a subcarrier (see Chapter 7) to a video broadcast; it is received in various cities and towns by a local FM radio station, which ignores the video, and remodulates the ticker data onto its own FM subcarrier signal, to be picked up by offices in the region. Receivers at the FM station will be VSAT-sized.

Dispatch is a service often used by utilities and repair technicians to inform service and maintenance crews of problems and assign them to jobs. It may also involve downloading to portable computers the information needed to facilitate the repair. Some such dispatch services are implemented terrestrially, often via a proprietary or shared network of radio towers. A less flexible system may involve field crews making telephone calls. Satellites have advantages for very wide area systems because of their large coverage areas.

Delivery tracking and confirmation, such as used by the various expedited-delivery services, allows system dispatchers, and even the senders and recipients, the ability to know in almost real time the whereabouts of a package. So far, such systems are implemented using terrestrial technologies.

The terminals for both kinds of services may be handheld "walkie-talkie" sized units.

3.4.8 Point-of-sale applications

The temporary glut of satellite capacity, and the resulting drop in price of the cost of space segment, led, among other things, to the rise of the satellite network connecting the stores of national chains. The first such major system was for Kmart, a large US-based merchandizing chain.

Figure 3.7 A VSAT used to connect point-of-sale data from a retail store to a corporate hub site for inventory control, credit card verification, and other intra-company data.

These systems are often referred to as VSATs, for very small aperture terminal, or microterminals (Fig. 3.7). They are basically point-of-sale terminals, possibly interconnected to a small computer, which can almost instantly communicate with, for example, a company headquarters. If a person buys an item at a store using a credit card, the computerized cash register updates the inventory levels in the store, sends the sale information to company headquarters, asks for authorization of the credit card, and receives authorization back at the store to complete the purchase. In the lodging industry, a motel may send a reservation request to another lodging. In the transportation industry, a gasoline pump will get authorization for a sale.

The terminals that do this are typically a satellite transceiver attached to a 1–2-m diameter dish located on the roof or beside the store. Datarates are typically in the few hundreds to few thousands of bits per second.

3.4.9 Internet and multimedia via satellite

To begin, we need to emphasize that "Internet" is not an application; it is a data-delivery service. Calling up a web page, sending e-mail, or downloading MP3 files are applications. However, we will use the rubric "Internet" to encompass all of these.

"Multimedia" is an indefinite, generic term encompassing the sending of a combination of video, audio, and data, usually directly to consumers. One important component may be a *return channel* from the consumer to the originator, for such purposes as ordering, polling, audience response, etc. Thus, some or all of the components of multimedia services may be carried using Internet protocols, over satellites or terrestrial links, or both.

Typically, as may you know from personal experience surfing the World Wide Web or using interactive television, the amount of information you send is rather small, whereas the amount you receive is large. To view a website, you need only send a few hundred bits of information: the address of the website (URL, or uniform resource locator); you get back perhaps megabits of text, sounds, and images. This asymmetry allows different kinds of systems to provide the service.

Hybrid Internet access using satellites is provided by splitting the incoming and outgoing data into separate datastreams carried by different technologies. Such a hybrid system requires a connection to the telephone system, and requests are routed through a local Internet service provider (ISP) over these landlines. The requested data (such as a webpage) are sent from the website to the ISP, which uplinks it to a satellite with your address on it. The satellite then broadcasts the data back to Earth to be picked up by your small dish and receive-only terminal. Several companies offer such services.

Full two-way Internet access, also called "return channel by satellite" RCS, uses satellite links in both directions. Your dish and terminal must be capable of sending and receiving, making the system more expensive. Such services are available for business and for residential access.

One of the major standards issues facing both satellite-delivered and cable-delivered multimedia services in the middle of this decade is which system to use to provide economical, reliable two-way communication between a service provider and the individual consumer. Two such standards have risen to the top as the major competitors. One or both will be built into the set-top boxes that connect a televison set to a multimedia service.

DVB-RCS, or Digital Video Broadcast-Return Channel by Satellite, is one such standard. The other is DOCSIS, the Digital Over Cable System Interface Specification, and its European version, EuroDOCSIS. The goal is to provide secure, reliable, and economical services to the user. Neither protocol yet has a critical mass of users, and as is common in the progress of hardware, there is the "chicken and egg" situation that the hardware does not become inexpensive unless adopted and manufactured in huge quantities, and it will not become popular enough to require manufacturing in large quantities until it becomes inexpensive.

It may well turn out that some providers will adopt one system, others providing another, thus again thwarting hopes of a global standard.

3.4.10 Remote monitoring and control

Many companies have valuable properties located in remote places. Examples include gas and oil pipelines, electrical substations, and meteorological monitoring instruments. In most such cases, the data demands are small, but the need is crucial for an industry.

Remote telemetry is the measurement of something from a distance. The trick is to get that information where it is needed. For example, the weather bureaus have mountaintop instruments and oceanic buoys periodically reporting on global conditions for better forecasting; natural resource companies get reports of flows through pipeline; environmental monitors may have instruments measuring snow depth in remote mountains so they can predict runoff and future water supplies. In many such cases, the dataflow is only one-way and very small. The terminals may be small, enclosed heliacal antennas with low-power transmitters.

SCADA is a two-way version, standing for supervisory control and data acquisition. An example might be a remote rural electrical substation, which can be monitored remotely, and commands sent to it to control switches, circuit breakers, etc. Pipeline monitoring may allow control as well. Terminals for these kinds of services are two-way, and are similar to VSATs.

3.4.11 Navigation, surveying, and fleet management

Lots of people want to know where they are, and some people want to know where other people or things are. There are terrestrially based navigation services, such as LORAN, Omega, and Decca, but to report a position back to some other location, another kind of link is required. Within cities or regions with terrestrial networks, this can be one way of doing the job. But for remote users, satellites provide a wider coverage.

There are a couple of companies providing such position determination and reporting services. Examples are OmniTRACS (Fig. 3.8) in the United States and the similar EutelTracs in Europe. These services can provide fleet dispatchers with low-speed data and messaging, and a positional accuracy of about 100 meters for their vehicles by using a pair of satellites, one for messaging and one for ranging. These services operate from a separate RDSS communications packages aboard Ku-band FSS satellites. For example, EutelTracs uses the Eutelsat II F4 satellite for messages to and from vehicles, and uses the Eutelsat W4 satellite for ranging. More than 20,000 vehicles in Europe use EutelTracs for fleet management.

There are two government-run global navigation systems. These provide position only to a receiver, which must somehow report it to somewhere else.

Whereas a decade ago, GPS and GLONASS were not considered to be a part of the commercial satellite industry, today they often are because most mobile users would like to know where they are. Civilian GPS applications already generate billions of dollars in sales of receivers and of specialized, value-added services. Within a few years, almost everything that moves will have GPS and/or GLONASS receivers aboard. This includes cars, trucks, golfcarts, bicycles, cellular telephones, boats, aircraft, and even hikers' backpacks. Within 10 years, unless you are spelunking or scuba diving, there will be almost no excuse not to know your position to amazing accuracy. Receivers for these services may be simple "patch" antennas or small helical antennas a few inches in size. Those transmitting information back to say, a

Figure 3.8 A mobile terminal used in the OmniTRACS fleet-tracking network. These cab keyboard units allow frequent e-mail communication between fleet dispatchers and over-the-road vehicles, as well as reporting of vehicle locations. (Photograph courtesy of OmniTRACS.)

fleet dispatcher, will be slightly larger, but still small enough to be hidden under a small dome or flat plastic plate on the roof of a truck or car.

The *Global Positioning System*, or *GPS*, is a system of two dozen Navstar satellites in 12-hour medium-altitude orbits in several planes. The satellites broadcast extremely precise timing signals. A user can receive signals from several satellites, determine distance from each satellite, and thereby determine position very accurately. Three, four, or more satellites need to be received simultaneously by a user for good positional accuracy. The GPS system operates on two frequencies in the L-band at 1.57542 GHz and 1.22760 GHz. See Fig. 3.9.

It should be noted that, unlike commercial RDSS systems that report positions to a hub, GPS user terminals are receive-only. For security, the military inventors did not want the users to be broadcasting signals giving away their positions. Thus, a

Figure 3.9 A receiver for the Global Positioning System (GPS) capable of finding the user's position to within about 10 meters anywhere in the world. This particular model, among others, also holds a database of major U.S. cities and highways.

civilian user who determines her position using GPS must find some other system if she wishes to report the position back to some central location. As an interesting aside, the precise timing signals from the Navstar satellites are also used worldwide to synchronize digital data flow over the Internet.

The military services, being concerned that would-be enemies could use the system against them, have devised a method called "selective availability" that slightly degrades the timing signals for civilian users, which consequently degrades the accuracy of position determination, and can turn off civilian availability of the system in times of emergency. This is to keep potential adversaries from using the system.

The original civilian accuracy, Standard Positioning Service (SPS), is on the order of a hundred meters horizontally and 156 meters vertically, with speed accuracy to within a few meters a second. Later, the U.S. government allowed an increase of the accuracy, so now ordinary users can fix their position with an uncertainty of around only 10 meters. This became possible because the U.S. military developed a technique of "selective deniability," which can degrade civilian accuracy in small regions without reducing the accuracy outside of combat zones.

GPS users have found another way to improve the accuracy of the degraded signals by a system called "differential GPS." It works this way: in many regions of the earth, fixed receiving stations of known location pick up the degraded satellite signal, and by knowing their own locations, can calculate how fuzzy the civilian timing signal is. These stations then broadcast to nearby GPS users a differential signal telling them how much to correct the position each user calculates for itself. Differential GPS can restore the accuracy of position and velocity determination to military accuracy, or even better. If a user can see many of the Navstar satellites simultaneously, and integrate these signals with differential corrections over a long time, the user can determine its position to an accuracy of millimeters. In fact, such systems are now being used to monitor the tiny motions of slippage along the San Andreas Fault in California.

Another accuracy improvement can be achieved by using the Wide Area Augmentation Service, WAAS. This is done by receiving auxiliary signals that are sent through commercial communications satellites, received by GPS receivers, and used to correct inaccuracies in the civilian position determination.

GLONASS (Global'naya Navigatsionnaya Sputnikovaya Sistema), operated by Russia, is a similar system for satellite-based navigation, with comparable accuracy. It operates in the L-band around 1.602 GHz, with each of the operational satellites at a slightly different frequency. There are 24 operational satellites, 8 in each of 3 circular 19,100 km high orbits inclined 64.8° with the orbits arranged 120° apart, with a period of 11 hours, 15 minutes. The spacing of the satellites is planned so that at least 5 satellites are simultaneously visible to user on the earth.

Like the GPS system, each satellite carries an extremely precise cesium atomic clock, synchronized with an even more accurate clock on the ground. Also like GPS, the GLONASS satellites provide two signals of different precision. One for civilian use, called "standard precision" (SP), allows a user to determine horizontal position within 57–70 meters. The "high precision" (HP) signal, for Russian military users, presumably allows a higher accuracy. And, again like GPS, differential corrections can allow even better accuracy for all users.

Future navigation systems are planned because some other countries are concerned by the fact that the GPS is under the control not only of the United States, but specifically of its Department of Defense. They are concerned that U.S. interests could lead to them being denied navigation services. Thus, several nations are planning totally civilian, international satellite-based navigation systems similar to GPS.

Galileo is a satellite-based navigation system now under construction by the European Space Agency. The initial launch is planned for 2005, and by 2008 the full 30-satellite system should be operational, allowing users positional accuracy of 1–2 meters.

The varied demands from billions of users worldwide is the reason why generic solutions to telecommunications problems seldom exist. One cannot answer a generic question, such as "Which is better, satellites or optical fibers?" or "Which should I use, TDM or FDM?" until you know the details of the users' requirements.

Always design a system by starting with the users' requirements and working toward specifying the technology, not the other way around.

Having now introduced the history and context of satellite telecommunications, it is time of begin explaining the technology underlying the industry.

Part 2

Technical Background

Chapter 4

Basic Definitions and Measurements

To begin, we must explain some basic terminology, most of which applies to all of telecommunications, not just the satellite side of it. We will also introduce the oft-used but often not understood measurement system of decibels.

4.1 Communications and networks

Telecommunications engineers often categorize the equipment into three broad categories: terminal hardware, switching hardware, and transmission hardware.

Terminal hardware is the equipment at each end of a telecommunications task. Depending on the task, it may be a telephone, a fax machine, an earthstation, a personal or mainframe computer, a television camera, a GPS navigation receiver, or your television set. Terminal equipment terminates the ends of the channels. In many systems, there must be some way of identifying each terminal; each must have an address. In telephone networks, these are telephone numbers; on the Internet, these are what are termed IP addresses.

Switching equipment controls the flow of the signals, and may not be required or present in all telecommunications tasks. Communications tasks which are *connection-oriented* establish, maintain, and disconnect fixed paths between users, such as in a telephone call. Switching is very important in such a system as the *public switched telephone network*, or *PSTN*. Such services are called *circuit-switched* because each telephone call is assigned a particular circuit for the duration of the call. On the other hand, *connectionless systems* route the signals over a mesh network of connections, often different paths for different parts of the same signal, such as for Internet traffic. Internet traffic is carried in bunches of bits called *packets*, and thus the connection is termed *packet-switched.* The devices that switch the packets are called *routers*. In contrast, a television broadcast usually involves no switching, since it is simply sent out in a wide range of directions for all to receive.

Transmission equipment carries the signals from one point to another, and typical real-world tasks use several sequential transmission methods to get the signal from the origin to the destination. Transmission equipment can be subdivided into two major types, those that conduct the signals through wires or fibers, and those that radiate the signal through air or space using radio or light beams.

Conducted transmission technologies are relatively more secure. However, cost is determined by distance, and they are subject to what is called the "last-mile" problem: since there must be a conductor the entire length of the channel without a single break, one sometimes finds a situation in which a signal has been carried across an ocean or continent, but then cannot get across river, a street, or someone's right-of-way. (Even in metric-system speak, no one calls it the "last kilometer problem!") For conducted signals, there must be a conductor—fiber, wire, or coaxial cable—to each terminal. For all but the simplest systems, there must be some sort of switching and routing control. Expansion of such a system is expensive because it requires running more conductors, with the attendant problem of getting rights-of-way. Breaks in the conductors can bring down all or part of a system (fiber-optic system operators live in continual fear of backhoe operators!). Conducted signals are also more physically secure from unauthorized interception. Figure 4.1 shows a typical optical fiber being laid in the ocean for communication.

Information-carrying waves conducted within a wire, cable, optical fiber, or waveguide are usually very contained, and have little chance to interfere with other signals. As long as adjacent signals are well shielded from one another, such "natural

Figure 4.1 Optical fiber cable is the foremost competitor to satellite communication, and is dominant on heavily-trunked routes, both over land and underseas between the continents. Here an optical repeater spliced into a transoceanic cable is lowered into the water from the cable-laying ship. (Photograph courtesy of Alcatel.)

resources" as spectrum can be used over and over again, so there is less competition for these resources.

Radiated transmission technologies, on the other hand, are less secure since they are usually broadcast widely. Their advantage is that, at least within the service range, the cost of sending a signal is not distance-dependent. The signals can often easily leap over topographical and political boundaries. There is not a fixed connection to each terminal, so expansion is easier. With a broadcast service, the number of receivers is irrelevant to the transmitter. Radiated waves, such as from terrestrial broadcast towers (Fig. 4.2), microwave link (Fig. 4.3), and satellites, however, spread far and wide, and frequently do cause interference among themselves. It is often difficult or impossible to shield one signal from another, meaning that spectrum often cannot be reused, and that one signal may significantly interfere with another. For this reason, there are more stringent technical regulatory constraints on radiated technologies than on conducted technologies. With few exceptions, transmitters, such as uplinking earthstations, must be licensed by telecommunication regulatory authorities. In many cases, receive-only terminals need not be licensed. Sometimes major receive-only earthstations may wish to become licensed so the authorities and other potential users will know where they are.

Communication satellites obviously fall into the radiated transmission category. Most satellites are pure transmission equipment, with little or no ability to control the flow or routing of the signals coming through them. This is why such satellites are sometimes called *bent-pipe satellites*, "mirrors in the sky," or "microwave towers

Figure 4.2 Terrestrial broadcast is a common and economical method of telecommunication, especially for broadcast services. On this one tower is a mix of antennas for several types of telecommunications services: cellular telephony, point-to-point microwave relay, and radio and television broadcast.

Figure 4.3 Microwave dishes such as these are common for terrestrial relay of data, voice, and video services, but are slowly being replaced by fiber optics for heavily-trunked routes. They are still valuable because they are easy to set up. In some frequency bands, they are also a major source of interference for nearby satellite dishes.

in the sky." Most satellites only receive the uplink, change the signal to the downlink frequency, amplify the signal strength, and send it to the ground. The satellite itself does not "know" if the signal is digital or analog, if it is a fax transmission or HDTV, and except for having some directionally beamed antennas, the satellite does not know where the uplink is coming from nor where the downlink is going to. Such satellites—the majority of them—are nothing but repeaters broadcasting their downlinks back to Earth.

A few satellites, those intended for what we might call telecomms traffic—that is, switched signals such as telephone calls and data exchanges, or for Internet traffic—do have some signal switching and routing capabilities, to varying degrees. These types of satellites, typified in geostationary orbit by the Intelsat satellites and in low orbits by the Iridium satellites, are said to have *on-board processing, OBP,* or *on-board switching, OBS.* We will explore the functions for bent-pipe and OBP satellites in Chapter 13.

4.2 Some definitions

Various characteristics of telecommunications connections go by specialized names and terminology. To speak with technical people, you need to know some of these.

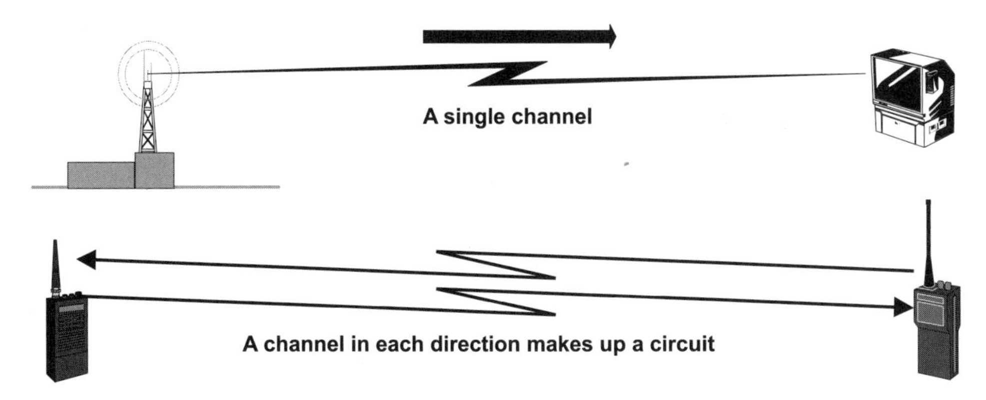

Figure 4.4 Channels and circuits. A channel is a one-way path from a sender to a receiver. Many telecommunications applications, notably telephone calls, involve two channels, making up a full circuit.

4.2.1 Channels and circuits

A channel is a communications path from a sender to a receiver. If there is another channel back to the sender, the two are connected by a *circuit*, or possibly more than one. A television broadcast is an example of a single-channel service, whereas a telephone call is a good example of a circuit, as there is a channel in each direction (see Fig. 4.4). In the telephone industry, each one-way channel is sometimes called a *half-circuit*. In the satellite industry, the term half-circuit is sometimes used to mean the two-way circuit between a single earthstation and a satellite, instead of the one-way path from a sender to a receiver. The satellite usage has two purposes: each earthstation's half-circuit allows it to monitor whether what it sent to the satellite got through, and from a legal and jurisdictional point of view, it often separates the roles of the originating station, the space segment provider, and the destination station.

Most of the channels through a satellite will, of course, be for the communications services of interest to customers. Alongside the communications channels, however, is frequently a channel or two devoted to the necessary "housekeeping" communications between technicians' earthstations. These are called *service channels* and are unknown and invisible to the general communications users. They allow technicians to talk to each other and thus coordinate operations of the system.

4.2.2 Direction of information flow

The term *simplex* describes a signal that travels only in one direction. For example, most paging systems are simplex: they send a signal to the person being paged but nothing is sent back to the person doing the paging. Simplex systems usually use only a single channel. Remote telemetry readings would be another example of a simplex connection.

Half-duplex, sometimes abbreviated HDX, is the term used to describe a system in which signals can go back and forth between two (or more) users, but only in one direction at a time. Amateur radio and citizens band radio, and some taxicab dispatches are like this. Half-duplex systems usually use one bidirectional channel.

Full-duplex, sometimes abbreviated FDX, describes a system that can carry signals in both directions at the same time. A telephone call is an example of a full duplex communication, but this may become half-duplex if a speakerphone is used.

4.2.3 Timeliness

If signals must be received as they are transmitted, they are said to be in *real-time*. Telephone calls and radio and television "live" programming are examples. Such systems are not tolerant of interruptions or long variable delays of the flow of information. Some systems tolerate fixed delays.

If there are some delays in communication, but they are small (only a few seconds) the system may be called *near real-time*. Examples are packet-switched applications, such as live chatting sent over the Internet. The amount of delay is sometimes referred to as the *latency*.

If a signal is held for minutes or longer at some point(s) along its path, the system may be called *store-and-forward*. E-mail, telegrams, and taped television shows are examples (although, of course, when the videotape is finally aired, the transmission must be in real time). A satellite example might be a low-Earth orbit satellite that receives some data while passing over the sender, stores it on board, and downlinks it when it is over the destination.

For digital communications, there is also the issue of synchronicity. The information, whatever it is, is expressed by a series of electrical pulses, to be explained more fully in Chapter 6. If these pulses are sent as a regular stream of data, controlled by some master clock, the communication is said to be *synchronous*. If the data is sent without such master control, it is called *asynchronous*.

4.2.4 Use of transmission facilities

If there is a permanent connection between or among users, these are often called *dedicated* channels or circuits. An example is a hard-wired office intercom system, with wires interconnecting every pair of offices.

If, on the other hand, a user makes actual use of the facilities only when necessary, this is said to be *demand assignment*. The telephone system is an example: you only have a telephone circuit when you lift the receiver. The major problem with hardwired dedicated systems is that the number of interconnections goes up as the square of the number of users, making large hardwired systems unfeasible. (For example, a hardwired system directly interconnecting only 100 users would require 4950 connections; 1000 users would require 499,500 wires; a million users would require 499,999,500,000 wires!)

4.2.5 Switching

Older transmission tasks, such as telex signals and telephone calls, usually connect the sender with the receiver by a circuit that is established for the duration of the transmission, often in response to some address (e.g., a telephone number) dialed by the originating party. The connection thus set up is called a *circuit-switched* connection (Fig. 4.5), and the connection is maintained until the call is ended. In contrast, many data transmissions (which can now include digital audio and video, not just "pure" numbers) make use of a different technique. At the sender's end, bunches of data bits are grouped together into packets, each identified by the "address" of the sender and receiver and with a serial number in the order it is sent. However, each packet may travel a different route to the intended destination, passing through many intermediate nodes—each of which examines the destination address and forwards it on. Such a system is called a *packet-switched* connection (Fig. 4.6). The two most common forms of this are the X.25 packet networks and TCP/IP, the Transmission Control Protocol/Internet Protocol that is rapidly becoming a primary way of data communication.

In a packet-switched transmission, the sender and receiver effectively have a circuit between them. That is, it appears that they are "hard-wired" together, but in reality the system is giving that illusion by properly handling the packetizing and

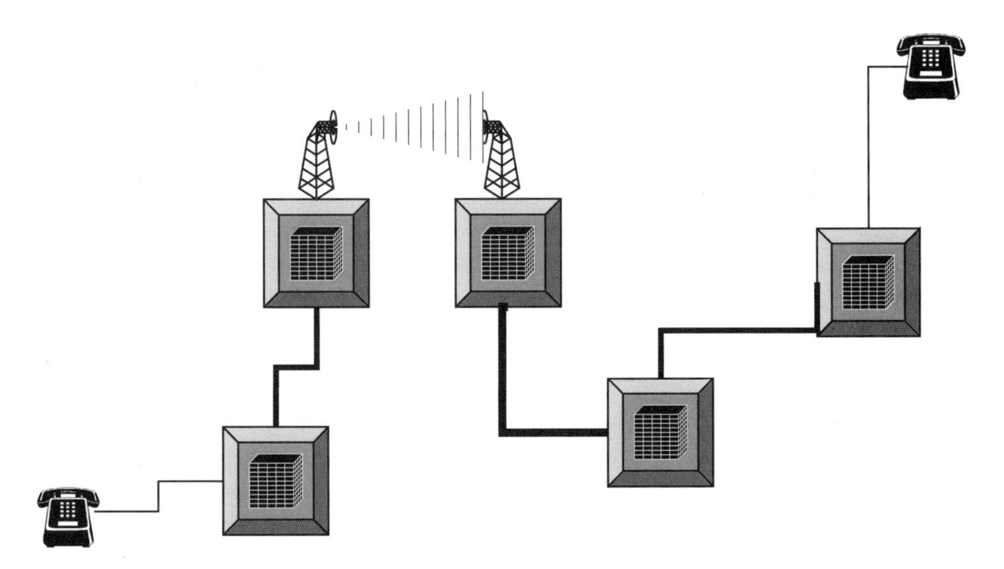

Figure 4.5 Telephone calls are an example of a circuit-switched technique: when you place a call, the PSTN establishes a circuit between you and the called telephone that is maintained for the duration of the call.

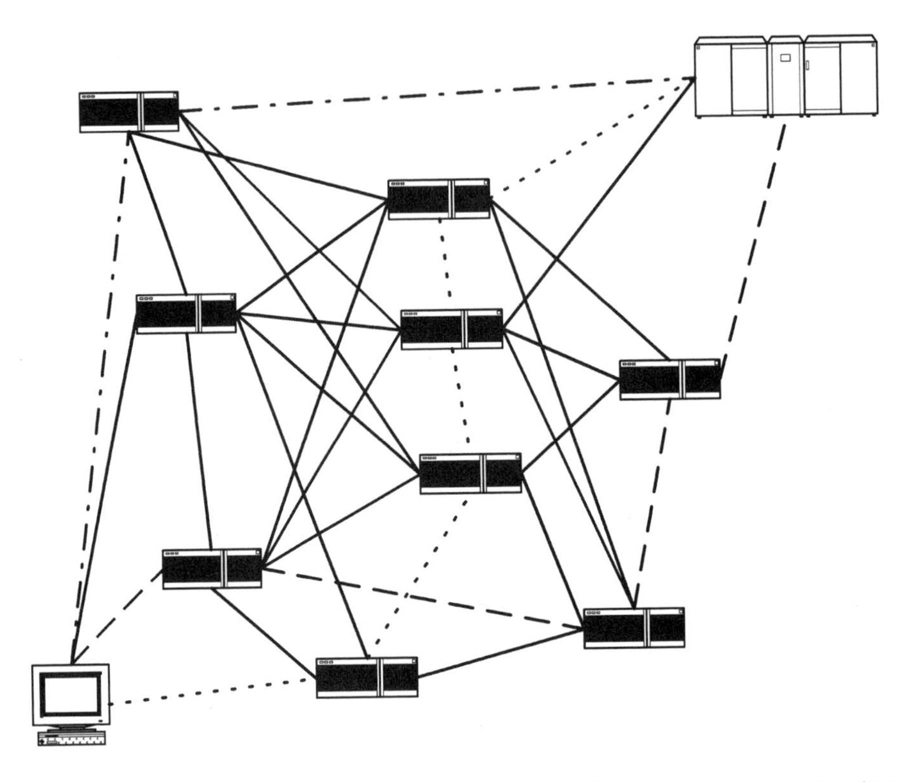

Figure 4.6 In connectionless, packet-switched networks, the datastream is broken into packets and numbered, and each may take a different path from sender to receiver, being reassembled in the correct order upon receipt. Such mesh networks are often more efficient, and are less failure-prone because of the multitude of possible paths for the data.

reassembly of the data. Thus, the two ends are connected by what is sometimes called a *virtual circuit*, one which appears to exist but actually does not in terms of permanent connections. Extending this idea further, if many users are interconnected by such a system, it may be called a *virtual network*.

4.2.6 Network geometry

The technical term for the geometrical arrangement of user interconnections is *network topology*. For a network of very few users, the connections may simply be a few wires among them. For larger numbers of users, the network topology becomes important. Some local area networks (LANs) in offices are *ring networks*, in which all communication goes around a big loop to which all of the users are attached. Another type is the *bus network*, arranged like a long backbone with offshoots connecting the users. Computer LANs and some cable television systems are like this. The telephone system is more hierarchical, like a *tree topology*, with large regional switching centers, somewhat smaller sectional centers, down to individual central offices serving individual users (see Fig. 4.7).

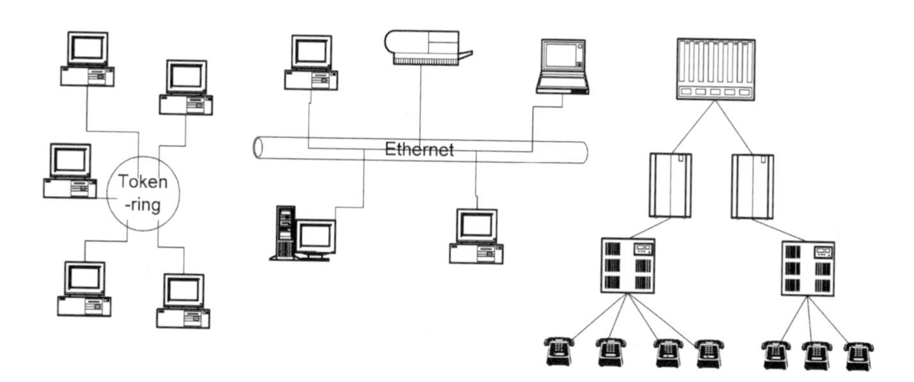

Figure 4.7 Ring, bus, and tree networks. While useful for such applications as office local area networks, cable television systems, or telephone systems, these network geometries are too rigidly hierarchical for implementation by satellite networks.

If there is one preeminently important location to be connected to many subsidiary sites, the topology, generically called *point-to-multipoint*, may also be called a *star topology* (Fig. 4.8). Any traffic between individual sites passes through the main hub. Some cable television systems are like this, as are teletraining programs, VSAT retail applications, and television broadcast systems.

If any user may contact any other user directly without going through an intermediary hub, this is called a *multipoint-to-multipoint*, or *mesh*, topology. For large numbers of users, the only realistic way of constructing a mesh network is to

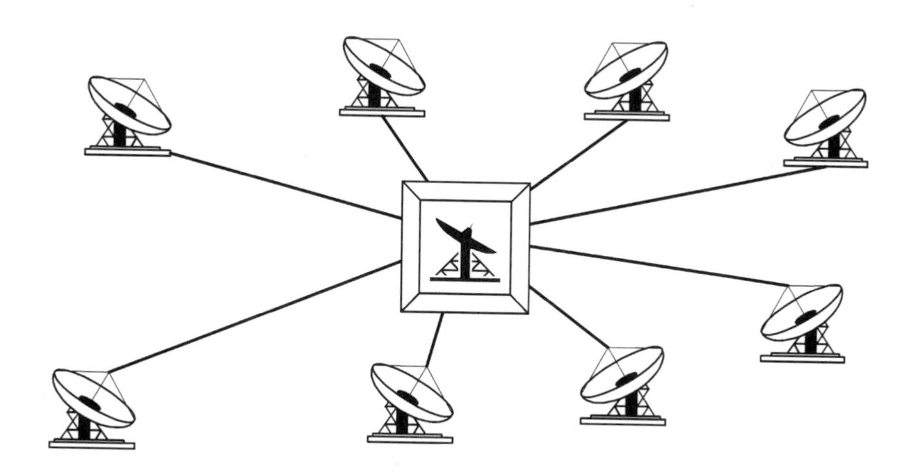

Figure 4.8 A star network, also called a point-to-multipoint network, typically has a central "hub" connected to remote "nodes" with which it communicates. If a node has traffic for another node, it must be routed through the hub. The flow of information can be one-way or two-way along the lines, and often is highly asymmetrical, with much more flowing in one direction than the other. Such networks are common for such applications as broadcast and cable video distribution, teleconferencing, and VSAT applications, and are well implemented by satellites.

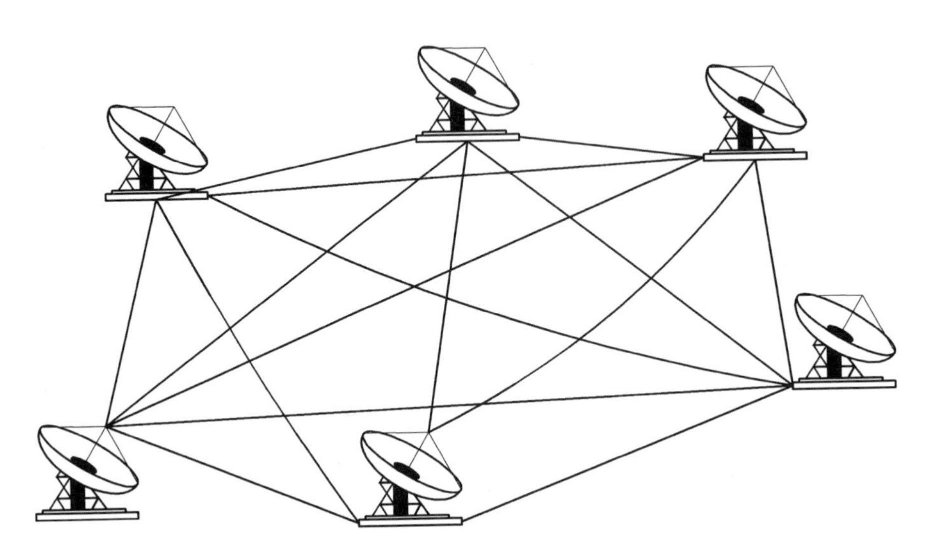

Figure 4.9 A mesh network, also called a multipoint-to-multipoint network, is characterized by the need for any individual node to get in direct contact with any other without going through some central hub. Large mesh networks are only possible using broadcast technologies because of the large number of wires or cables that would be required for a network with many nodes. The major satellite systems providing telecomms services are examples of such networks.

use some broadcast form of transmission, because the required number of wires grows too quickly. In the satellite industry, a good example is the Intelsat system of satellites and earthstations. Any earthstation may link through any satellite visible to it to any other earthstation also visible from the satellite (see Fig. 4.9).

The end user usually does not care about these details. As far as he is concerned, a telecommunications system is a black box: a signal goes in one end and comes out the other end, meeting financial and technical criteria for cost and quality. What goes on inside the black box is irrelevant except to the technical people responsible for its innards. All the user wants to do is perform some communications task and would do it by tin can and string if that would economically and technically serve the purpose.

4.3 Measurements: putting a number on it

Any technical endeavor requires measurements, and a technology that extends from a transistor on the surface of the earth to cislunar space requires many very large and very small ones. We will be dealing with very large distances and signal losses and transmission powers and almost incomprehensibly small received powers.

Our usual units of measurement are metric. It is common, for instance, to measure the size of a satellite dish in meters. Wavelengths common in the satellite arena are typically measured in centimeters or millimeters.

On the other hand, the distance between an earthstation and a satellite may be given in kilometers, statute miles, nautical miles, or even in terms of the radius of the earth (6378 kilometers, or 3962 statute miles). Since different references and organizations have their own preferred units, you need to be familiar with all of them.

The preferred metric system uses units and powers of units that come in groups of three. Each unit has a standard prefix letter. For example, for the unit of length we have the meter, and the next larger unit is 1000 (10^3) times that, or a kilometer. In Appendix E1 you can find a table of common prefixes and their values.

(A caution concerning the use of common names for large and small numbers: not all countries agree on the nomenclature. While in the United States and in the United Kingdom the terms "thousand" and "million" mean the same thing, terms for bigger numbers are not the same. In the U.S., a billion is a thousand millions, or 10^9; in the U.K., it is a million millions, or 10^{12}, and thus a thousand times bigger than an American billion. The number the U.S. calls a "billion" is called a "thousand million" in the U.K. and some other countries use the term "milliard" for the same thing. One exception is that the British now use the American-sized billion when referring to money, whereas they use the old usage for most everything else. Thus, be careful when using these big number words internationally.)

Another common unit measures the frequency of waves, and is called the hertz, symbolized by Hz. The old name for hertz is "cycle per second." (It was named in honor of German scientist Heinrich Hertz, who was the first person to show that electromagnetic waves could indeed travel through a vacuum.)

Yet another useful unit is the watt, abbreviated W, which is a unit of power, that is, energy per second. Transmitter powers and received signal powers are often measured in watts or its multiples.

Thus, for example, combining some units and prefixes, we have such measurements as the speed of light = 300,000 kilometers (km) per second; the frequency of the waves in your microwave oven of 2.43 gigahertz (GHz); the signal strength received by an earthstation may be around 10 picowatts (pW); or the time it takes to trace one line on an (NTSC) television screen is 63 microseconds (μs).

4.4 Decibels

You have probably seen the term decibel if you have had any exposure to telecommunications. It is the most common unit for measuring lots of things, especially power levels, but is frequently not really understood by nonengineers. An understanding of what decibels are and how they are used is very useful in order to have a grasp of the technical side of the business, and to understand what technical people are talking about. We will be using decibels often. They are commonly abbreviated dB (the preferred abbreviation), or DB, or db.

(By now you probably realize that a decibel is a tenth of a bel, named after Alexander Graham Bell of telephone fame, though we unfairly performed an "l-ectomy" on his surname!)

Decibels are very useful, for they allow us to simplify calculations and to express huge numbers and ratios, both big and small, in convenient terms. To do these things, you have to know about the two major characteristics of decibels:

First, measurements in decibels are *relative* measurements: they represent a *comparison* between two values. Unlike absolute units, such as in electricity where a measurement of 0 A would mean we have no current at all, a measurement of 0 dB does not mean no power or signal. It means whatever we want it to mean. Although arbitrary, it is very useful.

If this seems strange to you, consider another arbitrary, relative measurement scale you probably use every day: temperature. Let's use the Celsius temperature scale used by most of the world (although the principles apply to the Fahrenheit scale used in the United States and in a very few other metrically backward nations).

What is a temperature of 0°C? It simply is the temperature on this scale at which pure water freezes. Why? The entire answer is: "Because!" Because it is convenient; because it makes for a useful scale. And, please note, 0°C does not mean that there is no temperature! It is simply a useful starting point.

And what is 100°C? The boiling point of water. Why? Again: because it is convenient. This temperature scale is defined by dividing the temperature interval from freezing to boiling into 100 parts, thus setting the size of 1°C.

Note, also, that negative temperatures do not mean that temperature is "flowing the other way"; it merely means that the temperature is below the standard level we have set. The same is true for measurements in decibels.

Similarly, 0 dB is defined to be whatever we want, for convenience. A bit later we will see some common setting points for decibels. Negative decibel levels merely mean that the value being compared is below our zero-point standard.

A second property of measurements in decibels, one which often brings shudders to nontechnical people is that decibels are *logarithmic*. Just a seemingly small number of decibels can represent a very large number.

You probably have not had to bother with logarithms since your high school algebra class, when Ms. Harridan taught you about them and you memorized this stuff just long enough to get through the examination! But, contrary to the impression you may have had then (and maybe still), logarithms were not invented to bedevil students; they were invented to ease calculations. (The word even means "reckoning with numbers.") It is this property that allows us to use decibels to measure incomprehensibly big and small values somewhat understandably, and perform calculations with ease. For a mathematical refresher, consult Appendix E2.

A relatively small number of decibels can represent a huge number or a huge change. For example, let's take the case of a satellite link to your earthstation. Aboard the satellite are electronics called transponders, which receive the uplinked signal and send it back down to Earth. The final stage of the transponder is a radio transmitter, typically emitting a few tens of watts of power.

Can you guess what fraction—what proportion—of the total power sent by the downlink transmitter makes it into your earthstation dish: a tenth? A thousandth? A billionth?

Guess again, and think much smaller: a typical amount would be a thousandth (10^{-3})of a billionth (10^{-9}) of a billionth(10^{-9})! Another way to say that is that about 10^{-21} of the original signal makes it into your dish: $10^{-21} = 10^{-3} \times 10^{-9} \times 10^{-9}$. Now have you got a good gut-level understanding of just how big (or small) that is?

I don't either.

But we can say it in another, more understandable and useful way: The signal loses 210 dB in its travel from satellite to earthstation. Now, 210 is not an intimidating number; you could count that high easily, or drive that many miles in a few hours. So we can say this space loss, as it is called, is 210 dB. Note that 210 is 10 times the exponent of 21, and the minus sign indicates a loss in strength.

Getting ahead of ourselves a bit, we can take an example of how useful decibels are for calculations. Taking a simplified case, say we have a transponder transmitter that puts out 10 W with an antenna that increases that power by 100,000 times by beaming it into a narrow range, and that the signal then loses all but that thousandth of a billion of a billionth in the downlink to your earthstation.

To do that calculation in "real numbers" would go like this:

$$10\text{ W} \times 100{,}000 \div 1{,}000{,}000{,}000{,}000{,}000{,}000{,}000$$
$$= 0.\ 000\ 000\ 000\ 000\ 000\ 1\text{ W}$$

Doing the same calculation in decibels would go this way:

$$10\text{ dBW} + 50\text{ dBI} - 210\text{ dB} = -160\text{ dBW}$$

A lot easier! And you can imagine how much messier the "real number" calculation would be if we hadn't been using nice round numbers. (What we have just done is a very simplified calculation of what is called a link budget. The details of the calculation will explained later in Chapter 19.)

Measurements in decibels depend on comparing the ratio of two numbers, which may measure signal power, voltage, or even your salary and that of a colleague. These may be two independent measurement, or one may be some standard value, such as 1 W, to which the other value is being compared.

For the mathematical details and definition of decibels, see Appendix E2.

If you remember just two decibel values you will have an approximate guide to using and understanding decibels. Remember these:

$$3\text{ dB} \rightarrow \text{a factor of } 2$$
$$10\text{ dB} \rightarrow \text{a factor of } 10$$

So, decibels are relative. But when we wish to express something as an absolute measurement, we indicate what reference level we are using. For example, probably the most common unit measurement is power in decibel units, and the most common reference strength level is 1 watt. To indicate that we are using that as a reference level, we put the symbol W after dB: dBW. Thus a signal with a power of

0 dBW has a "real" power of 1W; a signal of 55 dBW has a power of 313,000 W; and a signal of –3 dBW has a power of 0.5W.

Other common measurements are dBm, in which a milliwatt is the reference level, and dBI, where I stands for "isotropic" (which means omnidirectional), and is used only with antennas (see Chapter 16).

Just about any change or comparison can be expressed in decibels. (Would you accept a 3 dB raise in salary?) Later on, we will be using decibels to measure bandwidth and temperature. In most cases, engineers convert to the decibel form simply by taking 10 times the logarithm of the value.

The important thing, as we will see when we reach the culmination of our task (the calculation of a link budget in Chapter 19) is that if we use decibels, all of the items in the calculation must be expressed in decibels. Alternatively—and much more messily—if we use real numbers, all values must be expressed in real numbers. You cannot mix real numbers and decibels in the same calculation.

4.4.1 Ups and downs of power

One final item concerning measurements. We will be dealing often with radio signals and how they change. Engineers often use specialized words to denote these power changes.

Common terms for measuring signal strength include *strength*, *amplitude*, *power*, and *flux density*. The term "density" usually means measurement "per something"; in this case, it refers to the amount of radio signal flowing though per square meter. For example, we might see a transponder rated as having a power of 34 dBW, or a signal flux density coming into an earthstation of –110 dBW/m^2.

Some oft-seen terms that technical people use to denote an increase in signal power are *amplification*, *increase*, and *gain*. For example, we might be told that an amplifier has a gain of a million times (60 dB) or that a dish has a gain of 33 dBI.

Customary words indicating signal decrease are *loss*, *drop*, and *attenuation*. For example, the space loss mentioned earlier is 210 dB, or a rain storm might cause a signal attenuation of 14 dB, or a slight error in pointing our dish may result in a signal drop of 1 dB.

Having seen how and why we measure things as we do, it is now time to go on and find out about these radio waves we use to communicate with our satellites.

Chapter 5

The Spectrum and Its Uses

Satellite telecommunication—and much other terrestrial telecommunication—uses radio waves to carry the information from source to destination. It is necessary, therefore, to understand their basic properties and the limitations on how these waves are generated, transmitted, and received.

Radio is just the lowest-frequency part of a phenomenon called electromagnetic waves. They are so-called because they are oscillating waves of electrical and magnetic fields traveling through space, and sometimes through other materials. One analogy is the waves produced on a string by shaking one end of it. Transmission through space is called *radiation*. (This should not be confused with the radiation caused by atomic and nuclear processes.)

The electromagnetic spectrum is the range of these waves, and is literally infinite. But because different waves behave differently and require differing technologies for their use, only certain parts of this spectrum are useful for communication. You have been exposed to such waves all your life, for they include visible light, the x rays used by your dentist, the infrared lamps that keep your food warm in restaurants, the ultraviolet rays that give you a suntan or burn, the microwaves that cook your popcorn, and the waves that bring you radio and television. As we have seen in Chapter 2, competition for spectrum is fierce.

5.1 Properties of waves

All of these different electromagnetic waves are similar and differ only in two major properties: frequency of vibration and power. (Other characteristics, such as how power changes with frequency, and a property called polarization, will be discussed a bit later.) When traveling through a vacuum—such as space—all of these waves behave the same way. It is when they encounter matter that the differences become important.

Different frequencies behave differently when they hit something. For instance, some will pass through air or water, such as visible light. Some are blocked by metals—for example, radio waves. Some pass through most anything, such as x rays. Some frequencies of waves can be made to travel through certain materials. For instance, lower-frequency waves such as radio waves will travel through metallic wires and remain within them, but cannot be carried by glass or plastics. Some radio waves will reflect off of layers in the atmosphere, allowing global shortwave radio. Higher-frequency waves, such as visible and infrared light, can be carried by glass and plastic, resulting in the fiber optic technologies used for communications. The very highest frequencies cannot be contained in anything.

5.2 The speed of light

A property common to all electromagnetic waves is that they travel at the same speed in a vacuum. This speed is usually called the "speed of light," but just as accurately could be called the "speed of radio." However, when these waves enter material objects, even such low-density things as the atmosphere or water, they slow down. The speed of light in a vacuum is the fastest speed in the universe—no material thing or signal can travel faster than this. Not even Congress can change this law of the universe. Depending on the units you want to use, the speed of light is

186,282.397 miles/second
299,792.458 kilometers/second
670,616,629.2 miles/hour
1,802,671,500,000 furlongs/fortnight
1 foot/nanosecond

This speed limit has an important effect on our ability to communicate over large distances. As we will explore in more detail in Chapter 10, geosynchronous satellites are all about 36,000 km, or 22,000 miles, above the surface of Earth. Thus, the round trip distance that a signal must travel from transmitting earthstation to satellite to receiving earthstation is twice this: 72,000 km or 44,000 miles if both stations are located at the subsatellite point directly below the satellite. In this case, a single one-way link between these earthstations will take the radio wave about 72,000 ÷ 300,000 = 238.6 ms, or about a quarter of a second (Fig. 5.1). If the two communicating earthstations are located on the edge of the earth just visible from the satellite, the distance is greater, and the maximum one-way transit time can be as much as 277.6 ms.

This is the origin of the infamous "satellite delay." This delay, also called *latency*, can be a problem for voice traffic and for some computer data protocols. In a two-way telephone call, if some of your own voice is reflected back to you from the other end of the line, you will hear your voice as an echo about half a second after you speak, which can be very annoying. In Chapter 17 we will look at how to get rid of the echo, but there is no way of getting rid of the delay. The latency can be a problem for data transmissions because some of the computer protocols require acknowledgments of receipt of data, and such acknowledgments are much slower over a long link.

Since, approximately speaking, all geosynchronous satellites are the same distance from every earthstation, and since nothing can go faster than the speed of light,

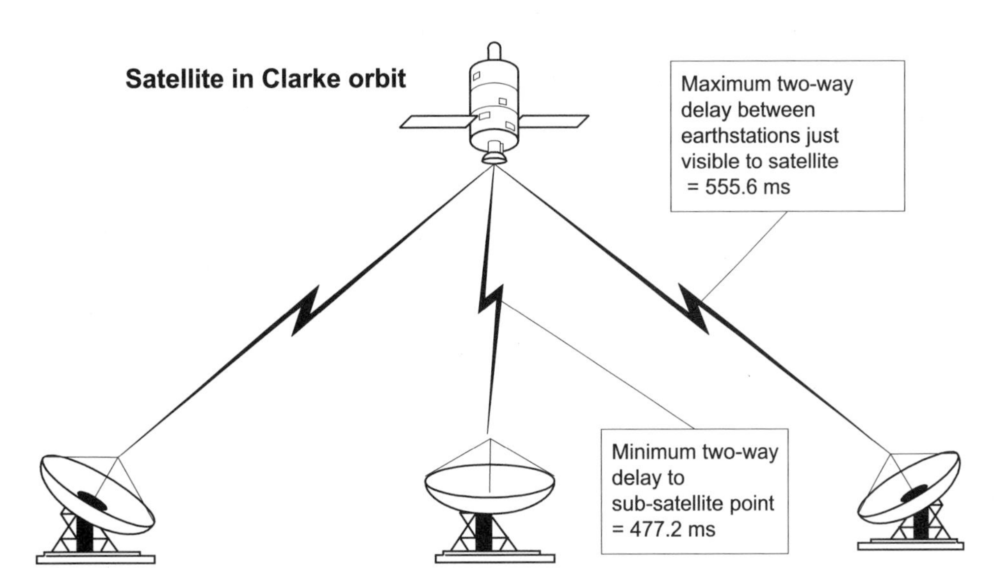

Figure 5.1 One of the drawbacks of communications through geostationary satellites is that it takes a measurable amount of time for a signal to go from an earthstation through the satellite to the receiving earthstation. The average two-way delay is about half a second. This can cause an annoying echo in voice channels, and problems with acknowledgment of data packets receipt for data protocols.

this delay is an ineluctable part of every communication with geosynchronous satellites. (One way to reduce the delay to insignificance is to use lower-orbit satellites.)

5.3 Inverse square law of radiation

Another property common to all electromagnetic waves is that, as you would expect, they get weaker with distance. We have the everyday experience that distant lights are dimmer than close ones. This is also a problem that greatly affects satellite communication.

Imagine a single point source of light (or radio, or x rays) surrounded by a sphere centered on the source. The total amount of light from the source will illuminate the entire inner surface of the sphere. On the inside surface of the sphere, the "signal strength" will be the total wattage of the light divided by the area of the surface, expressed in watts per square meter. If you remember high school solid geometry, you may recall that the surface area of a sphere is proportional to the square of the radius of the sphere. So, if you double the size of the sphere, the surface area to be covered doesn't just go up by a factor of two, but by a factor of two squared, or four; thus the light intensity per square meter has decreased by a factor of four (Fig. 5.2). A sphere three times as big as the first one will have nine times the area, so the illumination per square meter will be 1/9 of the first sphere, and so on. For this reason, we say that radiation obeys the inverse-square law: the intensity

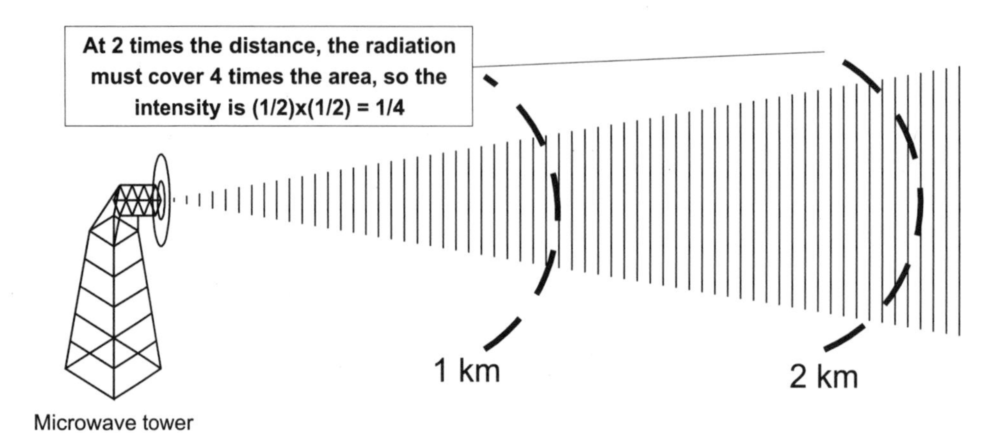

Figure 5.2 The inverse square law. A radiated signal spreads out as it leaves the source, with the loss proportional to the inverse square of the distance. Every doubling of distance causes another 6-dB drop in signal strength.

is proportional to $1/r^2$, where r is the distance from the source. Put in decibel terms, doubling the distance results in a signal decrease of 4 dB.

This property of radiation also has a large effect on satellite communication, since all geosynchronous satellites are roughly 36,000 km from any earthstation. This makes a beam from a satellite inherently weaker than, say, a signal from a nearby terrestrial television tower. Because satellite signals are typically much weaker than terrestrial signals, we will see in Chapter 8 that they require a greater range of frequencies (called bandwidth) than the comparable terrestrial signals.

The weakening of the strength of a radio signal as it travels through space is known as *space loss*. We will see how to calculate it precisely in Chapter 17. In the previous chapter, we saw just how very large the space loss is. To get ahead of ourselves a bit, take a specific case of a geostationary satellite directly over an earthstation on the equator—a distance of 35,878 km—sending a signal to the stations at the C-band downlink frequency of 4 GHz. Plugging these values into the formula from Chapter 17, we see that the space loss is 201 dB. Expressed another way, the received signal is only about 10^{-20}th of the transponder power coming out of the satellite!

5.4 Waves, wavelength, and frequency

Every wave has some *frequency*, the rate at which it oscillates, measured in *hertz*. If you imagine shaking a string, the faster you shake it, the closer together are the peaks along the string (Fig. 5.3). The peak-to-peak distance is called the *wavelength*. Note that there is an inverse relationship between frequency and wavelength: high frequencies are short wavelength waves; low frequencies are long wavelength.

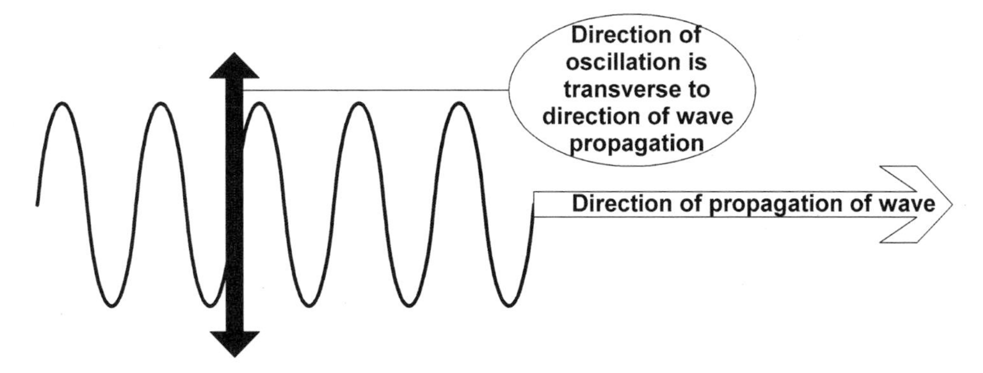

Figure 5.3 Transverse waves. Such waves, as on a string or a pond, the medium (the string or the water) moves perpendicular (transverse) to the direction the wave is travelling (down the string, or across the pond). The orientation of the plane of vibration is the property of polarization. (In contrast, sound waves are "longitudinal waves" and cannot be polarized.)

The laws of physics tell us that for any wave, the frequency times the wavelength is always equal to the speed of the wave, in our case the speed of light. To make things a bit easier, consider that in satellite communications we are usually dealing with frequencies in the range of billions of cycles per second, or gigahertz, abbreviated GHz, and wavelengths around the centimeter range, abbreviated by cm. Using these units, we arrive at a very simple pair of formulas for figuring out frequency and wavelength:

$$\textit{Frequency} = 30 \div \textit{Wavelength}$$

and

$$\textit{Wavelength} = 30 \div \textit{Frequency}$$

Thus, for example, the so-called C-band uplink signal from your earthstation to a satellite, which uses a frequency of approximately 6 GHz will have a wavelength of $30 \div 6$ or 5 cm, which is about 2 inches from peak to peak. The newer satellites using the 30 GHz Ka-band frequencies are dealing with wavelengths just 1 cm long.

So why do we care what the wavelength is? Because it controls the technology we use to transmit and receive the waves. For instance, have you ever seen a perforated earthstation dish, and wondered why the radio waves don't fall through the holes like rain does? The answer is that it is somewhat like a sieve: if the waves are much bigger than the hole size, they get reflected by the dish. Later, in Chapter 16, we will see that the power of a dish, and its directionality, are determined by the relative size of the dish and the waves it is working with. In Chapter 18, we will discover that the frequency greatly determines what effects the atmosphere has on our signals.

If you remember the high school version of the spectrum, you may remember the range of visible light, or colors, from the mnemonic fictitious person Roy G. Biv, which stood for the colors red, orange, yellow, green, blue, indigo, and violet. Beyond the visible spectrum, on either end, are frequencies invisible to the eye but useful for communization. Beyond violet—technically speaking, at higher frequencies—lie ultraviolet, then x rays, and the highest, gamma rays. These frequencies are little used in communications (Fig. 5.4).

Beyond red—toward lower frequencies—lie several invisible frequencies useful for telecommunications. Just below red in frequency is infrared, commonly used with optical fibers. Well below that lie radio waves, the lowest range of the electromagnetic spectrum. And they, too, spread over a wide band of frequencies.

5.5 Radio frequency bands

The earliest radio waves, produced by sparks, created waves of all frequencies, but gradually technology improved to allow control and limitation of the frequencies used. Around a hundred years ago, the first "wireless" signals used what we today would call "medium frequencies," abbreviated MF. As technology improved, we could use high frequencies, and low frequencies, which we called HF and LF. Technology improved further, so now we have a wide range of frequencies for our wide range of communication applications.

Table 5.1 presents some approximate frequency ranges with their corresponding wavelengths, names, and abbreviations.

Table 5.1 Radio Frequency Bands

Frequency	Frequency Band Name	Wavelength	Wavelength Band Name
<3000 Hz	ELF: Extremely Low Frequency	>100 km	Extremely Long Wave
3–30 kHz	VLF: Very Low Frequency	100–10 km	Very Long Wave
30–300 kHz	LF: Low Frequency	10–1 km	Long Wave
300–3000 kHz	MF: Medium Frequency	1–0.1 km	Medium Wave
3–30 MHz	HF: High Frequency	100–10 m	Short Wave
30–300 MHz	VHF: Very High Frequency	10–1 m	Very Short Wave
300–3000 MHz	UHF: Ultra High Frequency	1–0.1 m	Decimeter Wave
3–30 GHz	SHF: Super High Frequency	10–1 cm	Centimeter Wave
30–300 GHz	EHF: Extremely High Frequency	1–0.1 cm	Millimeter Wave
>300 GHz		<1 mm	Infrared

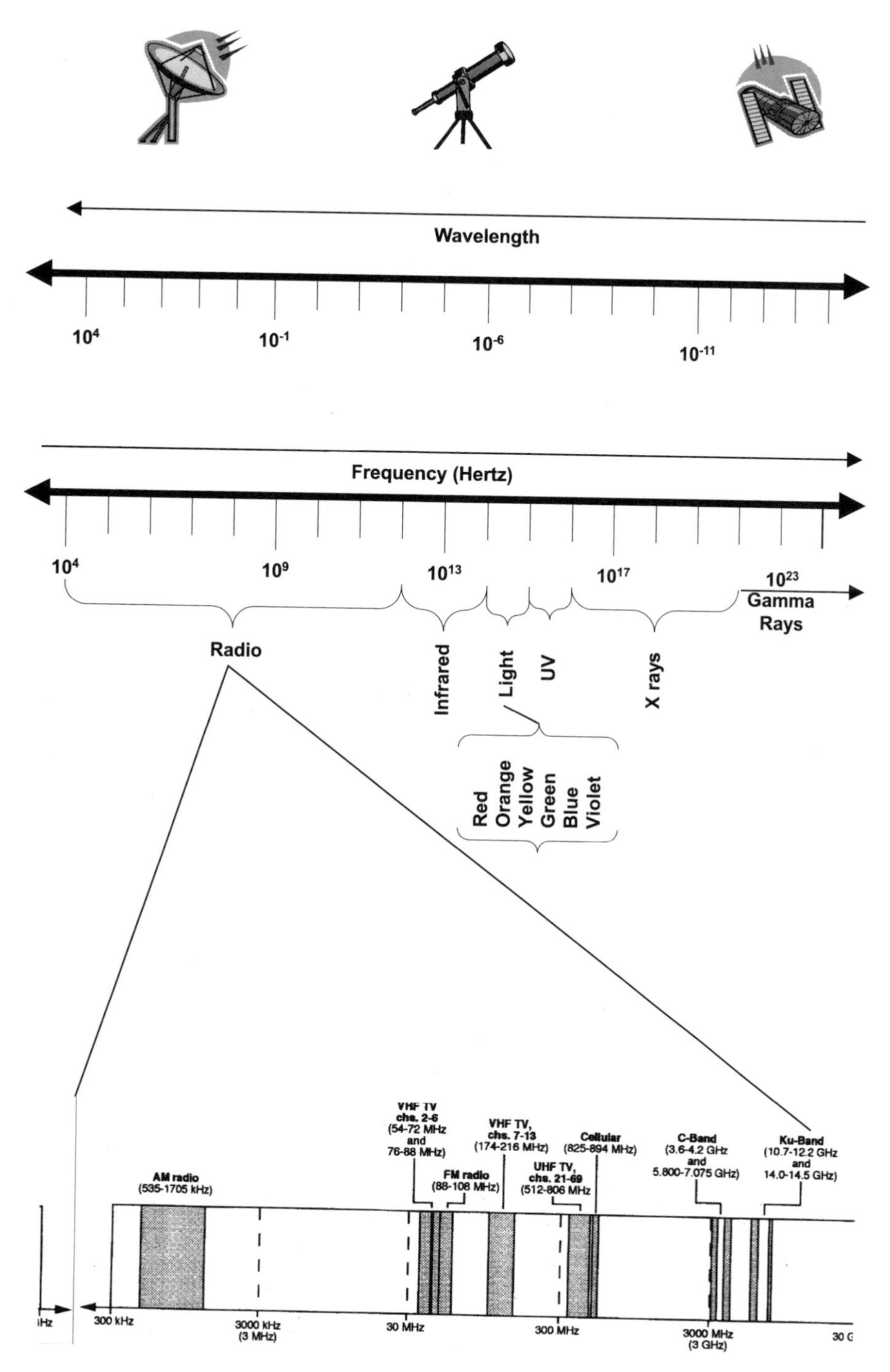

Figure 5.4 The electromagnetic spectrum. These oscillating electrical and magnetic fields cover literally an infinite range of frequencies, some of which are useful for communication. Different frequencies interact differently with matter, and so some waves are passed by some materials and blocked by others. The bottom part shows an expanded section of the radio spectrum of use to us for many types of telecommunication.

(I know one scientist who refers frequencies above 300 GHz as DHF, for "damned high frequency," but this is not official. Also, frequencies above about 300 GHz are not regulated by most telecommunications authorities.)

In the field of satellite communications, most of our applications occupy the upper several bands. VHF is used primarily for tracking and controlling satellites, and for the satellites called "Little LEOs" for mobile data applications. The lower part of the SHF range is much used for various mobile services, including the so-called "Big LEO" voice mobile services, and for navigational services. The upper parts of the SHF band are used for various fixed and broadcast services. EHF is thus far little developed, however many companies have recently filed for permission to use this band for high-speed data services because the lower bands are already congested.

Another common way of referring to frequency ranges is that of band designations using letters, such as C-band. These letters originated from early top-secret microwave research and are not in any particular order, nor do the letters stand for anything. Here are common bands and applications as used in satellite telecommunications:

VHF:	below 1 GHz	tracking and Little LEOs
L-band:	1–2 GHz	mobile services
S-band:	2–4 GHz	mobile and navigation services
C-band:	4–6 GHz	mostly fixed services
X-band:	7–8 GHz	government use in USA, Russia, Japan; various in other nations
Ku-band:	10–18 GHz	mostly fixed services
Ka-band:	20–36 GHz	fixed and mobile services, and intersatellite links
Q-band:	36–46 GHz	future highspeed data systems
V-band:	46–56 GHz	future highspeed data systems
W-band:	56–100 GHz	future highspeed data systems

The frequency ranges given here are approximate, not formal definitions. Confusingly, different users and different countries disagree on the exact boundaries of different bands.

C-band is sometimes called the 6/4 band; X-band is sometimes called the 8/7 band; Ku-band, in its various applications, may be called the 13/11 band, the 13-14/11-12 band, or the 18/12 band; the Ka-band may be referred to sometimes as the 30/20 band or the 40/20 band.

5.6 Frequency and bandwidth

With very few exceptions (lasers are one example), real-world signals occupy not just a single frequency, but a range of frequencies. This range is called the *bandwidth* of the signal. It is simply the highest frequency used minus the lowest frequency used.

The concept of bandwidth exists in all of communications. For example, your stereo sound system at home has a bandwidth, but it is called something else: frequency response. This is the full range of tones (frequencies) that your electronics can produce, from the lowest rumble of the longest pipes of a theater organ to the highest overtones of the violin and piccolo.

Furthermore, every application has some irreducible minimum range of frequencies it requires to communicate it with full quality. This range is called the *baseband* bandwidth of the signal. To carry the full orchestral range, your stereo system must handle the baseband bandwidth of sounds that the human ear can hear, which is approximately from 20 Hz on the low end to 20 kHz on the high end. When these sound waves are produced by an orchestra, they are waves in air, but when captured by microphones they become electrical waves in wires. If the electronics is not capable of carrying the full range of frequencies; that is, if the bandwidth of the stereo system is less than the baseband bandwidth of the music, then the sound you hear will not be at its full quality. Figure 5.5 illustrates the concept showing the "baseband bandwidths" of orchestral instruments.

You have probably experienced this frequently if you have ever been subjected to "music-on-hold" while waiting to talk to someone on the telephone. The bandwidth of the telephone system is only 4 kHz, because that is all that is required to carry a recognizable, understandable human voice. But it is not enough for music, so music-on-hold sounds poor, lacking all high and low frequencies.

On the other hand, it is not cost-effective to carry a signal in a system with a much higher bandwidth than is required, since it would cost more money but not

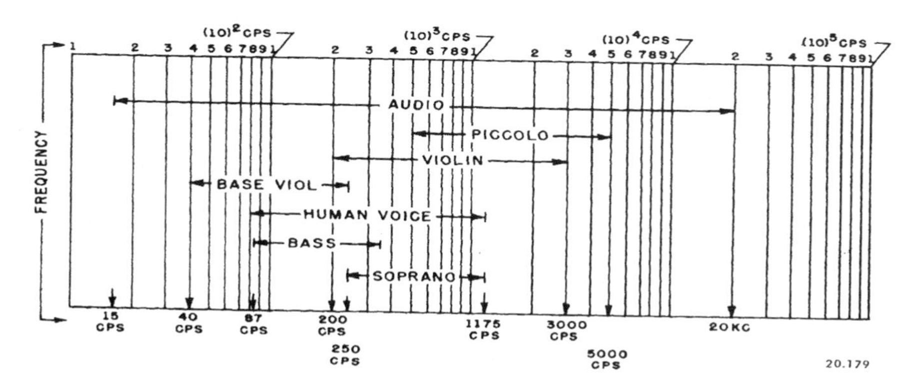

Figure 5.5 The concept of bandwidth. Using an acoustic analogy, each instrument of an orchestra has a characteristic range of frequencies it can produce: its baseband bandwidth. To reproduce all the tones of an orchestra, a sound system must have a bandwidth from 20 Hz to 20 kHz.

give better *useful* quality. For example, increasing the frequency response of your stereo out to 100 kHz would be expensive but wouldn't do you much good, since you can't hear that high (but any neighborhood dogs and porpoises might be impressed!). Since a good-quality voice can be carried in 4 kHz, that is all the telephone company uses.

Whereas a telephone voice requires a bandwidth of around 4 kHz, and "hi-fi" music (more officially called program audio) requires 15–20 kHz, a single NTSC (American) television channel requires 6 MHz. One can conclude that a television signal takes about 1000 times the bandwidth (1000 times the capacity) as a voice does, because there is more information in an image than in a sound. Thus, there is some telecommunications truth to the old adage that a picture is worth a thousand words!

Therefore, bandwidth is not only a measure of quantity, but also determines quality of the signal.

Thus, we have terms such as *narrowband*, *mediumband*, *wideband*, and *broadband,* but these terms are relative. Several narrowband signals, such as telephone calls, may be combined together into a broadband signal, but you should always keep in mind that terms like "narrow" and "broad" are relative. A bandwidth of 6 MHz is broadband in the telephone industry, but is a baseband signal for television stations.

Later when we calculate a link budget, we will need to include the bandwidth of the signal. To make the calculations easier, we will need to have all of the items that go into the link budget expressed in decibel form including bandwidth. For those interested in the details, Appendix E.3 gives the formula to convert bandwidth in hertz into its decibel equivalent. Just to give one example, a transponder with the typical bandwidth of 36 MHz could also be said to have a bandwidth of 75.56 dB-Hz.

In Chapter 7, we will see that the total information-carrying ability of a radio wave depends basically on two things: the power of the wave and the bandwidth of the signal.

5.6.1 More frequency terminology

The range of frequencies in the original signal is, as we saw above, the baseband bandwidth. But signals are almost never transmitted at these frequencies. For example, suppose you are being interviewed on a radio station. You speak with a range of vocal frequencies much like any other human: your baseband bandwidth. But to transmit your pearls of wisdom to the listening masses, the radio station must cause your voice to control the properties of the radio waves sent out from the station—a process called modulation, which we will explore more fully in Chapter 7.

The frequency of the radio station is called the *carrier frequency*, because it carries your signal to receivers within range of the station. The carrier frequency is always much, much higher than the baseband frequencies it is carrying, and has its own range of frequencies, its bandwidth. There is no fixed relationship between the frequency of the baseband signal and the frequency of the carrier, as long as the carrier is of higher frequency and the bandwidth of the carrier is at least as big as the bandwidth of the baseband signal (usually, it is much bigger). The specifics of the carrier frequency are not important for carrying your voice; you can be heard equally well on a station broadcasting at 650 kHz or at 1380 kHz (for example). It is the bandwidth and power that count, not the specific carrier frequency.

In the case of a signal linked through a satellite, your voice (or whatever the signal is) is sent to the satellite by the earthstation using a carrier frequency defined for uplinking in whatever band (C, Ku, etc.) is appropriate to the satellite, which is then turned around aboard the satellite and converted to the downlinking carrier frequency for that band. Your voice may have a bandwidth of 4 kHz, but by the time it is modulated onto the carrier wave, the frequencies carrying your voice may occupy a bandwidth of 45 kHz (determined by the modulation used), and that 45 kHz bandwidth of frequencies might be actually transmitted to the satellite at a frequency centered at 3.825 GHz.

To make a analogy, think of a long wall of a room. The span of the wall is like a range of frequencies. In the wall are some doors, each with some width. The width of a door determines how large a table you could roll into the room. But the position of the door along the wall does not control the capacity of the door. Thus the door width is like bandwidth, and the location of the door along the wall is like the frequency of the carrier wave. It is the width that counts, not the frequency.

5.6.2 Conversion and intermediate frequencies

In the example above, we had your 4 kHz voice being sent to a satellite at a carrier frequency of 3.825 GHz. Obviously, somewhere in the transmission chain of equipment there needs to be a change of frequency. Frequency change is often called *conversion* by engineers, and the devices that perform the task are called *converters*.

Electronic devices can be tricky things, and none is perfect. Most active electronic devices are not linear: they do not treat all frequencies the same. This is particularly true for amplifiers. When an electronic device is nonlinear, it produces some distortion in the signal, reducing its quality.

To minimize such effects, we often use *linearized* circuits that partially compensate for the deficiencies of the electronic devices. Another common technique is to perform frequency conversions in more than one step, for example taking your 4 kHz voice and converting it to some intermediate frequency, then converting that to the final carrier frequency. Such a technique is called *dual conversion*, and, not surprisingly, the frequency between baseband and carrier is called the *intermediate frequency*, or IF. Every radio and television uses intermediate frequencies for convenience.

You get several operational advantages by using an IF. For one, you minimize problems with very high-gain amplifiers by doing the conversion in two steps (a satellite transponder may have an overall gain of 10 billion times!). For another, you avoid distortions caused by interactions of the received signal with oscillators in the equipment. A third major advantage is that the IF provides a good standardized frequency that all intermediate stages of the transmission or reception chain of equipment can use. This promotes ease of interconnection and the economy of having standardized hardware.

In the satellite industry, an IF of 70 MHz is very common for use both aboard satellites and in earthstations, but it is not the only frequency in use.

5.7 Other wave properties

There are a couple of other properties of waves that are of interest and of use to us in telecommunications. One allows us to double our capacity, the other is of concern only under special conditions.

5.7.1 Polarization

Think back to our example of a wave produced on a string by wiggling it with your hand. You will note that the wave moves along the length of the string, but a given point on the string simply moves side to side. Because the string moves at right angles to the direction the wave is traveling, such waves are called *transverse waves*. And this produces a very useful property called *polarization*.

When you wiggle the string, you usually move your hand, and consequently the string, back and forth in a line or plane (Fig. 5.6). This makes all of the peaks on the string lie along a line. If someone is holding the other end of the string near eye level, he or she will see the string vibrate along a single line. If all of the vibration is along such a single line, the wave is said to be *polarized*.

If you vibrated the string in an up-and-down manner to send your "signal," you could just as easily have vibrated the string horizontally. If your receiver at the other end of the string consists of a sliding bar that can move only vertically, it will respond to your signal only if you vibrate the string up and down. On the other hand, if the receiver detects only horizontal waves, your could send a signal by wiggling the string right-to-left. In fact, if you had two sliding blocks at the far end, one of which moved only vertically, the other moving only horizontally, you could send two independent signals at the same time, over the same string, as long as one vibrated vertically and the other vibrated horizontally. Thus, on one "signal path" at the same time, you could double your transmission capacity.

Thus, polarization is a very useful technique for increasing telecommunications capacity. Another use for it is to minimize interference, for instance between adjacent satellites.

The technique of making all of the waves vibrate in one line is called *linear polarization*, and the orientation of the line of each wave is called the *plane of polarization*. There can only be two such planes, or lines, in use at a time, and these must be oriented exactly at right angles to one another. If they are, they are said to

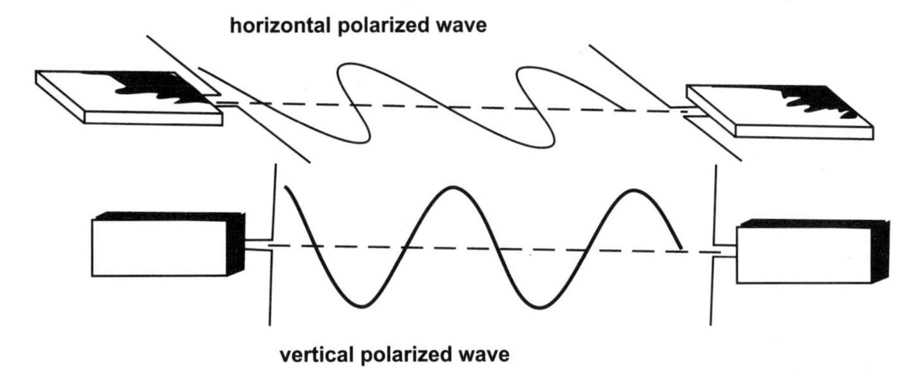

Figure 5.6 Transverse waves can be polarized: made to vibrate in only one plane. If waves carrying signals are made to vibrate in planes that are orthogonal (at right angles) to each other, they will be able to carry the signals without interference between them. Because of the right-angle constraint, only two such planes are possible, often called x and y. A receiver, looking toward the sender, would see the waves as a line of some orientation, which is why such polarization is called linear.

be *cross-polarized.* The absolute orientation of the planes is not important, only that they are arranged at right angles to each other.

Engineers can construct antennas that send and receive only polarized waves. In the satellite industry, the term "vertical polarization" is defined as waves vibrating perpendicular to the orbit of the satellite; horizontal polarization means waves vibrating in the plane of the orbit.

There is another way of polarizing waves. Instead of keeping the plane of polarization constant, the antenna can be made to cause the plane to twist in a regular manner, either clockwise or counterclockwise. This is called *circular polarization* (Fig. 5.7). Thus the waves "corkscrew" as they travel. Waves spinning clockwise (as seen by the sender, the same turning as when tightening a screw) are called *right-hand circularly polarized;* waves twisting counterclockwise are called *left-handed polarized.*

When you send out vertically polarized waves, the receiver gets vertically polarized waves; similarly for horizontal orientation. However, when you send out right-hand circularly polarized waves (abbreviated RHCP), the receiver picks up left-handed waves, since he or she is looking at the twisting waves from the other end. The opposite it true for LHCP: the receiver gets RHCP at that end.

For a particular path and frequency, you must use either linear or circular polarization for a single use, not both at the same time, for they interfere with one another. The technical term for the relative isolation, or independence of one polarized signal from another is *polarization discrimination.* Higher numbers are better and imply that there is less "crosstalk" between the two cross-polarized signals.

A common question is whether one method of polarization is better than the other. Like most other generic questions, the answer depends on the application you have in mind.

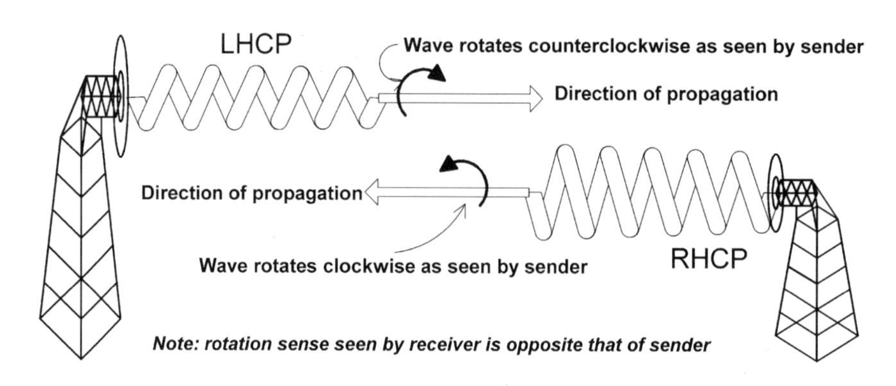

Figure 5.7 Circular polarization makes the transmitted waves rotate the plane of polarization. The two types are called left-handed and right-handed. A receiver looking toward the sender would see the waves corkscrewing clockwise or counterclockwise, respectively. The receiver receives the opposite "handedness" of the signal transmitted.

Linear polarization is simple, thus cheaper to do, and is the older technology. It has the advantage that it is less affected by atmospheric problems than circular polarization (see later in Chapter 18). For signals especially susceptible to the atmosphere, linear polarization may be preferable. But, because it requires the receiving antenna to be carefully oriented the same as the transmitting antenna, a slight tilt can reduce your signal. The complication comes at the earthstation because what is vertical and horizontal to the satellite is not vertical and horizontal relative to the ground at the earthstation.

With circular polarization, if the wave is spiraling in to you counterclockwise, the orientation of the transmitting and receiving antennas is not important as long as it is set to receive the proper handedness of the waves, either RCHP or LCHP. This can make installation of earthstations simpler, thus less expensive. For example, in the application called officially BSS-TV (Broadcast Satellite Service-Television), but popularly called DBS (Direct Broadcast Satellites), the specifications for the service dictate circular polarization, since your backyard dish will be installed by you or some other nonengineer, and polarization becomes one fewer technical detail to worry about.

5.7.2 The Doppler Effect

One property of waves we need to consider is called the Doppler Effect, named after Austrian physicist Christian Doppler (1805–1853).

It works like this. Think of a stone thrown into a pond, and the resulting ripples spreading out in all directions. Receivers on the edges of the pond all get a "signal" at the same frequency that was "sent" by the stone. But, if the stone was skipping across the pond as it made ripples, the ripples would be bunched up in the direction that the stone is traveling, and pulled apart on the opposite side.

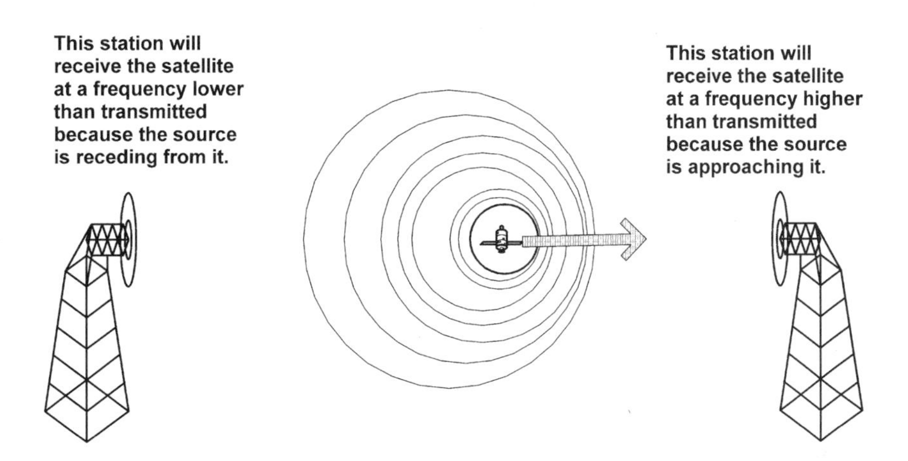

Figure 5.8 The Doppler effect. If a source of waves is in motion, the waves will be bunched up when received by a receiver toward which the source is moving (in the figure, on the right) resulting in receiving a higher frequency than transmitted. The opposite is true for a receiver from which source is moving away.

Therefore, receivers toward which the stone was moving would receive a signal at a shorter wavelength—and thus higher frequency—than the stone sent, while receivers from which the stone is moving away will receive a lower frequency.

In other words, if the transmitter and receiver are approaching or receding from one another, the frequency received will be different from the frequency transmitted (Fig. 5.8). The amount by which the frequency is Doppler-shifted depends on the speed of the waves and the relative speed of the transmitter and receiver. Higher speeds produce greater frequency shifts.

A very common example of a Doppler effect occurs when you stand near a railroad track with a train passing, blowing its horn. As the train approaches, the horn sounds at a high pitch; when just passing you, you very briefly hear the natural tone of the train's horn; and when it moves away, you hear a lower pitch.

Doppler shifts are only a very minor problem for communication with geostationary satellites because, by definition, they are (almost) stationary as seen from an earthstation. The newer low-orbit systems, however, are greatly affected. Consider that such satellites may be moving in orbit at speeds of 7,000 to 17,000 mph. Each is visible for only a few minutes during any particular telephone call. During part of the call, the satellite is approaching you, causing your handheld satphone to pick up a frequency higher than the satellite is sending; a few minutes later the satellite is over on the other side of the sky, moving away, so you receive a signal of lower frequency. Thus, equipment for such services must be what is often called *frequency-agile.* This, of course, complicates the electronics required in the satphone.

Having explained some of the ways waves behave, it is time to turn our attention to the two basic kinds of signals we use: analog and digital.

Chapter 6

Analog and Digital Signals

Telecommunications uses both analog and digital signals. We are most familiar with analog signals, because humans are analog devices. That is, the sensations we experience, such as sounds, images, tastes, and touches, vary greatly and in a continuous fashion. The term "continuous" means that they vary smoothly, and can have arbitrarily large or small changes. One sound can be a tiny bit louder than another; a leaf may be slightly greener than another or it might be brown. However, information is increasingly sent in a digital form because it is more reliable to do so and the quality received is much higher and more consistent (think of the contrast in quality between vinyl records and a CD). While the principles of digital transmission have long been known, the equipment to digitize and de-digitize analog signal for transmission was until recently prohibitively expensive for many uses, particularly for high-quality video.

Nevertheless, most signals we are interested in as humans, such as voices, music, and television, start out as analog signals and must be finally received as analog signals, whether or not they are carried between sender and receiver in an analog or digital form. Thus, we must understand the properties of both kinds of signals.

6.1 Analog signals

The very term analog means "similar," as when we say that things are analogous. In telecommunications, analog means that the electrical and electronic signals that represent the sensations we humans experience vary in a smooth manner in accordance with the sensation they represent. For example, when you speak, your voice is made of varying changes in tone (frequency) and volume (amplitude). If you speak into a telephone or microphone, the electrical voltages and currents your voice generates in the wires and circuits mimic the changes in frequency and amplitude of your voice. At one instant, you may be speaking such that the microphone generates exactly 1.00 volts (V); a short time later, it may be 0.93 V; and maybe later 1.01 V. Such an analog signal is an electrical image of your voice. This is similar to a television picture, where perhaps slightly different voltages and currents cause your television set to display those different colored leaves.

The important thing to remember about analog signals is that they vary continuously from "off" to whatever maximum value your voice or some circuit can

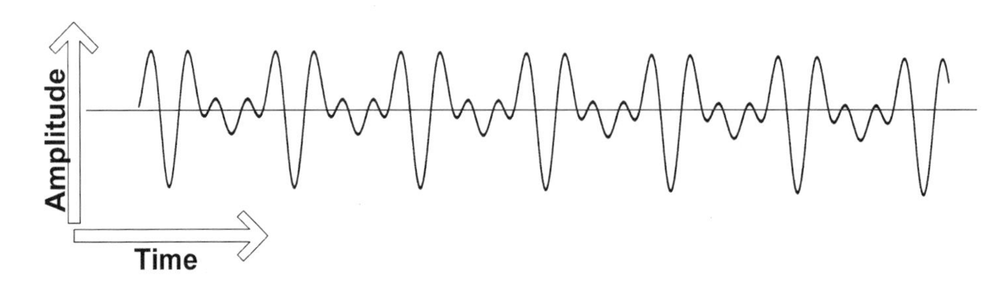

Figure 6.1 An analog signal. Analog signals vary continuously from zero strength to some maximum value, with no predetermined values. The signals we humans receive are all analog: sight, sound, taste, touch, smell.

produce. There are no fixed, predefined values of volume or color or voltage (Fig. 6.1).

6.2 Digital signals

In contrast, a digital signal can take on only fixed, predetermined values. Digital signals are sometimes said to use "discrete symbols," meaning that they are made up of fixed and distinct things which represent other, more common items. They are also sometimes said to be "quantized," meaning that they come in separate units. Digital signals are essentially codes that can represent analog values.

To see the basic difference between continuous and discrete things, think of the difference in measuring, say, a quantity of water and a quantity of bricks. You don't speak of "a water," you measure it by weight or by volume. You could have 1 kilogram of water, or 1.1 kg, or 0.99876674 kg, or any amount. In contrast, it would be convenient to measure bricks by number, and those numbers are always integers.

In telecommunications, the situation is similar. In a digital signal, we assign certain numbers, called symbols, singly and in groups to represent more common phenomena. While many people think of "digital" as a hot new technology, it is actually the very oldest form of telecommunication, in that it used discrete symbols to carry information.

Consider Morse code, named after Samuel F.B. Morse, its American inventor. (Other inventors in other nations almost simultaneously came up with similar ideas.) Even if you don't know Morse code (fewer and fewer people do—most are amateur radio operators), you know what it does: Morse code represents letters of the alphabet, numbers, and punctuation marks by combinations of electrical pulses of different lengths. These are the so-called symbols. We call the two kinds of pulses: "dot," symbolized by ·, and "dash" (or sometimes "dah"), symbolized by -. Thus in Morse code, · means "e," ·· means "i," ··· means "s," and ···· means "h," · - means "a," and so on.

To use Morse code in telecommunications, an operator presses down on a key causing current to flow for a long (-) or short (·) time. At the receiving end, these pulses of electricity show up as perhaps long and short lines on a papertape, or as long and short sounds.

The thing to note is that the symbols, - and ·, are fixed and discrete. There is no such thing as ¾ of a ·. Only dots and dashes. The meanings we give to them are arbitrary, but as long as everyone knows that ··· - - - ··· means "SOS" and that "SOS" is an appeal for help, then we can convey information using only pulses of electricity.

Using an arbitrary system of symbols to represent information may seem an unnecessary complication, but there is a reason, again exemplified by Morse code: it is a very robust communication method, and very resistant to noise. Morse code can be transmitted and successfully received over weak and noisy channels where a telephone call would be impossible.

6.2.1 Doing our bit for telecommunications

Today, telecommunications is inextricably interwoven with computation. Digital signals reign in both. For simplicity, the symbols we use are the most fundamental states you can have: on and off. Circuits are easy to switch on and off, and it is easier to tell if a circuit is on or off than to measure the exact voltage in a wire. These on and off pulses are usually represented by the symbols, or numbers, 1 and 0. Since there are only two possible states, or signals, this is said to be a binary code, so 1 and 0 are "binary digits" or *bits* (Fig. 6.2). (There are other possible number systems we could use, but we are not going to get into that.) For a refresher on the binary number system, see Appendix E.4.

One of the most common examples of this kind of system is in use in the personal computer you use every day. Since people don't deal well directly with pulses of electricity, but computers know nothing about such things as letters and punctuation, there has to be some sort of code that allows these human artifacts to be stored and represented in your computer. The code used is called the "American Standard Code for Information Interchange," better known by its abbreviation ASCII. Each letter and other symbol is represented by seven bits. In ASCII code,

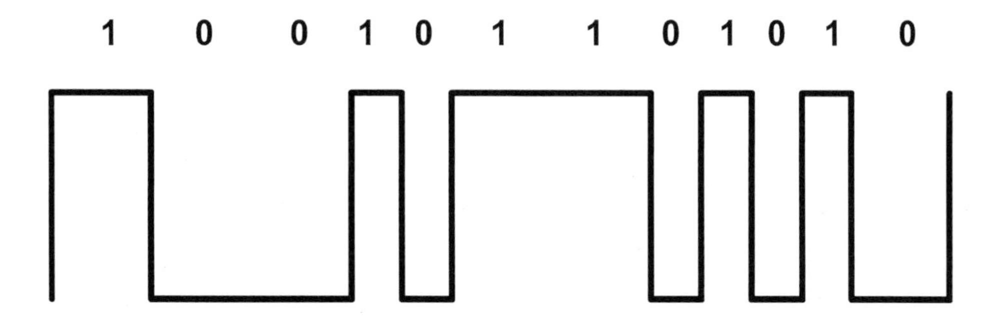

Figure 6.2 A digital signal. Such a signal encodes signal information into a series of pulses, called bits, which have only specified values. The most common ones are "on" and "off," often represented numerically as 1 and 0. A receiver detects only these values.

a = 1100001, A = 1000001, and % = 0100101. Thus, pulses of electricity carry human-understandable information.

As another example, consider your video monitor at your computer. Typically, you might be using a screen resolution of 800 horizontal pixels by 600 vertical pixels. Each pixel can be any color. The pixel color is made up by mixing different amounts of red, green, and blue, denoted by RGB. Each color can have any of 256 different intensities. Thus, each R, G, and B signal requires 8 bits, so each pixel requires 24 bits. And since there are 800×600 = 480,000 on the screen, a full screen requires 11.52 million bits, or 1.44 million bytes of data.

6.2.2 Bit players

Why do we use digital systems for transmission, and why is seemingly everything becoming digital, like your CD player and soon your radio and television? In the year 2000, about 80% of all satellite traffic was digital, and the percentage is rising. Why? Because digital systems are convenient and reliable.

For one thing, by converting an analog signal into a digital one, we can more easily quantify the amount of information to be transmitted. For example, how much information is in a telephone call, an e-mail memo, or a television picture? Simply by counting the bits, as in the example of a computer screen mentioned above, we can measure the "size" of our message and thus the demand it places on our telecommunication system. For example, here are some typical digital "sizes" of some typical messages:

Fire alarm: 1 bit
Memo: 3000 bits
One-page e-mail document, ASCII: 10 kb
One-page typed document, faxed: 500 kb
One minute uncompressed PCM telephone call: 3.84 Mb
One second NTSC television: 200 Mb
Black-and-white newspaper photo: 100 kb
Color magazine same photo: 2 Mb

Thus, we can measure the quantity of information in a signal by counting bits, and conclude that a television signal contains around 1000 times the amount of information as a typed document. We can also measure how fast information is transmitted, for which the unit is bits per second, b/s, and its multiples, such as kbps and Mbps.

6.3 Bring in the noise

Digital technologies are on the increase mainly because a digital signal is more robust and resistant to interference than an analog signal. While this has always been true, it has been expensive to do because sending a signal in digital form requires conversion from analog to digital, transmission of the digital signal, and conversion of that signal back to analog form for human use. What has made this possible recently is the great decline in the cost of computer memory and processing power.

The devices to do this are called *analog-to-digital converters* and *digital-to-analog converters*, sometimes written as ADC and DAC, or A/D and D/A. Increasingly, these devices are called coder-decoders, combined into the portmanteau term *codec*. Another common device is the modulator-demodulator, abbreviated to *modem*. All of these devices, whatever they are called, convert signals from one kind to another (and sometimes perform other related functions as well).

To see schematically how analog and digital signals differ during transmission, look at Fig. 6.3, which shows an analog signal. By itself, as transmitted, it is OK. However, in the real world, by the time it gets to your receiver, the signal is weaker and noise has been added. The noise comes from a variety of sources. Some are natural sources such as lightning, while many sources are man-made, such as sparks and power lines. In general, the noise is another analog signal added to the original.

Your radio receives both the radio station you want and the noise together, and the receiver *cannot tell the difference between them.* How could your radio know, for instance, that a sudden pulse is noise and not a cymbal crash in the music on the radio? It simply cannot. Thus, if the ratio of strengths of the signal you want and the noise you don't want is too low, then the radio station is not listenable, and you change stations. This *signal-to-noise ratio*, S/N or SNR, is a measurement of the quality of your received signal.

Suppose instead that the radio station transmits (and your radio set can receive) a digital version of the same program. Since it is a digital signal, it is made up of only discrete pulses. It is received along with the same noise as the analog signal. But, if you look at the schematic illustration in Fig. 6.4, you will see that the

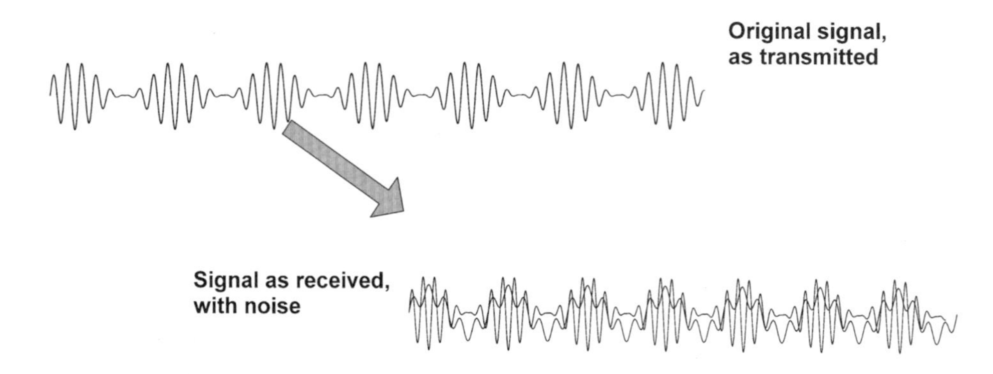

Figure 6.3 An analog signal may be of high quality as transmitted, but when received natural and manmade noise are also received. The major drawback of using analog signals is that the receiver cannot tell the difference between the intended signal and the unwanted noise, which results in reduced quality.

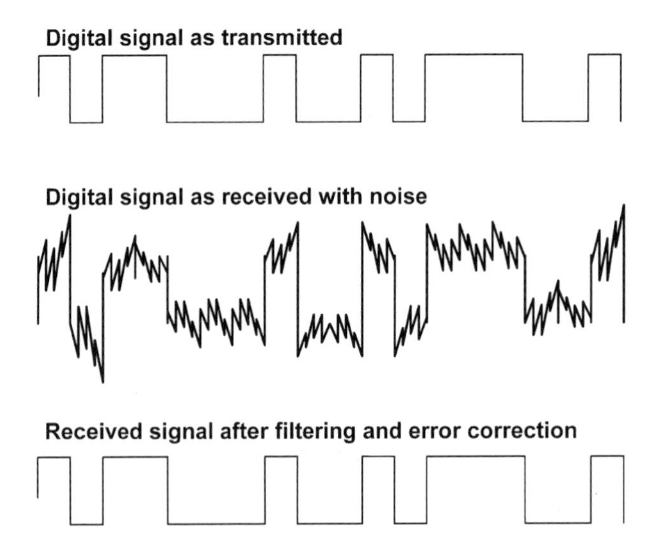

Figure 6.4 A digital signal consists only of bits. Even if these are received along with noise which distorts their original shape, if the noise is not too bad the receiver can still detect which are the bits, and filter them to regenerate their original form. This means that the receiver can tell the difference between signal and noise, resulting in a high-quality signal being received.

received signal still has identifiable pulses, even if they are noisy. It is the existence of the pulses, not their precise shape, that conveys the information. By running the noisy signal through a circuit that smooths out the pulses, your radio can reconstruct each and every pulse. Thus, your digital radio can tell the difference between signal and noise. The digits can then be converted back into a high fidelity audio signal and you will hear the radio station with a fidelity as good as if you were in the radio studio.

Now suppose, for instance, that you had to receive this radio station and amplify it to send it to a more distant listener. With an analog signal, when you receive the signal along with the noise, you unavoidably amplify the noise along with the signal, so you are relaying a degraded signal. Then more noise affects the signal you forwarded, so the distant listener gets a signal that is of even lower quality. Imagine repeating these steps many times, and what happens, for instance, to a telephone call sent along one of the old analog transatlantic cables.

However, if you are relaying a digital signal, you are removing any noise before retransmitting it onward, and thus you are relaying a signal that emerges from your relay transmitter equal in quality to the original signal. And so would any further relays. No matter how many relay stages there are, the final recipient gets a perfect signal.

For another analogy, imagine making a photocopy (which is an analog copy) of a document. It will not be as good as the original. Then make a copy of the copy, and a copy of the copy of the copy, and so forth. The quality keeps getting worse. Compare this situation with the case where you have your wordprocessor print out multiple (digital) copies of the same memo: each one is perfect.

Thus, we say that digital signals are not just amplified at each stage, as are analog signals, they are *regenerated.* This is the advantage of using digital transmission methods, and justifies the added equipment and expense of coding and decoding the signal for transmission.

Furthermore, the bits that represent the signal can be controlled and manipulated to make them more versatile. For instance, a signal might include not only a television picture, but some additional bits (you never see) that authorize your television set-top box to receive the movie. Or bits from different signals could be interleaved (carefully!) to allow many independent signals to be carried on a single channel.

The following is a summary of the major advantages of digital transmission:

- It is highly resistant to interference, because a receiver can tell the difference between desired signal and noise;
- Its signal quality is almost independent of distance and of the network topology (the geometrical arrangement of users);
- It allows better frequency sharing with other signals;
- It allows greater numbers of signals to be sent simultaneously when using a multiple-access technique;
- It allows the use of powerful data compression techniques, thereby allowing multiple digital channels to occupy the same bandwidth that would be occupied by a single analog signal, providing major economic benefits.

6.4 Digital compression

That last point is very important, and it works because much of the content of most signals is redundant, predictable, or does not change much. Here are a couple of examples.

As a simple example, suppose you want to fax this page to someone. (Don't: it's a violation of copyright laws.) Most of this page is white space, and on a typical typed page less than about 10% of the total area is characters (black). Yet, a fax machine has to scan every part of the page, line by line, and the lines are only 0.01 inches apart, both horizontally and vertically. Each picture element, or *pixel*, is a tiny square of the page 0.01 inches on a side. A single scan line on the page contains 850 (8.5×100) points. Thus, the total amount of information in a facsimile of a single 8½-by-11-inch page is $8.5 \times 100 \times 11.5 \times 100 = 935{,}000$ bits, even in the simplest case where the scan can detect only black and white, represented by 1 and 0. But a fax uses a compression technique. When it encounters a large string of the same color, such as a totally blank (white) line, instead of sending 850 0s it sends a special code that tells the receiving fax machine "the next 850 bits are white." This saves a lot of time and transmission capacity.

While compression can be, and is, used for signals carrying pure data, facsimile, and audio, its greatest benefits come in the video field because video is so much more demanding that anything else.

Recall that a television picture, while it appears to be moving, is, like a motion picture, made up of many still pictures, called frames, projected so quickly that your eye can't detect them individually. Each frame is made of hundreds of lines consisting of pixels. In North American television (called NTSC), there are 30 frames per second and 525 lines per frame (there are 25 frames per second, each of 625 lines, in the European PAL and SECAM methods).

Consider watching a golf tournament on television. Much of the time, the picture consists of three major items: a big stationary band of green grass at the bottom, a big stationary swath of blue sky at the top, and a very small white dot—the golfball—moving across them. From frame to frame (1/30 of a second), not much changes on the top or bottom. To save transmission capacity—i.e., to reduce the number of bits that have to be sent—why not just omit retransmitting the green and blue portions until they change, and only send the bits representing the white dot?

Actually, we can throw away even more than that. Since the motion of the white dot can be measured and thus predicted after a few frames, why not only transmit the bits for the dot infrequently, and let the receiver predict where the dot will be in the next frame for several frames in a row? Using these methods, one can avoid resending much of the picture, saving capacity.

In such a television picture, we can actually throw away over 98% of the information on the screen after it has been sent once! Then the bits that would have been used to send the unneeded information can be used to send an entirely different picture. Of course, when the picture of the golf tournament changes, and the camera sweeps across the crowd, almost every pixel on the screen changes from frame to frame, so those frames cannot be compressed very much.

So in one signal, a DBS operator might try to group some demanding events, such as sports, with some moderately demanding channels such as movies along with often-undemanding stations like the talking heads on a news broadcast. By using sophisticated digital compression carefully, the broadcaster can send 5 to 15 channels using only a single transponder aboard a satellite. This technique is basically what has made the DBS industry viable, especially in North America.

North American DBS systems allot 32 transponders to each DBS orbital location. Each was intended—20 years ago—to carry a single analog television signal. When the systems were planned, 32 channels was a lot, and would have been competitive with terrestrial cable television systems. However, by the time DBS was ready to get off the ground (for many years critics claimed it stood for "definitely behind schedule"), cable systems were providing customers with many dozen channels, and few consumers would have opted for DBS over cable if DBS only gave them 32 channels. But now, each of the 32 transponders can carry several digital channels, multiplexed together, so a DBS customer can get more than 150 digital channels using only 32 analog transponders.

Digital compression techniques go by names such as JPEG (for still images) and MPEG (for video). The technical details of how this is accomplished are very, very complicated and well beyond the scope of this introductory book. If you really want to know, consult some of the reference books in the bibliography.

There is one final item having to do with compression. The same word "compression" is also used for a technique to improve only analog signals, along with its opposite technique, called "expansion." The two words are often combined into another portmanteau word, "companding." Companding is a method of trying to maintain analog signal quality, and we will discuss its detail later in Chapter 8. You must keep in mind that there is no connection or similarity between digital compression and analog companding.

Having now provided the background on electromagnetic waves and the two basic types of signals, we are ready to explore how we use these to carry the vast and variegated signals that flow through the global telecommunications networks.

Chapter 7

Carrying Information on Waves

Telecommunications is the technology of carrying information by electrical and electronic signals. In the satellite communications sector of telecommunications, we utilize radio waves to carry the information, and to do that, we must somehow impress the information onto the radio wave.

Here we are using the word "information" in its most generic sense to mean whatever it is that we want to send. The information could simply be numbers from one computer to another, it could be a telephone call between two people, an audio or video broadcast to an entire continent, or a downloaded page from a World Wide Web server to your computer.

(This chapter, and the next, are among the longer and more technical chapters in this book. They contain a lot of information that you do not have to understand in detail to use satellite (or any other) telecommunication, but they will provide you with a fuller understanding of the subject. If you get over your head, skim over these chapters now, and come back later after reading later chapters that show how these techniques are used.)

7.1 Carrying information

The information we want to telecommunicate—numbers, words, sounds, images—must first be converted into some kind of electronic signal. The goal is to do this quickly, economically, and accurately, while striving for efficient use of the resources of power and of spectrum. As we saw in Chapter 3, different users and different services may have very different ideas about which of these criteria are most important. Thus, is it most important to know the desiderata of the intended services before trying to decide on the technology and detailed techniques to accomplish them.

Usually some intelligence we want to send (numbers, sounds, pictures) goes through many electronic manipulations during the journey from sender to receiver. Many times a single signal may be combined with other independent signals to promote efficiency of transmission. These electronic manipulations must be carefully planned and executed to achieve maximum fidelity of the signal.

The first task is to express the information we want to send, called the source, in an appropriate electronic form. For digital signals, this step is called *source coding*. This may simply convert the signal into an electronic form, or it may additionally encrypt the signal for security. This then produces what we earlier termed the baseband signal, a datastream representing a single telephone call, television program, or group of numbers for machine use. In addition, the particular service may wish to further encode the signal to provide for error-free transmission, and/or for security to ensure reception only by authorized receivers. Coding for error control and security may occur at several stages during a transmission—for example at the source and again at the earthstation. If the source-coded signal is further encoded for transmission to ensure accurate reception, this is called *channel coding*. This may occur in several places in the transmission chain between sender and receiver, and may again be for error control or for security or both. Figure 7.1 shows the steps in successfully telecommunicating information.

Next, this relatively low-power baseband signal must be transmitted in a manner that ensures reliable reception by the intended recipients. To do this, the baseband signal is used to control—the technical term is *modulate*—a relatively high-powered radio wave—the carrier—that will convey the information to the receiver. The details of the modulation technique will need to take into account the nature of the original signal, the characteristics of the sender and receiver, and the accuracy requirements of the particular application.

Finally, since a carrier is often capable of carrying many baseband signals, the baseband signals are often combined to use the system most efficiently by a process called *multiplexing*, to utilize the full capability power of the carrier wave. Again, the technical and economic needs of the particular application will dictate what technique(s) to use.

In the following sections, we will more fully explain some of the basic concepts of coding, modulation, and multiplexing, with the caveat that we are only skimming the surfaces of highly mathematical subjects. A detailed look at these methods is well beyond the scope of this book. For more details, you are referred to some of the massive tomes in the bibliography.

7.1.1 Coding

We have already seen the simple coding examples of Morse code, and ASCII, for digitizing text communications. More complicated signals, such as audio and especially video, require more complicated and faster techniques.

Much of the information we use today starts out as an analog signal: a telephone call, a television program, a sound recording. A few things begin as "pure" data: a facsimile transmission, a webpage, a Morse code signal, an e-mail, a position determination by a navigation system. As we have seen, information is increasingly transmitted by digital techniques, for all of the reasons explained in Chapter 6. However, this means that an extra step is required at both the sending and receiving ends of an analog application: making it a digital signal at the source and turning it back into a human-intelligible analog signal at the receiver.

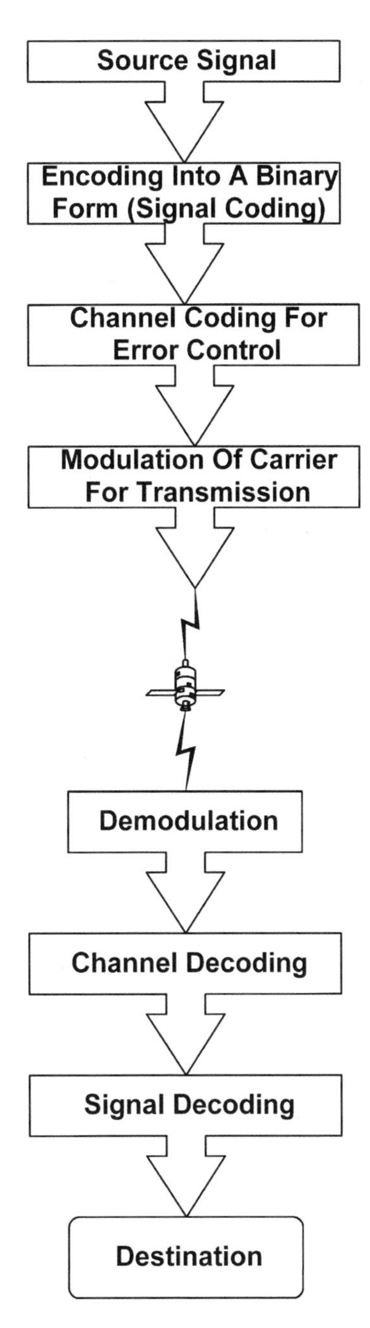

Figure 7.1 When information is transmitted, particularly digitally, it goes through many stages of coding and decoding. Some of these are designed to make an analog signal into a digital one, others to ensure error-free reception, and possibly still others to prevent unauthorized reception. One of the challenges for the telecommunications engineer is to make sure that all these processing steps do not interfere with one another, causing distortion of the signal.

To enable an analog signal to be transmitted digitally, we must first encode it into a digital signal. There are many ways to do this, depending on the information content and required transmission speed of the signal.

7.1.2 Pulse-code modulation

By far the most common technique for audio signals, especially telephone calls, is called *pulse-code modulation, PCM*. Its purpose is to convert a smoothly flowing audio signal into a digital stream of pulses. It is the way that some of the music you listen to is encoded, and the way that every telephone call is carried through the world's public switched telephone network. While your telephone call goes between your telephone and the local telephone company central office as an analog signal (unless you have special digital lines such as ISDN), one of the first things that happens to your voice when it reaches the printed-circuit card in the central office is that it is bandwidth-filtered down to the prescribed 4 kHz bandwidth, and then turned into digits. PCM has some advantages over regular frequency modulation for low *S/N* signals.

PCM works by sampling the audio signal. See Fig. 7.2 for a schematic illustration. Your voice is a squiggly waveform of voltages changing with time as you speak, and it contains frequencies up to 4 kHz. Every so often, the PCM circuit *samples*, that is, takes an instantaneous measurement of, the voltage of your voice. That voltage is simply a number. Now, instead of expressing that number of volts as an everyday decimal number, it is converted to its equivalent binary number, which as we have seen, consists only of 0s and 1s. For example, if at some instant your voice caused a voltage of 25 V, the corresponding measurement would be 00011001, which is the same as 25 in the decimal system.

Now two questions arise: How accurately must you measure the voltage, and how often must you measure it?

As for accuracy, trial and error has led to the conclusion that if you allow 256 different levels of voice power, from no sound at all to the maximum allowed in the wires, then you get a good representation of the human voice. To count up to 255 in the binary system (0 to 255 is 256 levels) requires 8 bits; so silence would be encoded as 00000000, while screaming into the telephone at a telemarketer would come out as 11111111.

Sampling the signal would seem to turn a continuous signal—your voice—into a choppy, discontinuous one. But as far back as 1928, scientist Harry Nyquist at Bell Laboratories showed that if you sample a changing signal at twice the rate it changes, you can later reconstruct the original signal with perfect fidelity. This 2-to-1 sampling rate is called the *Nyquist Rate*, or the *Nyquist Sampling Criterion*. Sampling at less than twice the original signal's maximum frequency is dubbed *sub-Nyquist*. To take your telephone call as an example, since the maximum frequency is 4 kHz, the Nyquist Rate would be 8 kHz, or 8,000 samples per second. Although this may seem fast to humans, to a computer it is coasting.

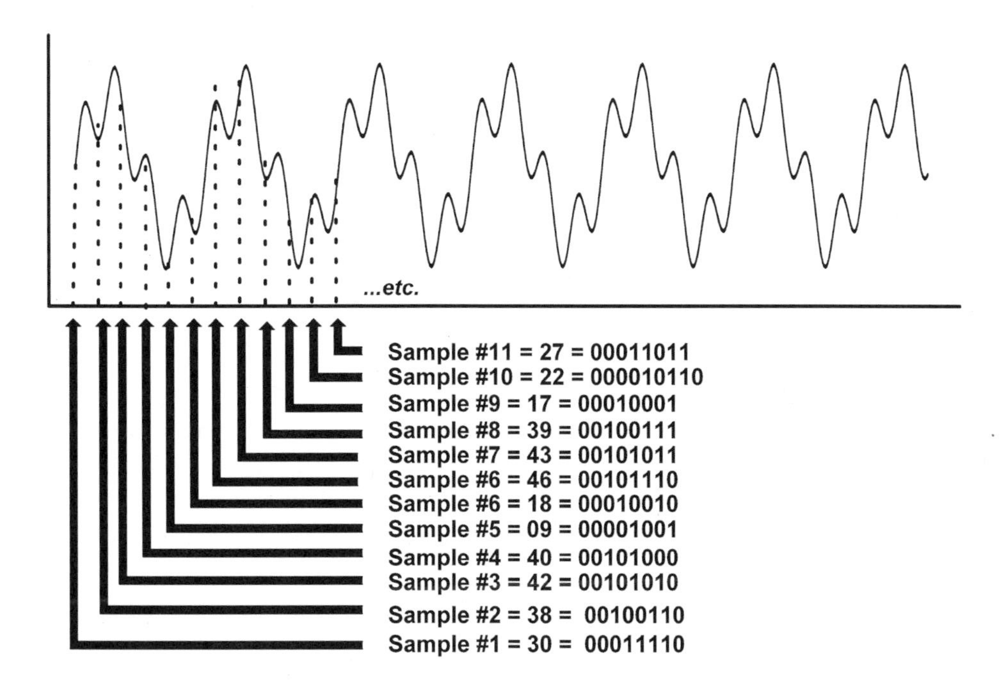

PCM works by periodically sampling an analog signal, measuring its amplitude at each sample, and then expressing that value in binary form, thus converting an analog signal into a digital signal.

Figure 7.2 Pulse-code modulation, one of the most common ways of turning an analog signal, such as a telephone call, into a digital signal for efficient transmission. The continuously-varying analog signal is measured ("sampled") at frequent intervals, and each such measured value is expressed as a binary number. For perfect fidelity, the sampling rate must be twice the maximum frequency contained in the source analog signal.

So 8,000 times each second, your voice is measured and each measurement uses 8 bits to represent the voltage. By simple mathematics, we thus see that an analog telephone call produces a digital telephone signal of $8 \times 8{,}000 = 64{,}000$ bits per second, or 64 kbps. You may find this number familiar, as it is considered the bitrate for a standard voice channel. (Some telephone systems use 7 bits per sample, with a corresponding bitrate of 56 kbps.) Because telephone technology came first and is so ubiquitous, the 64 kbps and 56 kbps channel speeds became global standard datarates.

To cut down on the number of bits per telephone call, we can either reduce the sampling rate—and consequently the quality of the voice—or reduce the number of bits per sample (one form of data compression) or both. Telephone companies, to maintain high quality, always sample at the Nyquist Rate, but often use compression. Since much information in your voice is redundant, there are times when you are not speaking, and you cannot change your voice very quickly, one clever technique is to transmit not the absolute voltage of your voice at a given instant, but only the value representing the change in voltage from the last sample. This is likely to be smaller than the absolute voltage, and will require fewer bits. Such techniques are called *adaptive differential PCM*, abbreviated ADPCM.

7.1.3 Encoding video signals

Video signals contain much more information, are much more complex, and require constant and very rapid transmission speeds compared to low-speed data signals or audio signals. Recall that the baseband bandwidth for a telephone call is 4 kHz, but for (NTSC) television transmission it is 6 MHz, over a thousand times as much. Thus, encoding an originally analog television program is much more complicated and demanding. Video signals very frequently contain an audio signal (the soundtrack), and often contain data as well, complicating the signal.

If you took an analog television signal and converted it to a digital signal by the simplest method, you would obtain a digital video signal requiring a bitstream of 100 to 200 million bits per second. That is a lot, and a huge demand on a telecommunications channel. Because of this, and because of the high amount of redundancy between the 25 or 30 frames of television each second, video is always coded with compression in mind, to reduce the actual bitrate needed to be carried by the satellite link. While there are many possible ways of digitizing a signal, the emerging global standard is a technique called MPEG-2. This encoding+compressing method is used by the two major digital television transmission schemes, DSS, used by DirecTV, and DVB (Digital Video Broadcast), used by almost all other digital television DTH services.

The details of MPEG are much too mathematical to go into here. You can just take it as a "black box" into which an analog television goes, and out of which comes a digital video signal with a datarate of a few megabits per second (the actual rate depends on how much compression you want).

7.1.4 Coding for error control

As we will explore more fully in the next chapter, when a telecommunications signal is sent, the receiver picks up not only the intended signal—greatly reduced in power due to space loss—but also receives noise from the environment. This noise can cause occasional bits to be received incorrectly: *errors*.

The measure of quality of a digital signal is called the *bit error rate, BER*. The BER is the average number of bits received in error as a fraction of the total number of bits sent. It is usually referred to as a power of 10, such as 10^{-4} or 10^{-8}. Confusingly, the same BER can also be stated as 10^4 or 10^8, so when you see a BER number, just ignore the sign of the exponent. Thus, a BER of 10^4 or 10^{-4} means that on the average, one bit out of 10,000 is received in error. (In the real world, errors often tend to come in bunches, called *error bursts*, but that is a detail we will not go into here.)

The number of errors that can be tolerated in a received signal depends on the particular application. The telephone companies long ago discovered by trial-and-error that a digitized telephone call can have a BER as high as 10^{-4} without the human ear hearing a degradation in the sound of the speaker's voice. Some military satellite systems, which need understandable but not necessarily perfect voice quality, allow as many as an error per every few hundred bits. Digital television reception may have error rates of 10^{-7} to 10^{-10} and not seem to degrade the picture. On the other hand, a bank transfer must be absolutely correct.

Coding requires equipment, and thus money, so error control is not costless. First, it takes time for the originating computer circuits in the transmission path to actually accomplish the coding, which will depend on the characteristics of the encoding computer. To catch errors, some kind of checking and correction procedure must be built into the communications protocol, and that takes time.

A receiver must first determine if there are errors. If there are, then some way must be provided to correct these to the accuracy needed. There are two basic techniques for error control: ARQing (pronounced "arcing"), and forward error correction.

ARQ stands for *automatic repeat request.* ARQing is done by adding to the information bits some extra bits used for checking. The receiver can use these to verify that all of the bits have been received correctly; then, if they have not, it can ask for the group of bits to be repeated. See Fig. 7.3.

The simplest version of ARQing is used to send bits around inside your personal computer and to its peripherals. It is called *parity.* It relies on the very simplicity of digital signals, that there are only two kinds of errors that can occur: a 1 is sent and a 0 is erroneously received, or vice versa. Take the ASCII code for example, that world standard coding that expresses text information and some control information (e.g., linefeed, carriage return) in a seven-bit "word." The word, of course, is just a string of 1s and 0s. Parity works by deciding beforehand that you want to add one extra bit to the data word, making up an eight-bit "byte," such that in every byte there is either an even number or an odd number of bits that are 1s. (It doesn't matter which scheme you choose.) So, for instance, if the sender sends an A, which in ASCII is 1000001, and the sender and receiver have agreed on even parity, the sender will append one extra bit to this code. Since the code for A already has two 1s in it, an even number, the sender appends a 0 to the word, and what is actually sent is the byte 10000010. Then if the receiver gets this and sees that there are an even number of 1s in the byte, it assumes that it was received correctly. So the receiver prunes the added 0 off the byte and correctly interprets the bits as representing A. It also sends a special signal, called an *acknowledgment,* or *ACK*, back to the sender to tell it that the data was received correctly. When the sender receives the ACK, it then sends the next bunch of data.

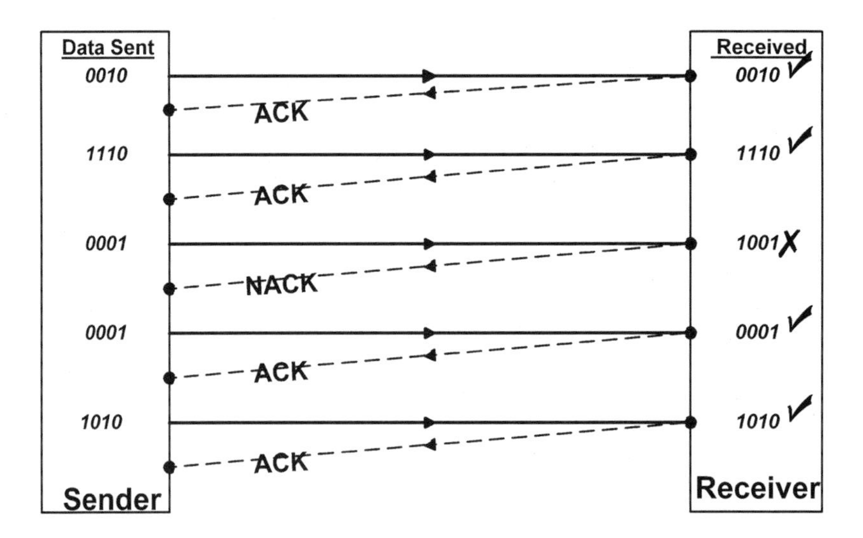

Figure 7.3 Automatic Repeat Request error correction works by having the receiver check each received data packet for errors. If the packet is error-free, the receiver sends an acknowledgment signal (ACK) back to the transmitter, which then sends the next packet. If the received packet is in error, the receiver sends a negative acknowledgment signal (NACK), whereupon the transmitter resends that packet. The longer the time it takes for the ACK-NACK signals to go between sender and receiver, the slower the communication rate.

But if some noise in the transmission path has caused an error, and, say, the received byte is 11000010, the receiver sees that there is a odd number of 1s and it has to be incorrect. Of course, any of the eight bits could be the one in error, and the receiver can't tell which one is the bad bit. So the receiver then automatically sends a *negative acknowledgment,* or *NACK*, signal back to the sender asking for the byte to be sent again. Hopefully it gets through error-free the next time.

There are several different levels of operation of ARQing. In the simplest, after each group of bits, the receiver checks them for errors, as described above. This is sometimes called stop-and-start ARQ, or half-duplex ARQ.

A slightly better method is for the sender to transmit continuously, and to number each packet of data in sequence. If an error is detected by the receiver, the NACK it sends back notes which packet was in error, and the sender backs up to that packet and begins sending again. This is called ARQing with pullback, and is somewhat more efficient.

Better yet, if the receiver can store incoming data after some error, the sender need only retransmit the single packet containing the error. This requires more buffering and storage at the receiving end, but increases overall efficiency. This method is called ARQing with selective retransmission.

You can see (at least) two problems with this simple method of error control. First, if two errors happen to the same byte, they will offset each other, so the receiver will count a correct parity and erroneously think the byte is correct. The second problem is that it takes time for the ARQ to report an error to the sender, and the sender has to stop transmission and resend the byte again. The longer the time this round-trip takes, the less efficient this method becomes. This technique works fine along the two-meter cable connecting your PC to your printer because the ARQing is done in microseconds. But for geostationary satellite links, where the round-trip takes about half a second, this really impedes communication. Half a second is almost eternity to a computer, and so the effective throughput on a noisy channel is low.

However, one advantage of ARQing is that, if you are willing to use sophisticated enough error-catching schemes and wait for the needed repeats, you can achieve almost perfect accuracy of reception.

7.1.5 Forward error correction

FEC is a powerful technique of error control that has all but replaced ARQing in long-distance telecommunications because it is more efficient. The scheme here is to add enough extra bits to the datastream being transmitted so that if an error is received, the receiver can not only detect that an error has occurred, but also correct it without asking the sender to send it again. It will not guarantee perfect accuracy, but can be configured to provide whatever accuracy is needed. It is not a matter of simply adding redundant bits on to the beginning or end of a group of data; instead, the data is multiplied by complicated formulas, or codes, that allow the receiver to detect not only that an error occurred, but which bit is in error. The simplicity of digital signals works to our advantage here because there are only two kinds of errors that can occur.

You can also see that FEC is required for a signal that cannot be interrupted to resend data to correct errors, such as in a digital audio or digital television broadcast. Each receiver needs to be able to correct the errors it receives.

There are many codes that can provide FEC. They often go by the names the people who thought them up: Reed-Solomon, Golay, Viterbi, and others. Some newer varieties are called turbocodes. Believe me, you do not want to know the details of the mathematics. Each has its strong points in terms of catching errors and difficulty of implementation. Although we have not discussed modulation yet, it is important to note that coding and modulation work together to provide efficient communications. In this context, "efficient" means achieving the error rate your applications require while transmitting the most bits per bandwidth and the lowest power.

Error control methods are rated by how much redundancy (and thus overhead) they provide to the information. Going back to the simple parity scheme, for every seven bits sent that actually contain information, you add an eighth parity bit. Thus, you effectively slow down your information throughput by about 1/7, or 14%, because all bits sent don't represent information. For instance, suppose you had a channel capable of carrying a thousand bits per second, 1 kbps. Without parity, you could send almost 143 seven-bit ASCII characters per second (1000÷7) over this channel, but some may be unknowingly received in error. Using parity, each seven-bit character encoding becomes an eight-bit byte, and thus you can only send 125 characters a second (1000÷8), a 14% decrease in effective channel capacity.

With FEC, the *coding rate* is expressed as a fraction, such as 1/2, 2/3, 3/4, 7/8, etc. A system using 1/2 FEC actually sends out two bits for every information bit, and enables good correction of errors, but it effectively slows your channel speed down to 1/2 its rated speed. An FEC of 3/4 does not catch as many errors, but it slows the effective speed down by only a third.

This has real-world business consequences. In the 1990s, some satellite system operators experienced shortages of space segment capacity because of failed satellites or failed launches of new replacements for older satellites. Most of their data customers were using 1/2 FEC. In order to increase the effective bandwidth available on those satellites on orbit, the operators asked their customers to shift to using 3/4 FEC. Thus, each transponder could carry more information, although the total real capacity of the transponder did not change. The increased possibility for errors in each channel was compensated for in other ways.

ARQing and FEC are useful for different kinds of tasks. The table below summarizes their characteristics.

Performance Issue	**ARQing**	**FEC**
Datarate	Low to medium	High to very high
Delay	High	Low
Good for real-time services	No	Yes
Complexity of decoding	Low	High
Requires return channel	Yes	No
Overall throughput	Low	High

7.1.6 Coding for security

Mathematical techniques similar to those used to guard against errors are used to guard against unauthorized reception, whether for corporate, economic, or national security reasons. It is particularly important for broadcast transmission technologies, such as satellites. A major advantage of geostationary satellites is that from a single satellite, 44% of the earth's surface can get your signal. A major disadvantage is that 44% of the earth's surface can get your signal, and you may not want everyone in the satellite footprint to get it.

Coding for security is just as complicated as coding for accuracy. Many companies have proprietary, or quasi-proprietary methods of keeping their communication (hopefully) only to those for whom it is intended. Entire multi-hundred page books are devoted to the subject.

Related to this is the topic of *conditional access*, which is the technique of uniquely addressing the communications stream so only intended recipients can receive it. The DTH television systems and DARS radio systems use such techniques to insure that only their paying customers get the programming. The difficulty and expense of encrypting and decrypting signals is dictated by just how important and valuable your signal is.

While we cannot go into detail about security techniques, it is important to realize that both coding for accuracy and coding for security are manipulations performed upon the original signal, perhaps several times before it is transmitted. Each such coding increases the total number of bits to be transmitted and received. Thus you pay a practical price for accuracy and security: you slow down the net amount of information you can send, because you are injecting all of these other bits into the datastream.

7.2 Modulation

Simply sending a constant radio signal conveys no information other than the simple fact that the radio station is on. To carry more information, and to utilize the communication method at maximal efficiency, changes must be made to the signal.

Modulation means change. In our case, it is the ways we change one or more of the properties of the radio carrier wave to make it carry information. The electronic device that modulates the signal information onto another higher-frequency wave, such as the IF, or which extracts the signal information from the IF, is called a *modem*, short for modulator-demodulator. In the case of multiple information signals carried on multiple carriers, a separate modem is needed for each carrier. They are located at the earthstation.

Taking a more personal human example, you modulate (change) the volume and pitch of the vibrations of the air coming from your lungs to form sounds that make up speech in order to send the information in your mind to that of another. These air pressure modulations reach the listener's ears and modulate the position of the eardrum, which in turn modulates electrical signals reaching the listener's brain, which are then interpreted as information.

Electronically speaking, we modulate one or more properties of radio waves to make it carry information. The three properties of a wave that we have to work with are its amplitude, or strength; its frequency, or wavelength; and its phase. See Fig. 7.4.

We have already discussed the concepts of amplitude; it is just the power of the wave. If the signal is a sound reaching your ears, its amplitude is the amount of air pressure; if that audio signal is in wires, the amplitude might be measured as the voltage or current in the circuits; if the signal is a radio wave from earthstation to satellite, the amplitude might be measured in watts per square meter of the radio wave.

Frequency, or equivalently, wavelength, has also been discussed. It is simply the rate at which the signal changes with time. Frequency modulation works by converting changes in amplitude to changes in frequency.

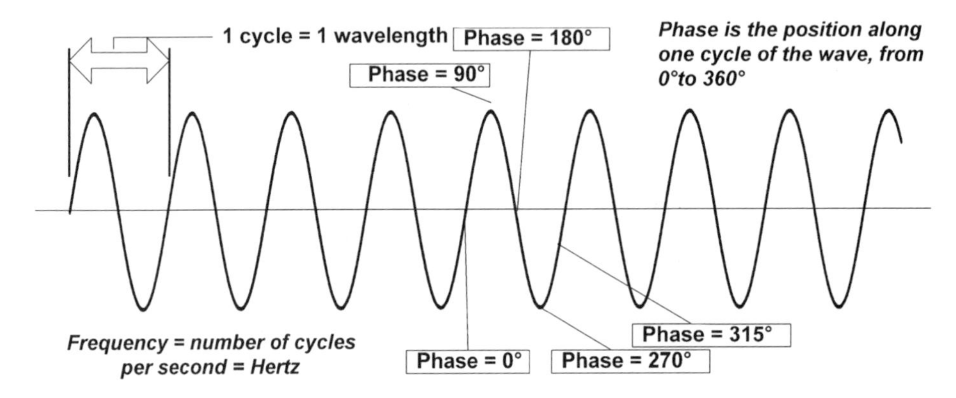

Figure 7.4 Properties of waves. Simple waves have a frequency, the number of waves per second. One hertz is a wave per second. The distance between the crests of the wave is the wavelength. There is an inverse relationship between frequency and wavelength. The term phase describes the position along one wave at any instant, from 0° to 360°. Complex waves can be thought of and analyzed as combinations of such simple waves.

Phase is a term that describes where the wave is at a certain time along one cycle. At the beginning of a cycle, or wave, the phase is said to be 0°; a quarter of a cycle later it is 90°; halfway through a cycle the phase is 180°; and three-quarters of the way along the phase is 270°. At the end of one cycle, the phase is 360°, equivalent to 0°, as on a circle. Phase modulation works by converting changes from 0s to 1s in the data into changes in the shape of the wave.

The radio wave that transports the information is called the *carrier wave*, and in some cases there may be several related carriers. If there is one major relatively wideband carrier, it may be called the main carrier. This is most common in sending video signals. Sometimes there are narrower-band carriers sent alongside of the main video carrier. These smaller signals are called *subcarriers*. In most television broadcasts, there is the main video carrier, and at least one subcarrier that carries the associated audio (soundtrack) for that television signal. If the bandwidth of the channel permits, other completely unrelated subcarriers may be used within the channel to carry such things as audio and low-speed data (Fig. 7.5). In the satellite industry, each analog television signal carried on a transponder is typically accompanied by several such subcarriers, and some transponder owners may make more revenue from selling or renting these subcarriers than they may get from their main television signal.

One important point to keep in mind is that the radio carrier itself is by its very nature an analog signal. The electrical and magnetic waves described in Chapter 5 oscillate continuously. These carrier waves may then be modulated by either analog or digital signals that they are intended to carry. When we talk about a transmitted analog or digital signal, what we really mean is that the carrier is analog modulated or digitally modulated. The carrier wave itself is always an analog signal of vibrating electrical and magnetic fields in space.

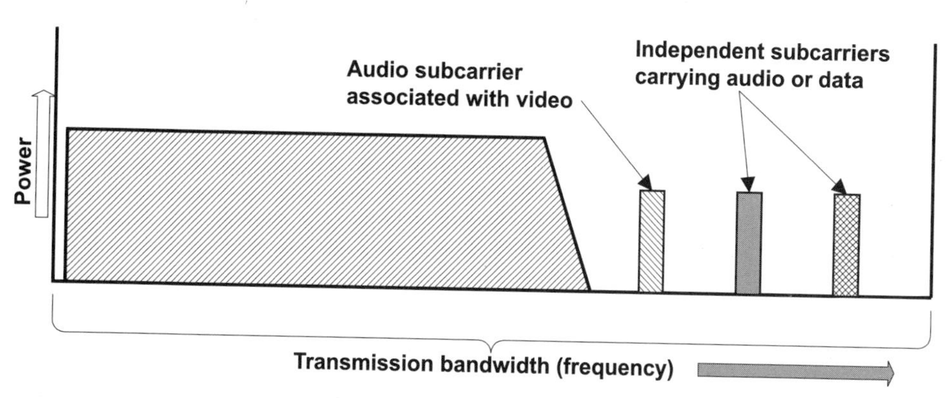

Figure 7.5 Often real-world signals are composed of several parts. In particular, a television signal consists of a main "video carrier" that takes up most of the frequencies of the television channel, and (at least one) "audio subcarrier" that holds the soundtrack associated with the picture. To avoid interference with adjacent television channels, there is vacant bandwidth beyond the audio subcarrier. With careful control of powers and bandwidths, this otherwise-wasted spectrum can be used to carry low-demand services, such as audio or low-speed data in other small-frequency bandwidths called subcarriers.

7.2.1 Analog modulation

In the case of an analog signal modulating an (always analog) carrier, we usually have two methods of modulation at our disposal: we can change the power or change the frequency of the carrier in response to changes in the signal input.

Changing the power, or amplitude, of the carrier is called *amplitude modulation, AM*. You are familiar with AM radio stations. Figure 7.6 shows a wiggly line that might be the power output of an AM radio station. Notice that it is a continuous signal, but its amplitude changes. The changes are controlled by the baseband signal your voice produces by talking into a microphone in the studio, producing corresponding analogous changes in the voltage or current in the wires. These electrical changes, in turn, control the power—the amplitude—of the radio station. If the maximum output power of the radio station is 50 kW, then you are controlling that much power with merely the sound of your voice.

Since the power of the radio station goes up and down in response to whether you speak loudly or softly, we immediately see the major drawback of AM. The desired signal from the radio station competes with natural and man-made noise coming into the listener's antenna. So when you speak softly there is more likelihood that your immortal words will be drowned by interference. AM radio does not have a very good *dynamic range*—the range between the softest and loudest sounds possible—because the soft sounds produce weak signals that get swamped by noise. As we saw in the last chapter, the receiver cannot tell if amplitude changes are the result of the original signal or of noise.

An alternative is to change not the amplitude of the radio station, but only its frequency; the radio station's power remains permanently at maximum. This is

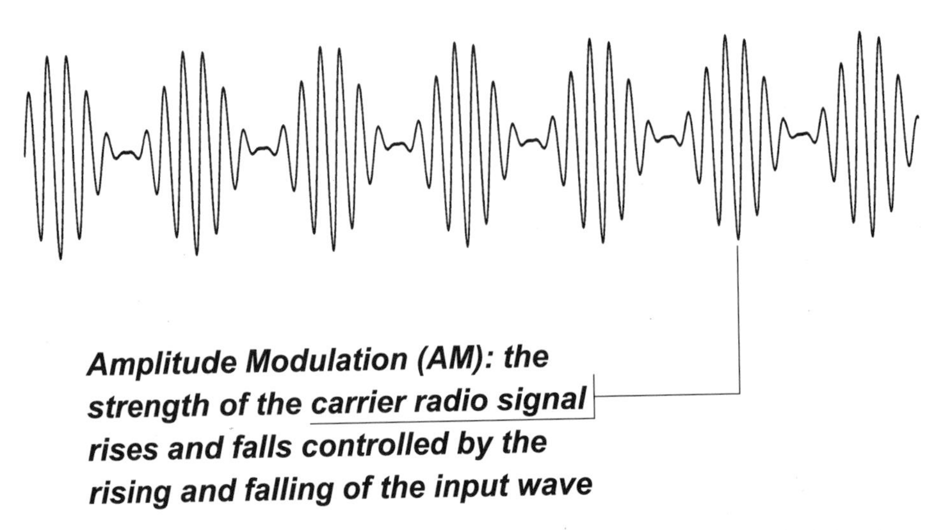

Figure 7.6 Amplitude modulation. In this scheme, the strength of the original signal controls the high power of the carrier wave, thus modulating its amplitude. The fidelity of analog AM signals is not very high because the signal can be variably reduced and noise can enter the channel on the way to the receiver.

called *frequency modulation, FM*. At an FM station, when your voice in the microphone gets louder, the frequency of the radio station increases slightly; when you speak softly, the frequency drops a little, but the strength of the station stays the same. Most sources of noise are amplitude changes, so if we build an FM radio receiver to detect only changes in frequency and not changes in amplitude, it will be resistant to noise and have a better dynamic range. This is why FM stations sound better than AM stations; indeed, once they were referred to as the "good music stations." FM is the most common modulation scheme used when satellites carry analog signals.

In exchange for this better quality, FM stations must occupy a broader band of radio frequencies, in other words, they take up more spectrum. This is a kind of prelude to a general principle we will explore more fully in Chapter 8: that bandwidth is one of only two parameters that together control the quality and quantity of information transmitted. The other is signal power.

For emphasis, let us note that in both examples above, an analog signal—the radio carrier—was modulated by another, much lower frequency analog signal, your voice. But, we increasingly want to take advantage of the benefits of digital transmission techniques, so we use a digital signal to modulate the analog carrier.

7.2.2 Digital modulation

Remember that in a digital transmission, the carrier itself remains an analog signal. It is the modulating signal that is digital. Thus, if the modulating signal represents something that was initially analog in nature, such as a sound or picture, it must have first been encoded. It may have been encoded several times, for such purposes as compression, error control, and security.

The simplest digital signal consists of a string of 1s and 0s. How these control the carrier defines the three basic means of digital modulation. See Fig. 7.7.

In the simplest form, similar to AM radio, the 1s and 0s control the strength of the carrier. However, instead of producing a range of carrier strengths, an incoming 1 results in the carrier at full strength, while a 0 results in no carrier at all—in other words, on and off. This can also be called amplitude modulation, but is often called, to emphasize its digital character, *amplitude shift keying, ASK.* Cable modems typically use 256QAM, or 256-phase quadrature amplitude modulation. Some satellite transmissions use 16QAM.

Morse code is a simple example of ASK, except in this case, we have two durations of "on" called dot and dash. It sometimes comes as a surprise to today's digitally conscious people that the very first telecommunications method was digital!

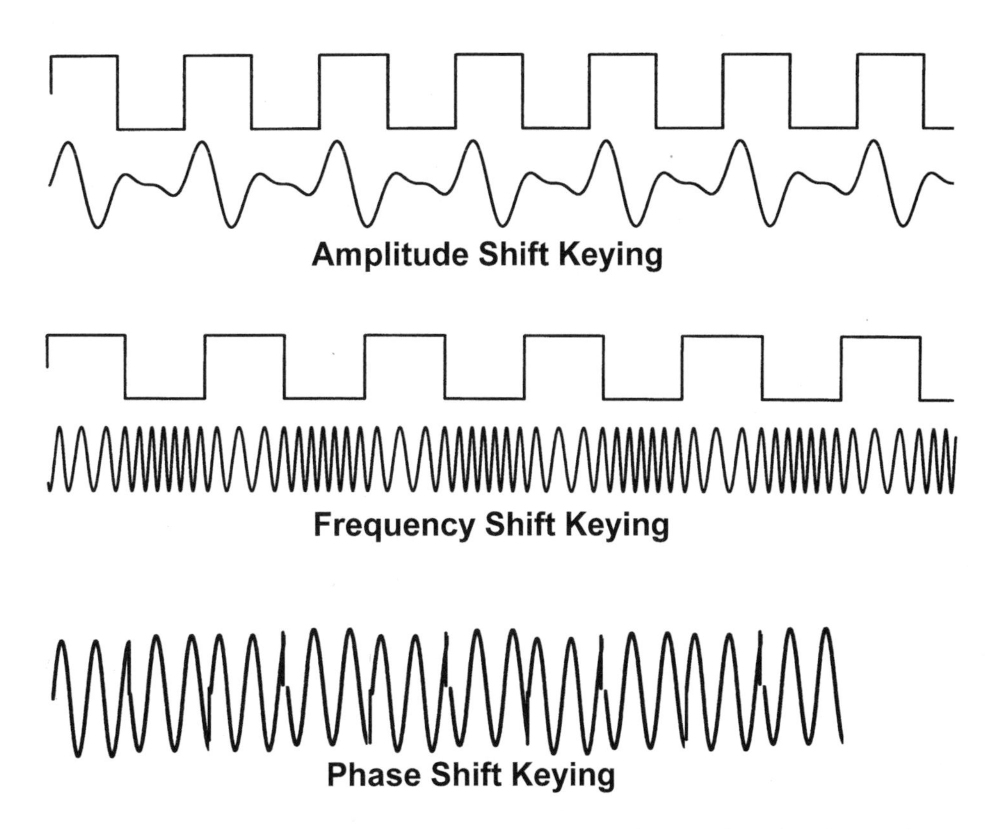

Figure 7.7 Digital modulation. In this scheme, the bits of the original signal may change various properties of the carrier wave (which is always an analog signal). In amplitude modulation, more commonly called amplitude shift keying (ASK), different strengths of the carrier represent the 1s and 0s of the signal. In frequency modulation, or frequency shift keying (FSK), different frequencies represent the bits. In phase modulation, or phase shift keying (PSK), when the signal changes from a 0 to a 1, or vice versa, the phase of the carrier shifts.

If the 0s and 1s of the incoming digital signal control the frequency but not the strength of the carrier, we get digital frequency modulation. To emphasize the digital nature, this technique is more commonly termed *frequency shift keying, FSK*. In FSK, for example, a 0 may be represented by a low tone, and 1 by a higher-pitched tone. If the receiver of the signal is sensitive to only these two tones, it will reject any noise of different tones and thus receive a higher-quality signal.

As a real-world example, this is how the modem connected to your personal computer acts as an interface between a device that works with pulses of electricity, and a telephone system that carries the frequencies of the human voice. Recall, *modem* is a contraction of *mo*dulator-*dem*odulator. It converts the 0s and 1s from your computer into tones the telephone system can handle.

The third wave property we can alter is the phase. While this is harder to do and to receive, it provides many benefits. *Phase modulation, PM*, more commonly called *phase-shift keying, PSK*, is the most spectrally efficient way of sending digital information. With PSK, the carrier always remains at the same amplitude and frequency. When the incoming signal changes from a 0 to a 1, or vice versa, the phase of the carrier shifts. PSK is very resistant to interference, since little in nature can produce such a phase shift. Phase modulation is the most spectrum-efficient way of sending digital signals.

Recall that phase refers to the distance along a wave's cycle. We can choose how far we shift the wave when the transition from 0 to 1 or from 1 to 0 comes along. In the simplest case, we can shift the wave half a wavelength; in other words, if the wave was going up when the bit arrives, the wave is altered to make it now go down, or vice versa. Since there are only two such changes possible with a half-cycle shift, this is called *binary phase-shift keying*, or *BPSK*. If we shift by a quarter of a cycle, there are four possible shifts, and so this is called *quadrature* (or *quaternary*) *phase-shift keying*, or *QPSK*. There are many other variants, including 8PSK, 12PSK, etc., but BPSK and QPSK are the most common in satellite transmissions because of the amount of noise inherent in the link. For satellites, we can expect usage of 8PSK to grow to increase throughput. (In contrast, because of less noise in the system, fiber optic systems often use 64- or 128-PSK.)

To provide even higher speeds over channels with limited bandwidth, various combinations of ASK, FSK, and PSK can be used. This is the way, for instance, that telephone modems have been able to increase in speed from their original 50 and 110 bps speeds, which now seem glacially slow, to the newer V.90 modems capable of 56 kbps. They accomplish this by a complex combination of modulation techniques (and can reach even higher effective speeds by using data compression).

One common combination modulation techniques is called ASK/PSK, in which both the amplitude and phase of the carrier wave are varied. This is also known as QAM, *quadrature amplitude modulation*, and comes in several levels, such as 16-QAM, 32-QAM, etc, indicating how many different amplitude/phase combinations are used.

7.2.2.1 Energy dispersal

The effect of any modulation scheme is to spread the intelligence you wish to transmit across a range of frequencies of the carrier radio wave. Because the nature of the intelligence (datastream, telephone call, television signal, etc.) changes from application to application, and changes with time during a transmission, the modulated signal must be kept within a useful range of powers so as not to interfere with other signals. This is accomplished by a technique called *energy dispersal*, which evens out the power distribution. It reduces the possibility of interference, both with other signals and from intermodulation distortion. Most transmitters use some sort of energy dispersal scheme, with the details dependent on the nature of the signal.

7.2.3 Modulation, forward error correction, and throughput

Modulation and coding work together to produce a reliable signal, within the allowed bandwidth and economic constraints of the particular application. Since satellite bandwidth is limited, typically to 500 MHz for FSS satellites, one prime consideration is conservation of bandwidth, or, put operationally, getting as many bits as possible into a given band. The number of bits per hertz of bandwidth is referred to by engineers as the *packing rate* or *packing ratio*.

As an example, the following table shows some representative digital modulation methods, and the number of bits they allow per hertz for several FEC coding rates. It should be kept in mind that one other important parameter is not mentioned here: power. Higher degrees of modulation (for example, 8-PSK versus BPSK) typically require more power while conserving bandwidth.

PSK Level	FEC Rate	Theoretical bits/Hz	Typical bits/Hz
BPSK	none	1.0	0.80
	1/2	0.5	0.4
	2/3	0.67	0.53
QPSK	none	2.0	1.6
	1/2	1.0	0.80
	2/3	1.33	1.07
	3/4	1.50	1.20
8-PSK	none	3.00	2.40
	1/2	1.50	1.20
	2/3	2.00	1.60
	3/4	2.25	1.80

7.2.4 Spread spectrum

Spread spectrum is a coding and modulation technique that allows many independent signals to use the entire bandwidth of a transponder simultaneously. It is highly mathematical, and relies on uniquely coding each signal in such a way that the codes used for each one do not interact with each other. The encoded individual signals are simply added together at the transmitter, intentionally spread over a wide bandwidth much broader than the individual signals, and then sent. A receiver picks up this combined signal, and to extract the one datastream it wants, must know the specific code used to encode the data.

Because this kind of technique allows many signals to share a communications channel, it is often referred to as *Code Division Multiple Access,* or *CDMA*. We will discuss this more in Chapter 20.

There are two generic kinds of spread spectrum techniques. The first, called *direct sequence*, spreads the combined signal over a wide range of bandwidth simultaneously. The other, termed *frequency hopping*, uses a narrower-band signal that is "hopped" around over several frequencies in a predetermined manner within the assigned channel bandwidth.

Spread spectrum communications are highly secure, since the encoding codes must be known, and highly resistant to interference and jamming. They were originally developed for military secure communications services, but are now used in the commercial sector for mobile telephones and satellite communications.

7.3 Multiplexing

So far, we have seen how a single signal may be carried in a telecommunications channel. Each signal begins as a single baseband signal. Usually the transmission technology—wire, coaxial cable, transponder—is physically capable of carrying more than one such signal. To use that wire, cable, or transponder to its full cost-effectiveness, we need to combine those baseband signals with others to fill up the information-carrying ability of the channel.

The technique for doing this is termed *multiplexing*. Engineers often shorten the term multiplexing to *mux*. It combines many narrower-band signals together to fill the useable bandwidth of the channel. (Sometimes, of course, most often with video, a single signal fills the capacity, so no multiplexing is done. It is most common with audio and low-bitrate datastreams.)

An everyday analogy may help explain multiplexing. Consider the water and sewage systems connecting an urban home to the municipal water supply system and the sewage system. Leaving your home is typically a 4-inch-diameter sewer pipe. You can easily understand that it would be much more expensive and troublesome to run an individual pipe directly from your home to the sewage plant. Instead, to save money (not to mention reducing wear on the street) your home's drain pipe feeds into a larger neighborhood pipe capable of carrying much more, and this may in turn feed with other neighborhood pipes into even larger conduits. In "telecomspeak", all of your sewage is multiplexed together until the capacity of the largest conduit is reached.

On the other end, the water coming into your house from a municipal water system is typically supplied through a two-inch pipe, but your home does not have a dedicated pipe running from the pumping station to the house. A huge water main pipe leaves the pumping station, and the stream of water is repeatedly "demultiplexed" and split off into smaller and smaller pipes, finally serving your house.

In the electronic world, there are three physical properties we can subdivide in order to allow several independent signals to use the same pathway: space, frequency, and time. Subdivision techniques can also be combined to allow higher levels of multiplexing.

Before getting into these in detail, let us get ahead of ourselves a bit by mentioning that these same techniques, and one additional one, will come up again in Chapter 20 under the heading of multiple access. The goal of both is to maximize communication.

The difference between multiplexing and multiple access is not so much a difference of what is done, as it is one of where in the system it is done and by whom. For example, suppose you and 23 officemates simultaneously needed to make telephones calls to Paris. The telecommunication company your call goes through does not have wires dedicated just to your calls, at least for most of the path. First, however, there is a multiple-access choice by you and your colleagues, as you each choose which telephone set to use. Each provides identical access to the telecommunications system.

As your 24 different telephone calls reach the local central office, each is filtered and digitized, and 24 baseband signals are produced. For efficiency, however, the telephone network multiplexes your calls together onto a single pair of wires or a single cable. Whereas you and the others chose and had control over which telephone set you used, you have no control (and, if properly done, won't even have knowledge) of the fact that all of your signals have been multiplexed together into a broadband signal. This broadband signal is carried somehow to a gateway earthstation, which is linking to a satellite, but so are perhaps thousands of other earthstations linking to the same satellite. Thus your earthstation, and all of the others, must work out some multiple-access arrangement so all can use the satellite without causing mutual interference. When your calls get to France, they are *demultiplexed* (*demuxed*) back into 24 individual calls so that 24 telephones ring in Paris.

In a somewhat overly simplistic summary, we can say that combining signals is called multiplexing when the system does it automatically, and is called multiple access when the transmitting and receiving entities have some control and selection of the process.

7.3.1 Space-division multiplexing

When each signal is given some individual path to follow, this is generically called *space-division multiplexing, SDM*. A simple example is that of many pairs of wires—each pair carrying a single telephone call—along a single cable duct.

A good terrestrial example is your cell phone. The cell you are in is covered by one frequency; to avoid interference, each adjacent cell uses a different frequency. But in cells across town not adjacent to yours and thus not liable to cause or suffer interference, the same frequency used in your cell can be reused. This conserves valuable spectrum.

In Chapter 20, we will see that the satellite multiple-access version of this technique is multiple radio beams.

7.3.2 Frequency-division multiplexing

Frequency division works by assigning each independent signal its own unique band of frequencies to use. This is rather like a chorus with basses, tenors, altos, and sopranos each singing a different message. A similar technique is used in the fiber optics industry to send independent datastreams down a single fiber, each stream using a different color (wavelength) of infrared light.

Sometimes we need to multiplex together signals of differing characteristics of power and bandwidth. In such a case, sometimes called generic FDM, if we have an allocated bandwidth for transmission (of a transponder or coaxial cable, for instance) we need to fit the disparate signals into the allowed bandwidth with care and make sure that there are *guardbands* of vacant frequencies between them to minimize interference among all of the signals. If one carrier leaves and another wants to use the channel, we must be sure that it will fit in without causing interference to itself or other signals. Such a generic FDM use is seen in Fig. 7.8.

In many applications, however, all of the baseband signals are identical in power and bandwidth usage. Telephone calls and low-speed datastreams are examples. In such a case, we can evenly spread them out over the allowed bandwidth. Basically, we subdivide the transponder or cable capacity into a number of identical power-bandwidth channels. Since all signals are identical, it does not matter which channel a carrier is assigned to. Such a scheme is called *single carrier per channel*, or *SCPC*. See Fig. 7.9.

Another way to multiplex telephone calls that all use the same basic frequencies is for the multiplexing system to alter them to combine them for transmission, and the receiver to impose the opposite alteration. For example, suppose we have four telephone callers, A, B, C, and D, who place calls over a system that is capable of carrying all four calls along a single pair of wires. Each caller's signal occupies a baseband bandwidth of 0 to 4 kHz. In order to combine them to efficiently fill the bandwidth of the wires, the multiplexing system may work this way: Caller A's call is carried just as it is generated, using the frequencies from 0 to 4 kHz; B's call has

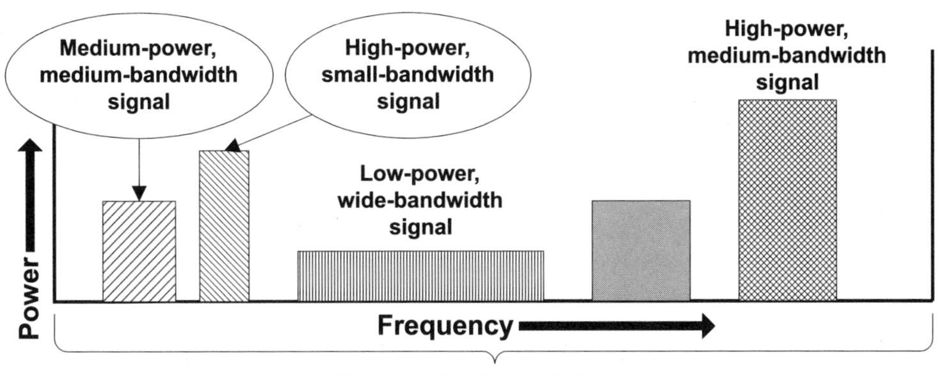

Figure 7.8 Frequency multiplexing places signals of differing power and bandwidth requirements within some allowed transmission bandwidth such that they do not interfere with each other.

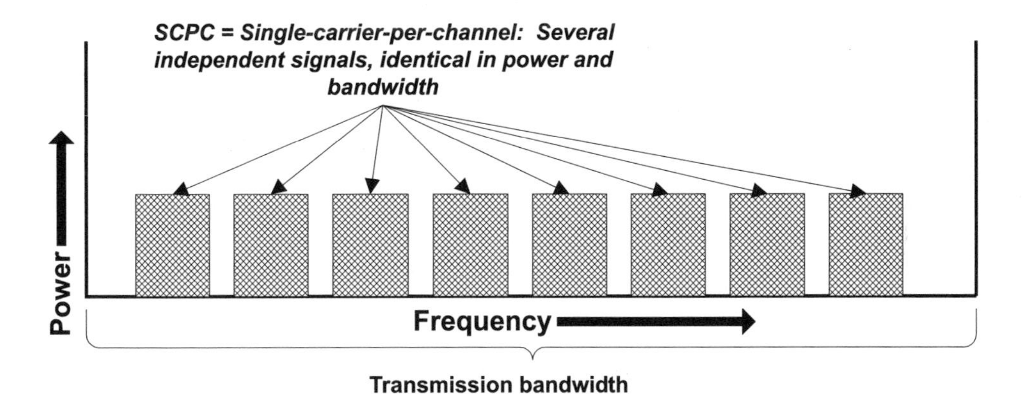

Figure 7.9 Single carrier per channel. Often we wish to transmit simultaneously many similar signals, such as telephone calls, within one wide bandwidth. Since each signal (call) has the same power and bandwidth requirements, they are easily spaced within the transmission bandwidth.

every one of its frequencies boosted by 4 kHz and is "stacked" on top of A's call; similarly C's call is raised by 8 kHz and put on top of B's, and D's is increased by 12 kHz and put at the top of the stack of frequencies. Thus, the four calls are carried along the wires as a "broadband" signal with a frequency range of 0–16 kHz. At the receiving end, the respective amounts are subtracted from each caller's signal, resulting in four independent baseband telephone calls again.

(In real-world telephone systems, more calls are actually multiplexed together. In the U.S. telephone system the lowest-order grouping is 24 calls; in Europe it is 30 calls.)

If those multiplexed calls in the example above need to be further multiplexed when they go into a still-higher capacity transmission system, the entire block of four calls can be increased in frequency and stacked on top of another similar block, and so on. Many hierarchical levels of multiplexing are possible and in use, the highest level combining thousands of simultaneous calls.

Looking ahead a bit, note that most communications satellites are inherently frequency-multiplexed. For example, each Fixed Service Satellite (FSS) is typically allotted a total bandwidth of 500 MHz. It is impractical to build a single circuit on the satellite with this large a bandwidth. Instead, most satellites have this assigned bandwidth subdivided into narrower bands, each assigned to a circuit path through the satellite called a transponder. Some typical bandwidths for individual transponders are 36 MHz, 27 MHz, 54 MHz, and others, chosen to suit the requirements of the traffic to be carried on a particular satellite. If a particular transponder of a satellite is required to carry a group of signals each with a bandwidth smaller than that of the transponder, the transponder bandwidth can be further subdivided—multiplexed—to provide the services desired.

As a matter of peripheral interest, the satellite industry's biggest competitor uses frequency-division multiplexing in its optical fibers, but since optical engineers prefer to refer to wavelengths, rather than to frequencies, they call it *wavelength-division multiplexing*, WDM. (Recall that frequency and wavelength are inversely related; see Chapter 5.) When the wavelengths (or frequencies) of the light through the fiber are spaced closely together, it is called *dense wavelength-division multiplexing*, DWDM. Fiber communications systems can operate at high speeds with multiple sources and destinations because they do not have to worry about three things of concern when operating satellites: shortage of frequencies, beam interconnections, and the long latency of satellite links.

7.3.3 Time-division multiplexing

TDM works by allocating different time slots to different signals. While it can be used for analog signals, it is most common for digital signals, where each signal is made up of a sequence of bits. The bits are sent in groups that may be called packets, words, frames, etc. To avoid interference between adjacent packets, small intervals called *guardtimes* are inserted between packets.

Consider the same example of four telephone calls. Now each caller's signal is a stream of bits. In a TDM system, a bit (or possible group of bits) is taken sequentially from each caller. See Fig. 7.10. At the receiving end, they are reassembled into single calls once more.

You can immediately see the need for some organization to this kind of system. It is necessary that the receiving end know which bits belong to A's call, which to B's, etc. Thus, there must be *synchronization*. In general, in a TDM system, somewhere there is a master clock ticking along uniformly, and this is the drummer to which all of the signals and equipment march. This need for tight synchronization

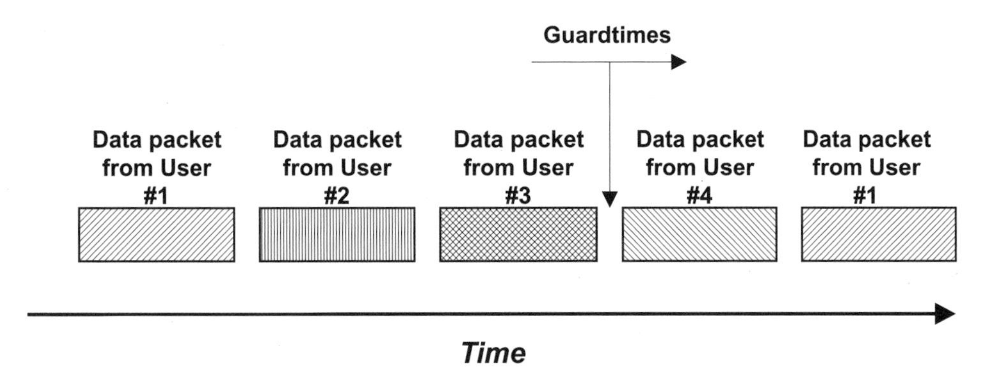

Figure 7.10 Time-division multiplexing allots time slots for each user to send data, with small "guardtimes" between data packets.

complicates TDM systems as compared to FDM systems, but allows higher speeds and thus more efficient telecommunications. (Of course, the system must operate fast enough that, again like a motion picture, the bits can all be recombined to produce the sensation of a continuous telephone call to the user.)

7.3.4 On demand

If all of the users of a multiplexed system have similar requirements for telecommunication, each user may be permanently assigned a frequency channel on an FDM system or a fixed time slot on a TDM system. Often, however, different users have different requirements, and it would be a waste of capacity to permanently allocate a channel to a user who was not always using it.

To increase efficiency, the system can adopt a *demand assignment* protocol, sometimes called *DAMA*, for *demand-assigned multiple access*. This allocates a portion of the system's capacity to each user as he needs it. See Fig. 7.11.

There are two basic ways of allocating the capacity. A system using *contention* allots capacity on a first-come, first-served basis. Using a classroom analogy, whoever raises a hand first gets to speak first. An alternative is *polling*, in which the system asks each user if it needs to communicate. In the classroom, the instructor asks each student in turn if he or she has a question.

Thus, we can see that the digital compression system we discussed in Chapter 6 is a time-division demand-assignment multiplex system. In this case, the system assesses the instantaneous needs of each digital television signal and allocates that channel an appropriate number of bits per second. As the needs of each channel change from millisecond to millisecond, the allocation to each channel varies accordingly.

Such a system operates on the statistics of the distribution of bits (demand) for each channel, and is sometimes called *statistical multiplexing*, which engineers like to shorten to *statmux*. Since each user is allocated capacity on the basis

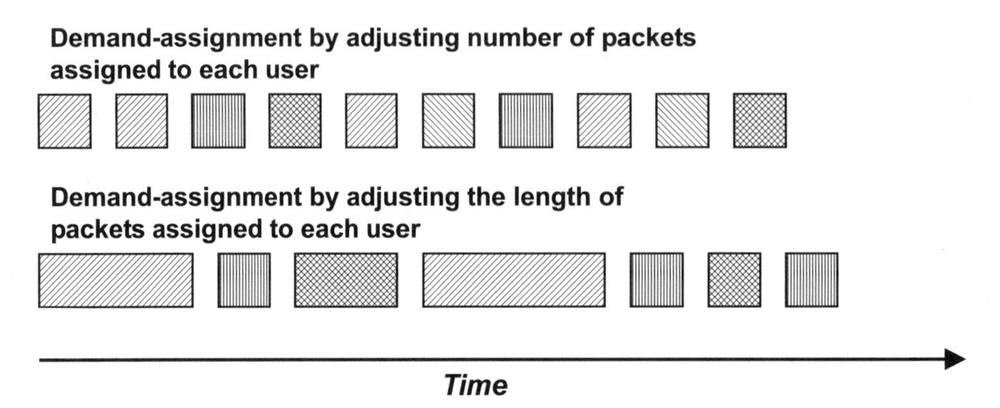

Figure 7.11 Demand assignment allows heavy users more capacity than light users, making for more efficient communication. In a TDM system, heavy users can either be assigned to use timeslots more frequently, or to use longer timeslots, than the light users.

of instantaneous need, if several users happen to simultaneously need high capacity, the system may not be able to allocate all of the capacity that each one needs at once. This can produce delays or drop-outs in data that may (or may not) be unacceptable, depending on the details of the user's need and the data transmission protocol in use.

Generic TDM and statmux thus have several relative advantages and disadvantages. For TDM, these are

- ☺ It is not protocol sensitive, since each user gets a known time slot;
- ☺ It is good for time-sensitive transmissions, where delays are not acceptable;
- ☺ Being simpler, it is usually less expensive than statmux; but
- ☹ It makes less efficient use of the carrier capacity.

On the other hand, the relative advantages of statistical multiplexing are:

- ☺ It allows higher overall speeds;
- ☺ It makes more efficient use of the system; but
- ☹ It can introduce data delays; and
- ☹ It is more expensive.

7.4 Networks and protocols

The arrangement of the network interconnecting users, and the selection of the particular techniques used to communicate, are critical to providing a useful, economical service. The network interconnections, as we saw in Chapter 4, are referred to as the network topology. The particular selections of hardware, connections, modulations, and codings come under the heading of communications protocols.

In the beginnings of satellite communication, television and telephony dominated the traffic, and both were carried using analog signals and using frequency-division multiple access (which will be covered in more detail in Chapter 20). Data was then a very small part of the total traffic. Video and voice are typically characterized by having fixed channels and fixed rates of transmission. Even when carried digitally, such applications used so-called *synchronous* data transmissions.

Today, most voice traffic and increasingly much of the television is digital, with data being the fastest-growing segment of the business. Over the past several decades, efficient data protocols have been developed for terrestrial networks and have been adapted for use over satellite links. However, some protocols need substantial adaptation to function well over the links to geostationary satellites because of the inherent and ineluctable delay due to the distance out to the Clarke orbit. Furthermore, the new demands for data require constantly changing datarates, typified by "bursty" transmissions. The older synchronous protocols are not suited to such demands. The explosive expansion in demand for Internet services, in particular, has led to new ways to use terrestrial protocols. The newer protocols do not carry the bits at constant rates, and are said to be *asynchronous*. In these, the data bits—whatever they represent—are grouped into bunches called packets, of either fixed or variable size.

It should be emphasized that "digital" is not an application itself, it is a way of transmitting the information of an application. The applications are such things as audio, voice, television, point-of-sale data, Internet traffic, etc. As we have seen, some of these begin and end as analog signals, others begin as data and sometimes end up as analog (for example, downloading a webpage).

During the last half of the 1990s, the Internet, previously a government/academe network, became a public information service, to the great benefit of society. Demands for increased communication have soared. One of the problems that has arisen is that of adapting the Internet, whose protocols are a decade old and assume error-prone, low-datarate channels that can connect with each other quickly, to the satellite characteristics of highly-accurate, high-speed channels that have significant delay. This delay, often called *latency*, of course arises from the distance of the geostationary satellites from the earth. This has meant that engineers have had to come up with new techniques, protocols, to efficiently carry Internet traffic via satellite.

7.4.1 Internet via satellite

The Internet uses and is defined by a pair of protocols called *Transmission Control Protocol, TCP* and *Internet Protocol, IP*. Since these are used together, they are usually referred to in the combined abbreviation TCP/IP. These define how the data is partitioned and carried, and contain techniques for error control since the original Internet was designed to work on noisy, error-prone mesh networks. TCP/IP is a connection-oriented protocol meaning that it relies on getting acknowledgments of each data packet sent out.

Error control in TCP/IP uses an ARQing kind of system, in which one node of the packet network sends a packet out to the next node, where the packet is checked for errors. If none are found, the receiving node sends an acknowledgment back to the sender. This all takes time, which is minute when the terrestrial nodes are a few miles or tens of miles apart. However, when the link is over a 44,460-mile satellite connection, these delays limit the speed at which Internet packets can be transmitted. (Recall that this is the minimum one-way link length; for satellites close to the horizon, the pathlength is greater.) Using the bare-bones TCP/IP over a satellite link limits the throughput to less than 1 Mbps.

There are techniques that can get around this. Since the original TCP/IP was planned for error-prone and possible nonfunctional links between nodes, and since satellite links are highly reliable, the technique fools the transmitter into thinking that receiver has successfully received each packet. This is called *spoofing*, and can greatly increase the throughput.

Another possibility is to "encapsulate" the Internet data into another form. One of the most-used methods involves packing the Internet packets together to form the same kinds of information packets used in *DVB*, the *digital video broadcast* standard used worldwide for digital television. One advantage is that DVB receivers are common and relatively inexpensive.

Another technique, borrowed form terrestrial networks, is to use a technique called *asynchronous transfer mode*, or ATM (don't confuse this with automated teller machines!) ATM encapsulates the data bits into fixed-length "cells" for transmission. Each cell contains both the desired data and some "header" bit that allows for identification of the data, not only as to source and destination, but also its nature, timeliness requirements, etc. Thus, ATM can accommodate a wide range of types of traffic vaguely described by the currently popular term "multimedia."

Several companies have invented ATM or DVB satellite internetworking equipment to make the protocols work. Using such devices allows satellites to play a major role in the global expansion of Internet traffic.

Another technique is to use an alternate Internet transmission protocol called *user datagram protocol*, or *UDP*. This is a connectionless system with no acknowledgments and thus can move data at higher speeds, but must rely on forward error correction to catch errors.

Having covered the technologies and protocols used to send signals, it is time to consider the last item that limits our ability to communicate: noise.

Chapter 8

Signal Flow, Quality, and Noise

Whether as users or providers of telecommunications services, we are concerned with the quality of the signal. In general, there is a correlation between quality and cost, and each user and provider must decide the relative importance of each.

For a human-related signal, such as a telephone call or television picture, there is first the receiver's subjective perceived impression of quality. This is imprecise and relative, and varies with the experience, expectations, and expertise of the receiver. Engineers attempt to find more objective numerical measurements that hopefully correlate with the users' subjective opinions.

If there were nothing to interfere with our intended signal, we could in theory send a perfectly good signal at very low power. In the real world, however, we are faced with numerous sources of interference, to which we give the generic term *noise*. The signal is whatever it is that we wish to transmit, and the noise is anything that we do not want. Note that "noise" is *any* unwanted signal competing with the one that you want. Even a brilliant speech or beautiful music is noise if it interferes with what you are doing by leaking into your signal.

8.1 Analog signal quality: signal-to-noise ratio

Thus, a common measurement of the quality of a analog signal transmission is called the *signal-to-noise ratio*, often abbreviated to *SNR* or *S/N*. It is important to remember that it is the *ratio* that is important, not just the absolute size of either the signal or the noise. *S/N* is usually measured in decibels.

At the source of a transmission, say a telephone call or television broadcast, the originating signal has some inherent SNR, determined by many factors. For instance, a voice signal coming from a radio station's soundproof studio will have a better initial SNR than a telephone call made from a noisy hotel lobby telephone. During transmission to the receiver, the signal will unavoidably suffer some degradation in quality, that is, its SNR will decrease. The goal of the telecommunication engineer is to minimize that degradation. When the radio wave carrying your signal is received by the end user, it is turned back into a useable signal, and the received quality will again be measured by its SNR.

Of course, the received signal can never be better than the quality of the transmitted signal.

8.2 Digital signal quality: bit error rate

If the signal to be sent is a digital one, we have an objective measurement of the quality of the signal, which is simply how accurately it is received. Presumably, the original digital signal is of perfect quality: it is just a bunch of bits that may be just numbers (say for a spreadsheet) or may represent something else, such as a PCM-encoded voice. Again, during transmission to the receiver, the signal will unavoidably suffer some degradation in quality.

At the receiving end, the usual measurement of quality of a digital signal is the *bit error rate*, or *BER*. This is a fraction—usually a power of 10—that describes the average percentage of errors.

8.3 Quality during transmission: carrier-to-noise ratio

Often the greatest amount of the degradation occurs during the transmission stage(s) of the channel. When the signal is carried on a radio wave, called the carrier, such as in satellite communication, the measurement of the quality of the signal during transmission over the links to the satellite is called the *carrier-to-noise ratio*, abbreviated *CNR*, or more commonly, *C/N*. *C/N* is also measured in decibels. Sometimes we use C/N_O, *the carrier-to-noise-density ratio* instead. This is the same as CNR, but measured per unit bandwidth.

C/N, or C/N_O, is the "holy grail" of the telecommunication engineer. The bigger the *C/N*, the better. All else being equal, the received quality is totally dependent on the carrier radio wave's *C/N*.

The final SNR of a signal and the BER of a digital signal will both depend primarily on the quality (value) of the *C/N*, and will also depend on the details of the modulation and other manipulations to which the signal has been subjected. Improvements in final signal quality due to the details of the coding and modulation schemes used on the signal are known as *coding gain*.

A more useful measurement than *C/N* for digital transmissions—though completely equivalent—is a quantity called the *energy-per-bit-to-noise-density ratio*, symbolized by E_b/N_O. (A few people actually acronymize this and pronounce it as "EB-noh.") Since the bits are really pulses of energy, the higher the bitrate, the shorter the pulse, and the less energy each pulse contains. Thus, you can see that higher-speed data is more susceptible to noise. A formula for converting between *C/N* and E_b/N_O is given in Appendix E.11.

8.4 Improving signal quality

Given that there is no such thing as a perfect communications channel, and that signals will always be somewhat decreased in quality, what can you do about it to minimize the degradation? That depends on the nature of the signal and the requirements for the particular signal.

8.4.1 Companding analog signals

Recall that for an analog signal, we often use a matched pair of techniques called *compressing* and *expanding*, which are combined into one word, *companding*. This technique is basically just a matter of precompensating for known or anticipated defects in the transmission path, and essentially involves boosting the signals most likely to suffer interference at the sending end and correspondingly reducing them again at the receiving end. (Note that this kind of compression has no connection with the digital compression mentioned in earlier chapters.)

Look at Fig. 8.1. The lines represent some of the signal strengths in, for example, a radio program that you want to transmit. Near the bottom, but above the level of the faintest sounds in the original signal, is a noise level. You can find out what this is simply by listening to the receiver when there is nothing being transmitted. This noise would swamp the lowest-level sounds—the whispers and pianissimo passages—and thus reduce the range of sounds received. To precompensate for this, we compress the transmitted signal, raising the softest sounds to a strength above what we anticipate the noise will be. At the receiver, they are expanded, and in the process, the strength of the noise is reduced, too, preserving the quality of our program.

Companding is used only for analog signals, and is most often known by some trademarked terms such as Dolby™, dbx™, etc. In the tape-recording business, it is called NAB equalization, and in the FM radio side of the industry, it is called preemphasis. All are examples of companding.

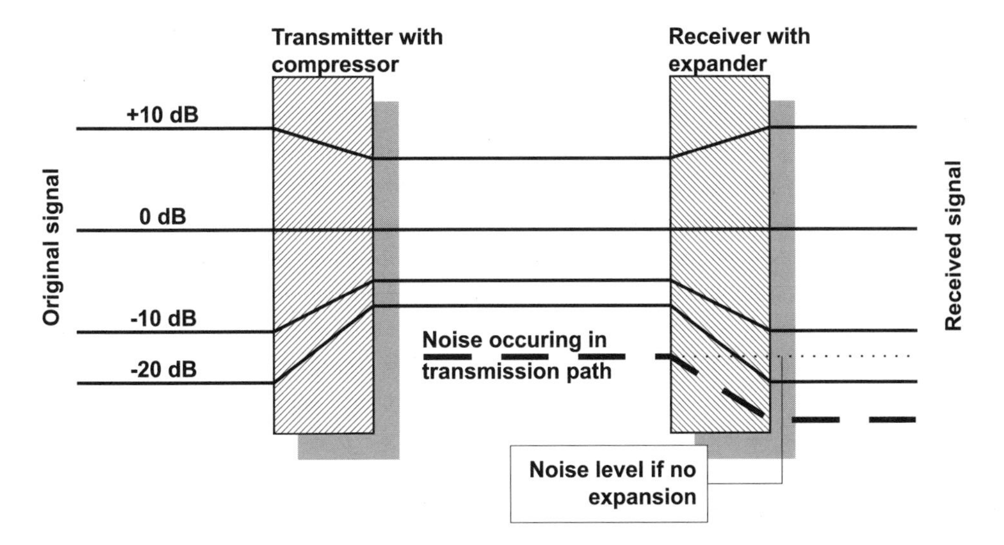

Figure 8.1 Companding improves the received signal-to-noise ratio and maintains a broad dynamic range by precompensating for noise in the transmission channel. During transmission, low-level signals are boosted to powers higher than known noise levels, and then appropriately lowered at the receiver. There are many variants of companding, usually known by the trademarked name of a particular technique for a particular purpose.

8.4.2 Error correction for digital signals

Maintaining the quality of digital signals is done by the error-checking and control techniques discussed in Chapter 6. The two choices are ARQing and forward error correction. As mentioned there, you must know what the user requirement is for accuracy in order to make the proper trade-off between quality and cost. As with analog signals, these quality criteria often are based on empirical subjective quality estimates by the end users.

8.5 The communications circuit and the cocktail circuit

Consider that you do the same sort of thing when you wish to converse in the presence of noise. To see this, let us invoke a human analogy that we will use several times throughout this book to illustrate some concepts.

The analogy is cocktail party. Let us suppose that you have been invited to a huge reception at the Embassy of the Duchy of Grand Fenwick. You wish to converse with your companions. The air in the ballroom can be thought of as the communications medium.

If the ballroom is fairly empty, there is not much background noise, the originating *S/N* is pretty good, so you and your colleagues converse normally. But as more people join the party, the noise level goes up and the *S/N* of your conversation goes down. To precompensate for this, you raise your voice slightly so your companions can hear you over the crowd. When the room is very noisy, you talk even more loudly. If you talked that loudly in a quiet room, your friends would think you were strangely shouting at them, but when you speak loudly in a noisy room, they compensate for this by realizing what you are doing. You are "compressing" and they are correspondingly "expanding" your speech to allow your conversation to take place. Eventually you reach the point where little communication is possible.

We will return to our cocktail party in later chapters.

8.6 Noise figure and noise temperature

Now that we know that noise determines our ability to communicate, it is time to ask just what noise is. The short answer is that it is anything besides your signal. Noise comes from many places.

Many sources of noise are natural. Unfortunately, everything in the universe, from the atoms to the galaxies, produces electromagnetic noise. For natural sources, the amount of noise produced, and the peak frequencies produced, depend primarily on one physical parameter: temperature. For this reason, telecommunication engineers often refer to the amount of noise by a number called the *noise temperature*, sometimes symbolized by NT or N_T. (Noise temperature is measured in degrees on the Kelvin temperature scale.) If the noise is measured instead in decibels, it is referred to as *noise figure, NF.* You can convert back and forth between *NT* and *NF* using the formula given in Appendix E.5, or, more conveniently, using a chart like Fig. 8.2.

Natural noise can come from the sun and moon, from water vapor in the atmosphere, from mountain peaks that may happen to be within the field of view of your earthstation antenna, or from almost anything within view of your antenna.

Note that the wider the bandwidth you receive, the more natural noise enters your receiver, interfering with your desired signal. The range of frequencies through which you pick up this interference is called the *noise bandwidth.*

In addition, there are many sources of man-made noise (Fig. 8.3). Some are external to your electronics, others are internal. External sources of interference include nearby satellites (which may be near the satellite you are aiming at and thus within the edges of your antenna's beam), any nearby electronic equipment, power

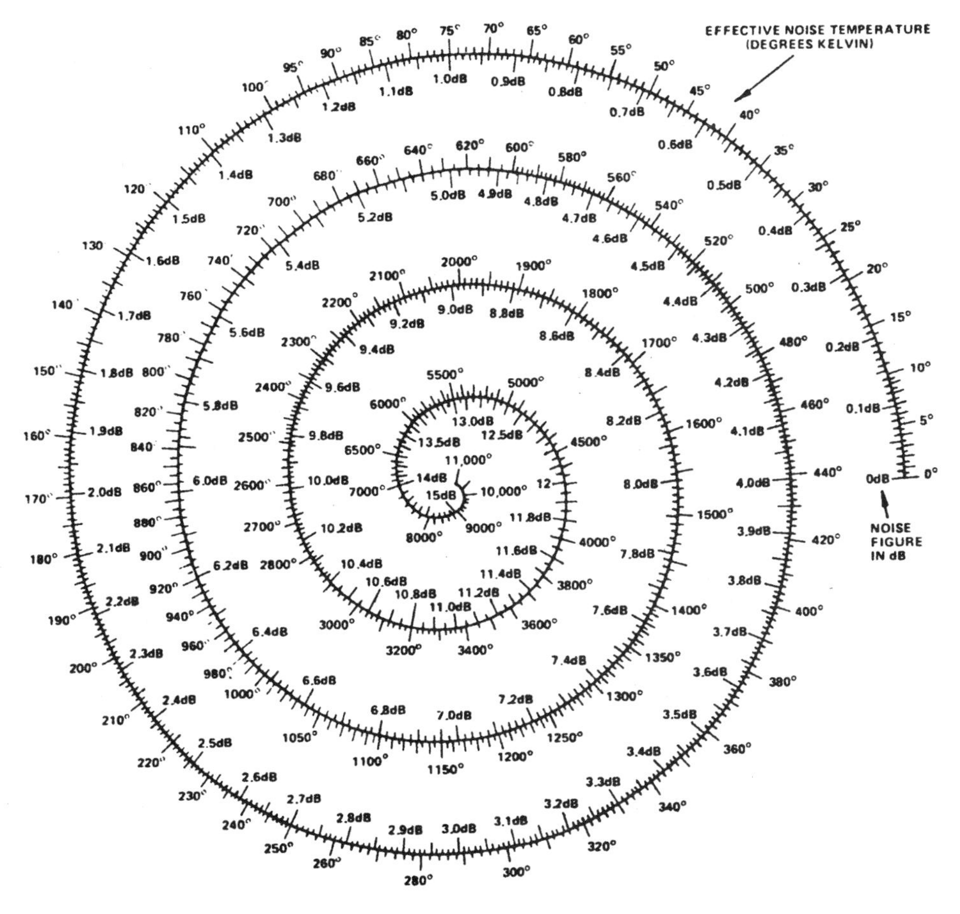

Figure 8.2 Noise figure and noise temperature are two equivalent ways of measuring noise levels. The noise temperature (NT) is measured in degrees Kelvin (K) more often called just Kelvins. The Kelvin temperature scale begins with 0 at absolute zero, and has degrees the same size as the Celsius scale. Noise figure (NF) is measured in decibels. In telecommunications, the lower the better. (Image courtesy of Andrew Corporation.)

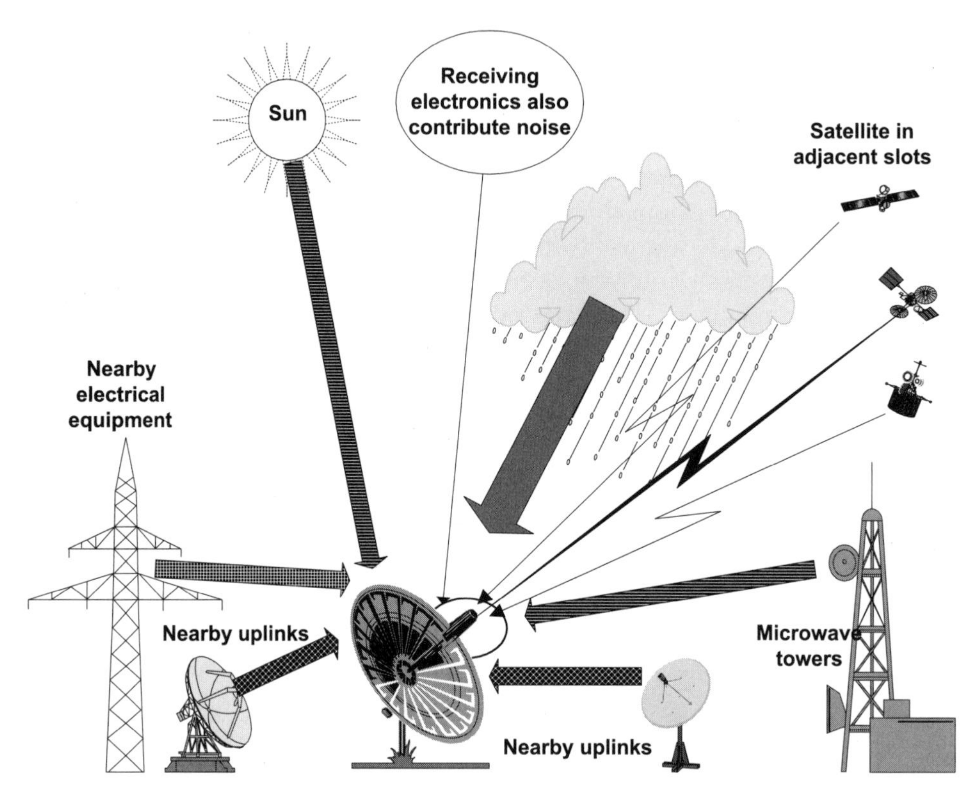

Figure 8.3 Sources of electromagnetic noise abound. Some are natural, since everything in the universe radiates some electromagnetic noise. Some noise is produced by the unintentional spillover of signals from other electronic devices. There is more interference in an urban environment than a rural one, which is why teleports and gateway stations are often located outside cities, or well shielded if within cities.

lines, nearby earthstations, and (at some frequencies worst of all) nearby microwave relay towers. Many complex factors control the amount of noise interference you will receive from external sources, including their frequency and spectrum, how strong these sources are, how close, their directionality, polarization, shielding, and how susceptible your system is to noise. The process of minimizing the mutual interference of various transmitters and receivers is called *frequency coordination,* an often contentious technical and sometimes political issue.

Noise can also be produced within your own electronics. The components that make up your amplifiers and other components generate their own noise that can interfere with the received signal. The typical amount of energy coming into the antenna from a satellite is only around a few trillionths of a watt, so it does not take too much noise to interfere with it. For this reason, the first active electronic hardware the signal encounters is called a *low-noise amplifier*, or *LNA*. The name implies that this piece of equipment should amplify the very weak received signal while contributing very little competing noise of its own. We will consider LNAs and their kin more fully in Chapter 17.

8.6.1 Intermodulation noise

Various signal processing operations within the electronic components, particularly amplifiers, can also produce internal noise, both in earthstations and in the satellite's electronics. For some applications, this can be the major source of noise, and thus the limiting factor for communications capacity.

The culprit here is often a phenomenon called *intermodulation distortion, IM*. It occurs naturally in all amplifiers that are used to carry more than one frequency simultaneously, such as satellite transponders and even your home stereo system. Using the latter as an example, your stereo amplifier may be able to put out 1000 W, but if you turned the volume up that loud the sound would be unlistenable due to distortion. This is because when the amplifier gain is near its maximum (or minimum, but that is less of a problem), all of the simultaneously present frequencies in the signal mutually interfere with one another, producing distortion, a form of noise that competes with the desired signal. (If you look at the system specifications that came with your stereo system, you will see mention of its frequency response (aka bandwidth) at some level of intermodulation distortion, IM.)

This means that if you are dealing with a range of frequencies, you cannot use your amplifier at full power or you will create your own noise. To avoid this, we reduce the amplification of the system (turn it down) until the distortion is minimized. The technical term for "turn it down" is *backoff*. Backoffs in the 6 to 10 dB range are typical when you are using a frequency-division-multiplexing or multiple-access method, because you are putting many different frequencies through the system simultaneously. Realize that a backoff of 6 dB means that we have to turn down the gain of the amplifier to only one-quarter of its full power to avoid IM noise! (We will see later that one advantage of time-division multiplexing or multiple access is that only a single signal is going through the system, meaning that we do not have to backoff and can use the full-rated power of the equipment.)

The throughput characteristics of an amplifier can be improved, and thus the amount of backoff needed reduced, by using a *linearizer*. This is a wideband filter circuit located before the input to the amplifier that precompensates for the lack of linearity of the amplifier. Thus, you might read about a transmitter employing linearized traveling wave tube amplifiers (TWTs).

8.6.2 Satellite-to-satellite interference

Another often major source of noise is from satellites in orbital locations close to the one that you are linking to. The "beam" of the antenna, whether that of the earthstation on the ground looking at the satellite, or that of the antenna on the satellite looking at the earth, is not a perfectly cylindrical shape, like that of a laser. Instead, the beam is actually a cone-shaped range of directions. (In Chapter 16 we will see that the breadth of the cone depends on the size of the dish and the frequency in use.) When an antenna is receiving, even though it may be pointed exactly at the antenna sending to it, it is slightly sensitive to signals coming from the side. Thus, as seen by an earthstation, satellites too close to yours will send a small part of their signal into your receiving antenna, causing *downlink interference*. Conversely, while an antenna is transmitting to your satellite, inevitably a small portion of the signal spreads out and is received by the antennas of satellites orbiting near yours, causing *uplink interference* in them.

8.6.3 Terrestrial microwave interference

For earthstations located in urban areas, particularly those operating in the C-band, the biggest source of interference is likely to be other nearby earthstations and terrestrial point-to-point microwave towers. Terrestrial C-band microwave services share the exact same frequencies with the C-band satellites. The interference problem is less at Ku-band frequencies and above since the spectrum bands are not shared identically as they are in C-band.

Microwave towers are typically not more than 25 miles apart, and they transmit with powers on the order of tens to hundreds of milliwatts. Your earthstation is trying to pick up a few trillionths of a watt from a satellite, and the satellite signal simply cannot compete with the one from a nearby microwave tower. Similarly, your earthstation may be sending out the equivalent of several million watts, even billions of watts, while a nearby microwave antenna is trying to detect a few microwatts from the next tower.

So each of you causes interference to the other. Some solutions to reduce terrestrial interference include shielding the earthstations and microwave towers with metal fences or concrete walls.

8.7 The limit on capacity

In 1948, a researcher at Bell Laboratories, Claude Shannon, found that there is a theoretical maximum capacity for any telecommunications circuit. This is often called *Shannon's Law*. In real-world applications, we usually cannot come close to Shannon's theoretical limit. Shannon's Law says that the maximum possible bitrate for a signal is a function of only two things: the bandwidth of the signal and the C/N of the signal. You need not remember this in detail, but for the record, the mathematical statement of Shannon's Law can be found in Appendix E.6.

What you should remember is: for a given quality of signal, there is essentially a trade-off between bandwidth and power (assuming noise is a constant).

One analogous way to think of this is to imagine that the information we want to send is like water flowing through a pipe. What determines the amount of water that can flow is not simply the height of the pipe nor simply the width of the pipe, but both together which produce the cross-sectional area of the pipe. In this analogy, the height of the pipe is like power, and the width of the pipe is like bandwidth. See Fig. 8.4.

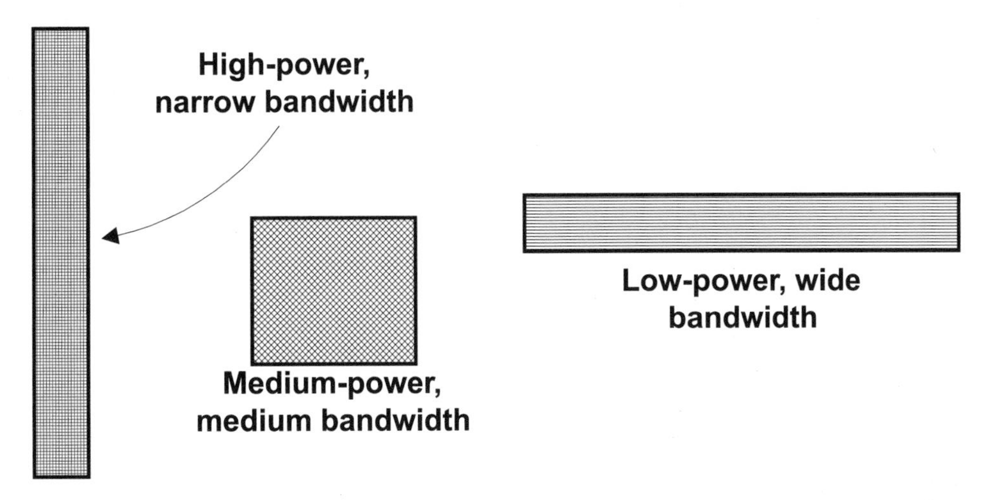

Figure 8.4 Shannon's Law. Flow of information in a telecommunications channel is somewhat like water flow in a pipe. What counts is the cross-sectional area of the pipe, not its shape. In telecommunications, the "dimensions" of the pipe are bandwidth and power, so there can be a trade-off between them. Because satellites are distant and thus their signals are received at lower power, they must use a larger bandwidth than comparable nearby terrestrial transmissions.

This implies that if our signal is weak, we need to transmit a wider bandwidth to maintain quality. This is an important point in satellite links. Since geostationary satellites are distant, their signals are weak, compared to a terrestrial broadcast. Therefore, in general, the same signal at the same quality from a satellite must be of greater bandwidth.

Taking a specific example, a terrestrially broadcast analog television channel occupies a bandwidth of 6 MHz and produces a good picture on your television set within the service area of the television station, which is typically a few tens of kilometers in radius. A couple of decades ago, when satellites were less powerful than they are today, sending the same quality analog picture via satellite required a bandwidth about six times greater, or 36 MHz. (This is why the most common bandwidth for satellite transponders is 36 MHz.) Today, we can get a similar quality analog television signal in about half that bandwidth. Nevertheless, since all signals to and from geostationary satellites are inherently weak, they must be correspondingly of greater bandwidth.

8.8 Digital and analog systems' response to noise

Finally, let us examine how digital and analog signals are affected by noise. In a typical application, there is a range of acceptable signal quality, from a "perfect" signal to one that is marginally useful. Thus, there is some threshold of useful signal quality.

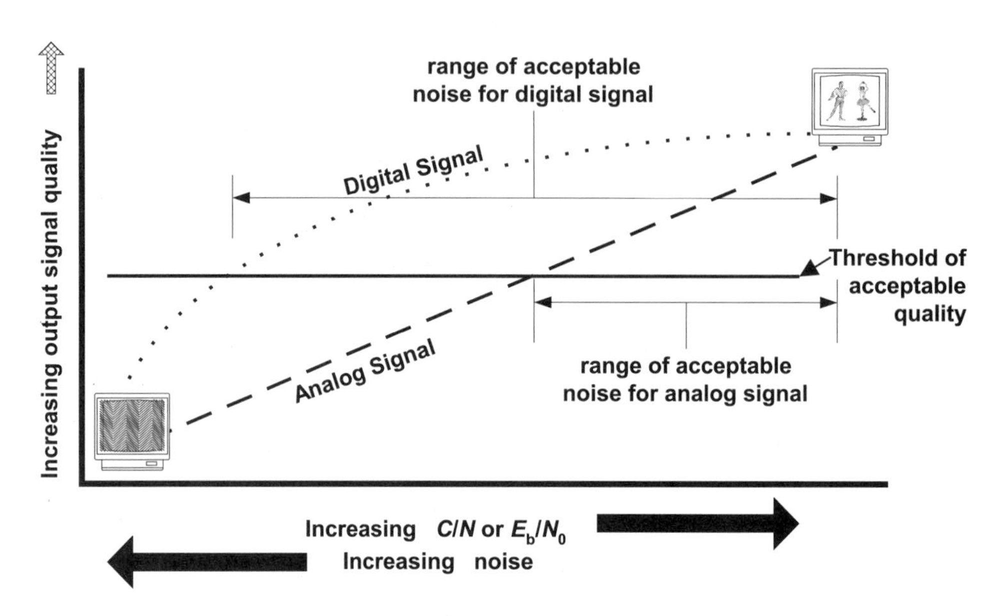

Figure 8.5 Digital signals behave better in a noisy environment because receivers can tell the difference between signal and noise until the noise level reaches some high level. In any transmission, there is usually some threshold of acceptable quality. In a noisy environment, digital signals remain above that threshold over a wider range of noise level than do analog signals.

Consider Fig. 8.5, which shows schematically the relationship of carrier quality with received signal quality for both kinds of signals.

At the top right, we have a very high *C/N*, and consequently a very good quality signal, such as an excellent television picture. With an analog signal, moving to the left in the chart as *C/N* gets worse, the picture gradually degrades in an approximately linear proportion to the degradation of the carrier. At some point, the picture quality drops below the threshold of desired quality, and the signal becomes practically useless. The range of the values of *C/N* within which you have an acceptable picture is shown in the figure.

But with a digital signal, since (up to a point) the receiver can detect and correct received errors as *C/N* (or E_b/N_O) decreases, the quality of the television picture stays almost perfect. The range of *C/N* over which the picture is acceptable is much greater than that for an analog carrier. Finally, at some level of *C/N*, the carrier quality drops below a point at which errors can be corrected, and the picture quality drops off suddenly.

Thus, as long as the digital carrier is above the threshold, you get a perfect picture; once below the threshold, no picture at all. In contrast, an analog picture slowly deteriorates as the signal gets worse. This ability of a digital transmission to hold output quality over a wider range of carrier quality (i.e., over a wider range of noise) is another reason for the trend toward digital transmission.

A good everyday example of this difference in how analog and digital systems respond to decreasing C/N can be seen on your television screen. If you receive over-the-air (analog) signals, you have observed that the farther you are from the television tower, the worse your picture quality is. If, instead, you have satellite-delivered digital television, most of the time your picture is perfect. Even when a rain storm comes over your house, until the rain gets very heavy, the picture remains perfect because the receiver can correct errors. When the rain becomes so heavy that the C/N is near the receiver's correction threshold, the picture gets a bit "blocky", and just a little more heavy rain removes the picture from your screen altogether until the rain lightens up.

We have now covered the basics of how radio waves are used to convey information. If you thought that this section of the book was over your head, turn the page. It is time to turn our attention to the satellites themselves.

Part 3

The Space Segment

Chapter 9

The Space Environment

Most people know that space is a vacuum. Some therefore assume that it should be a rather benign environment for satellites. There is no wind, nothing rusts, and there is sun all of the time. Although there is indeed little material in space, what little there is, along with the immaterial flow of radiation in space, makes for an environment that is challenging for spacecraft engineers to design a complicated piece of hardware that will last a decade or more with no repairs. Quality control and reliability are at a premium, since no one makes "house calls" in the Clarke orbit! This is a primary reason for the high cost of a satellite system.

There are five important environmental factors that influence the design and operation of a satellite, whether used for communications or for some other purpose. These space environmental factors are

- gravity;
- vacuum;
- radiation;
- meteoroids; and
- space debris.

The relative importance of each may depend on the details of the satellite's orbit, as we will see later in this chapter.

9.1 A matter of some gravity

All of Newton's Laws still apply in space.

It is not true that there is no gravity in space. Everything that has mass is attracted to everything else that has mass by a gravitational force that, as Sir Isaac Newton showed, is proportional to their masses and inversely proportional to the square of the distance between them.

All material bodies have mass, whether on the earth or in space. This mass results in inertia, and thus, it takes a force to change the motion of something. What a satellite does not have in space is weight, for weight comes from resisting gravity. You have weight because the floor keeps you from falling, but if you jump off a cliff you will be weightless (briefly!).

Figure 9.1 Satellite testing. During manufacture, satellites are constructed of the best tested parts and then the whole system is repeatedly tested. Here, for example, the HotBird-3 satellite, manufactured by Astrium, undergoes an electronics test in a chamber designed to absorb stray signals. (Photograph courtesy of EADS Astrium.)

Since satellites in space do not have weight, they can be constructed with structures that would be flimsy on Earth, incapable of supporting their own weight. In particular, satellites may use huge, relatively thin solar panels to get energy, or large, mesh unfurlable antennas that have little more strength than a big umbrella. With no weight, such components will work well in space.

However, when satellites with such systems are being designed and tested (and retested and retested and....), special support devices like cranes and buoyant balloons must be used to hold them up. Even worse, during the stages of launch from Earth into orbit, the satellite payload is stressed with forces five to six times the force of gravity. Because of this, such things as large solar panels or antennas must be designed to be collapsible. Once the major propulsive maneuvers of launch are over, they can then be deployed, extended, unfurled, and put into operation. Figure 9.1 shows a typical commsat being tested in a special test chamber.

9.2 High vacuum

Space is almost empty, but what little there is left is important. Even in the best laboratory vacuum chambers, we cannot remove enough of the air to create a vacuum as "hard" as that which exists just a few hundred miles above the surface of Earth. When satellites are tested during construction and before launch, we use vacuum chambers to get as close to space conditions as we can.

This is important because many materials change their properties in a vacuum, compared to those in an atmosphere. For example, glass is a bit more brittle in space, because there are no air molecules to help strengthen its surface. Some seemingly solid materials turn directly from a solid to a gas and evaporate when exposed to hard vacuum in a process called *sublimation*. Some types of rubber and plastics sublime, and so are useless as spacecraft components. Further, bearings must be sealed or the lubricants will leak out.

There are charged particles in space, and as the satellite orbits, it runs into them and they may accumulate on its surface. This builds up a static electric charge, much like you do on a dry winter's day when you walk across a carpet. Different parts of a satellite may charge up differently. Since there is no air or water or other fluid surrounding the satellite, there is nothing to dissipate this charge, and it continues to build up unless the satellite has been designed to get rid of it. If the static electric voltage gets too high, you can have an electrical spark—miniature lightning—streak across or through the satellite's components, often causing major damage. Such sparks can cause electrical upsets to the circuitry, melt components, blow fuses, or cause breaks in wires. The larger the satellite, the worse the problem. Satellites whose orbits carry them through the heavy radiation regions surrounding Earth (called the Van Allen Belts), the medium-altitude orbits, are more subject to these kinds of problems than satellites in very low orbits or in Clarke orbit.

Another consequence of the lack of a surrounding atmosphere is that nothing evens out the heating load on the satellite. The only mechanism by which a satellite can get rid of heat is radiation from its surface, which is a rather inefficient process. One part of the satellite is facing the sun, picking up a lot of heat energy. Just a few meters away, the other side is facing the cold of space. This produces mechanical stresses on the satellite's components that can warp structures. Parts of many satellites may be covered with white reflecting cloth or plastic, or the common gold-plated plastic foil. (This is done not to literally "gold-plate" the satellite and raise its cost, but because gold film is an excellent reflector of heat radiation.) Some satellites use moveable louvers that can be turned to allow absorption or radiation of heat.

In addition, the electronics and mechanical devices aboard the satellite generate their own heat as they operate. Without atmosphere, something must get rid of this heat by carrying it to the satellite's surface where it can radiated away into space. On larger satellites, you find such things as heat pipes, which are closed convection tubes that collect heat at one end and release it at the other end.

Thermal control and regulation of a satellite is a major design consideration. The problem gets worse as the satellite gets bigger and consumes more electrical power. Satellite manufacturers plan ever larger satellites with 10 to 25 times the power consumption of typical satellites of only a decade ago. Thermal problems increase commensurately.

And while most of the time the problem is to get rid of excess heat, there are times in which a satellite is in the shadow of the earth (in eclipse) and must not be allowed to get too cold or the batteries will weaken and onboard fuel may freeze. Thus, designers of the satellite must anticipate both extremes.

9.3 Radiation in space

Radiation comes in two forms, and most people are confused about the differences and effects of each. One type is called "non-ionizing" and primarily produces heat; the other is called "ionizing" and has more severe effects. Both forms significantly influence the design and thus the cost of a satellite.

The first form of radiation is electromagnetic, i.e., light and its relatives, the same kinds of waves we are using for communications. The largest single effect is from the sun and is in the form of heat. Every square meter of spacecraft facing the sun is being heated at a rate of about 1400 W. That is about the same as the heating power of a clothes iron on its highest setting. All of this goes into heating the satellite, and must be dealt with, as we saw in the previous section. The parts of the sun's radiation that contribute the most to heating are in the visible and infrared regions of the spectrum.

Ultraviolet light from the sun has another deleterious effect: it slowly decreases the ability of the solar panels to convert sunlight into electricity. Since all of the power used by the satellite comes ultimately from these solar panels, this effect slowly reduces the electrical capacity of the satellite over its lifetime. Most communication satellites use silicon solar cells in their solar panels, and these can lose about a couple percent of their output each year. Thus, at the end of its lifetime in orbit, the solar panels are supplying quite a bit less than they were when the satellite went into orbit. Some recent satellites use solar cells based on gallium arsenide, which is more resistant to the space radiation.

Whichever is used, satellite designers must put aboard solar panels large enough so that even at the end of operational lifetime they provide enough power for satellite operations. This means that at the beginning of its life, a satellite has too much power. It also means that the solar panels must be larger, hence more massive, than they would be if their output didn't decrease. And this brings up an important point about satellite construction in general.

Design of a satellite is usually a "zero-sum game" as far as mass is concerned. You usually design a satellite to some maximum mass limit, such as what an anticipated launch vehicle can carry into orbit. Thus, everything put into the satellite is a trade-off of mass for some function. A few more kilograms that need to go into solar panels, for instance, means a few less kilograms of fuel that the satellite can carry in its tanks for stationkeeping, which in turn means a shorter operational lifetime (and hence revenue) for the satellite.

There is a second effect of sunlight on satellites. Sunlight actually produces a tiny force on the satellites, but because of the large area of the solar panels, this can add up. As the satellite orbits Earth, half the day it is moving toward the sun, and solar light pressure slows the satellite a bit; during the other half of the orbit, the sun speeds up the satellite. This has the small effect of changing the orbit of the satellite. Such effects are corrected, along with other, much larger, perturbations by firing the motors of the reaction control system aboard the satellite.

9.3.1 Cosmic rays

The other form of radiation consists of atomic particles. These are pieces of atoms whizzing through space at speeds often close to the speed of light. Most of these are protons, some are electrons, and much rarer are heavier ions. They have the ability to penetrate many materials. When they hit or pass through material, they can ionize atoms by stripping off their electrons.

Electrons, in particular, affect the solar panels in much the same way that ultraviolet light does, reducing their output. They also contribute to electrostatic charging of the satellite as mentioned above.

Protons are a major source of trouble. They can burrow into the computer circuits that control the satellite and actually give commands to the computers, causing them to malfunction in many ways. In extreme cases, they can cause total failure of the satellite.

The earth's magnetic field traps subatomic particles in wide volumes around the planet. These are called the Van Allen Belts (Fig. 9.2). Satellites that pass through these regions regularly, such as some of the MEO satellites, are subject to particularly intense radiation.

For this reason, all satellites must use what are called *radiation-hardened* (or "rad-hard") electronic components. These are expensive to make and increase the cost of the satellite. There are only a handful of companies in the world that make space-qualified radiation-resistant chips, and because of their actual and potential uses in military hardware, their sale is restricted. Many of these radiation-hardened

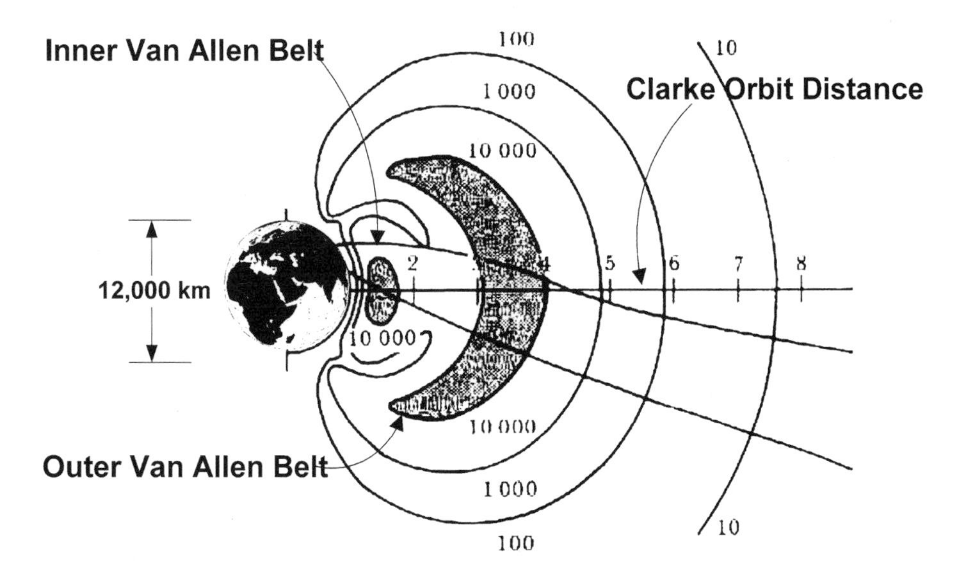

Figure 9.2 The Van Allen radiation belts around Earth. There are two major regions of charged particles trapped in Earth's magnetic field. GSO satellites orbit outside the outer belt, and LEO satellites below the inner belt. Some MEO satellites may swing through the belts during each orbit. The numbers on the curved lines are densities of charged particles.

components are treated for export purposes as "munitions" for which permission to export must be obtained.

It is not practical to harden computer components against the most intense cosmic rays because the shielding would increase the satellite's mass too much. As a consequence, we have seen (and will doubtless see again) a few cases in which satellites have been damaged by especially intense bursts of radiation. For example, in 1994, the Canadian satellites Anik E1 and Anik E2 were hit by a burst of particles spewed from the sun in a small explosion. Both satellites became non-operational due to electronic failures. Anik E1 was restored to service in a few hours; however, Anik E2 had had its onboard stabilization system totally fried, making it unable to stay pointed at Earth. After much work, Telesat engineers devised a way of keeping it operational by frequently beaming commands from a satellite control earthstation.

Satellite design engineers do take some steps to try to ameliorate the effects of radiation. Filters and surge protectors can prevent some radiation-induced problems in the circuits. Satellites that orbit at medium distances and consequently pass through the Van Allen radiation belts can be equipped with glass covers for their solar panels to reduce radiation degradation.

Solar storms wax and wane in a cycle that parallels the number of sunspots on the sun's surface. This cycle has an average period of 11 years, but peaks have come as frequently as 8 years apart and as infrequently as 16 years apart. Apart from these semiregular cycles, a solar flare occasionally sends out a blast of particles. If the blast happens to spray the region around Earth, all satellites in orbit are at risk. (On the bright side, we also get colorful displays of the aurora!)

9.4 Meteor-oids, -ites, and -s

There is also some solid material in space. In the space environment of satellites, we are concerned with the smallest of this material, which we call *meteoroids*. If a meteoroid happens to encounter the earth and burns up in the atmosphere, the streak of light is called a meteor; if it survives its fiery plunge and hits Earth, it becomes a meteorite.

The meteoroids range in size from miles across—which are very, very rare—down to microscopic bits of cosmic dust. Millimeter-sized particles that are the size of beach sand are fairly common. If you wonder why we are concerned with a grain of sand, the answer is that it may be moving through space at very high speeds. Even a small object traveling fast can do damage, since the energy of motion increases as the square of the speed. (That is why a few-gram lump of lead becomes a deadly bullet when fired from a gun at a thousand feet per second.) A 1-gram meteoroid moving at a typical speed of 40 km/sec would have the same energy as a small car speeding at 90 mph. If it hit a satellite, it would be the same energy as if a bomb went off beside the satellite!

Some of these grains are bits chipped off of asteroids by collisions or grains of rock released as the icy cores of comets disintegrate with age. Some may be left over from the formation of the solar system. As Earth orbits the sun, it sweeps up some of this material. It is estimated that between 10 and 400 tons of meteoroids fall to Earth every day. The concentration is greatest close to Earth, and lower at the higher Clarke orbit. Several times each year, Earth predictably crosses the highly elongated orbits of defunct comets that have strewn dust and rocks along their paths. These concentrations of meteoroids are seen on Earth as meteor showers. On a typical dark night, a single observer on Earth can see 5 to 10 meteors an hour; during a meteor shower this may rise to 50 to 60 a hour.

In the late 1990s, a potential meteor shower problem worried satellite operators, customers, and their insurance companies. One particular meteor shower is known to greatly increase the meteoroid flux. Whereas in normal years the so-called Leonid shower, which peaks on November 17 each year, produces maybe 50 meteors per hour, once every 32 or 33 years Earth encounters a dense clump of left-over comet material, and the observed rate of meteors jump enormously for a few hours. In 1966, for several hours observers counted upwards of 1000 meteors per minute! But in 1966, there were very few satellites in space to worry about. There were lower peaks in 1998, 1999, and 2000, which had no damaging effects on satellites. However, there is nothing one can do to prevent it or ameliorate the potential impact—pun intended—other than be prepared to launch replacement satellites if any now on-orbit become hors de combat.

Usually the effects of a satellite being hit by a typical meteoroid are small. Over the typical 10–15 year lifetime of a satellite in Clarke orbit, you can expect a few pinpricks of less than a millimeter or so in the outer surface of the structure, and a slight "sandblasting" that will also decrease the efficiency of the solar panels somewhat. But impacts are a statistical phenomenon: the effects will depend on how large the satellite is, what orbit it is in, and how long it is in operation. The more satellites in orbit, the greater the overall threat. The new constellations in low- and medium-orbit are at greater risk than those in Clarke orbit.

9.5 Space debris

Naturally occurring space junk is a minor problem, but we have added to it. The old comic strip character Pogo is noted for several sayings, the most famous which applies here: "We have met the enemy and he is us." In this case, the detritus from about half a century of the space age has become a collision risk. From the beginning of the Space Age on October 4, 1957, to the end of the twentieth century, we have launched thousands of rockets, each of which results in dozens to hundreds of pieces placed into space. If there is an explosion of a launcher in space, that can add many times that number of objects. For instance, in 1994, a Pegasus vehicle exploded in orbit, releasing an estimated 300,000 pieces, of which 700 were big enough to be tracked. Figure 9.3 shows a plot of the trackable

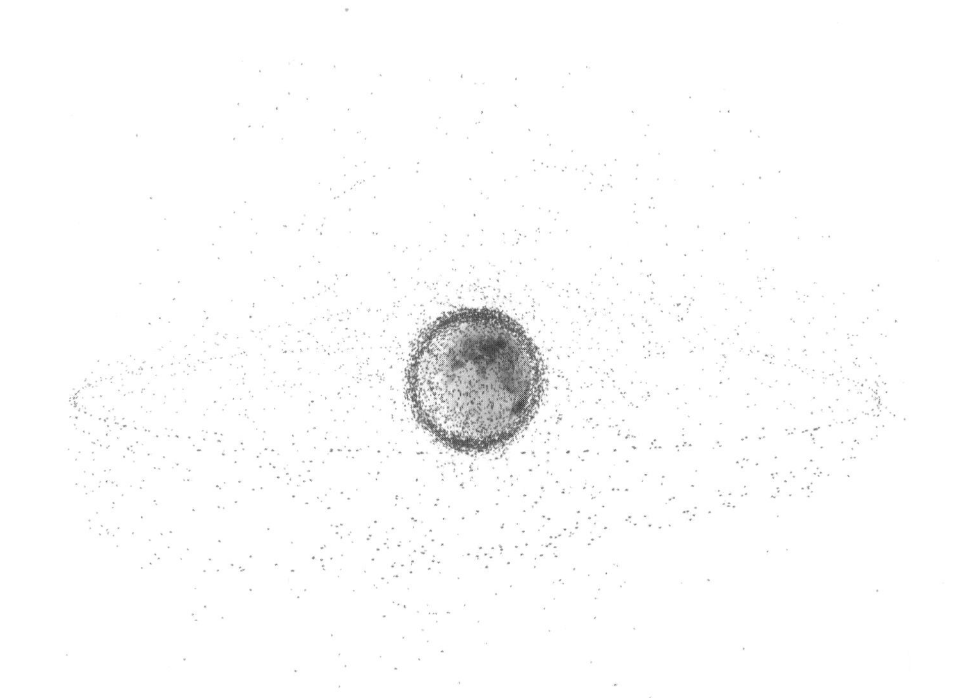

Figure 9.3 The distribution of trackable space objects near Earth, as of April 2003. Approximately 9000 objects are shown,of which only about 900 are working satellites. The rest are retired satellites, parts of satellites and launchers, and debris from collisions and explosions. In addition to all these, there are probably millions of bits too small to track. (Image courtesy of Analytical Graphics Inc.)

objects in orbit as of April 2003. Tens of thousands to millions of other pieces are too small to be tracked.

Only fairly large objects—more than a few centimeters in size—can be tracked by the military's space surveillance systems. Even so, at any one time, there are usually 8,000 to 10,000 objects in the database. There are an estimated 35,000 orbiting objects the size of a marble or bigger, and probably millions of tiny pieces of debris. A 1999 study estimated that 4 million pounds of space junk are in lower orbits, and other objects that are too small to be tracked amounted to over 100,000 pieces larger than 1 centimeter in diameter.

The concentration is highest in the lower orbits. However, material in low orbit, a few hundred kilometers up, is slowly pulled down by air resistance and eventually falls out of orbit. The time this takes depends on the density of the piece and its altitude. Between 1957 and 2000, about 17,000 pieces of space debris reentered Earth's atmosphere.

Anything higher than a few hundred kilometers is up there for good, practically speaking. Worse yet, collisions inevitably occur, and two pieces of debris colliding can produce a spray of thousands of particles. Much of this debris seems to be from a dozen intentional explosions. In the 1980s, the U.S. and Soviet Union tested their abilities to destroy each other's satellites by sending up bombs to get next to a test target and explode. So we have literally been hoisted with our own petard!

The rest of the debris is material resulting from satellite launches and launch failures. In some cases, the upper stages of launch vehicles were not empty of fuel when their engines were shut down. Months to years later, the fuel in these "time-bomb" upper stages exploded and produced a myriad of space shrapnel. (Today we have learned to dump residual fuel or burn the engines to completion.) Other debris is from such simple things as the springs that push satellites out of the launchers, nuts, bolts, and even chips of paint. After almost every mission, the space shuttle returns with several tiny pits in its windshield caused by hitting chips of paint at speeds of around 25,000 km/hr. Some years ago, a French military satellite was destroyed by collision with an Ariane rocket upper stage. Two small satellites that were recovered after less than a year in low orbit each had a couple of dozen visible impact craters on their surfaces.

The space debris problem is growing. With the hundreds of launches planned over the next decade, especially for the newer low- and medium-orbit systems, it will only get worse, and the systems most at risk are exactly these nongeostationary systems. Only with experience will we gain a knowledge of just how serious the risk is.

And we don't know what to do about it. There is no way to catch something traveling that fast, coming from all directions. If you come up with a method, you can make a fortune!

The details of the space environment in which a satellite must function will depend on the particular orbit it is in. The situation for the geostationary satellites in the Clarke orbit is different from those orbiting only a few hundred miles above the earth's surface. This brings us to the topic of orbits in the next chapter.

Chapter 10

Orbits

Communications satellites are in orbit around the Earth. The properties of a satellite's orbit control its visibility from Earth and the details of how we link to it. We need, therefore, to have a basic understanding of the properties and relative advantages and disadvantages of different kinds of orbits.

The force of gravity is constantly acting on the satellite. A satellite stays in orbit because the downward force of gravity pulling the satellite toward Earth counterbalances the momentum of the satellite imparted to it by the launch vehicle and which continues even after the rocket has stopped thrusting. Without this gravitational pull, the satellite would fly off into space. Without friction or drag from air resistance, an object will theoretically remain in orbit forever. Recall that the moon has been orbiting the earth for billions of years without anyone pushing it around!

In addition to the pull of the Earth (and we will see later that even that is not uniform), the Sun and Moon pull the satellite in various directions. This causes what are called *perturbations*, which slightly pull the satellite out of its intended orbit and orbital slot. This drifting must be corrected for the satellite to operate properly, and so an important subsystem of most satellites keeps the satellite in place. This need to control the satellite adds to its mass, complexity, and cost, as we will see later in this chapter.

About four centuries ago, an astronomer in Prague named Johannes Kepler used accurate years-long records of planet observations to arrive at an empirical understanding of how planets—and everything else—orbit. About 70 years later, Isaac Newton used newly formulated principles of physics and mathematics to show not only that Kepler was right, but to demonstrate the physical basis of satellites' behavior.

10.1 Kepler's Laws

Kepler found that three laws describe the fundamental properties of orbiting objects.

Here are Kepler's Laws as they apply to our communications satellites:

1. *Closed orbits have the shape of an ellipse, and the ellipse lies in a single plane.*

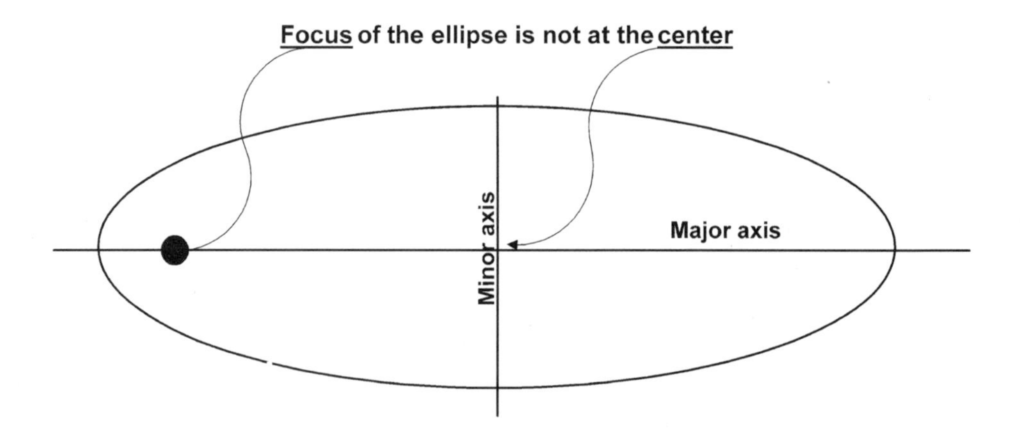

Figure 10.1 Properties of an ellipse. Closed orbits are ellipses, of which a circle is just a special case in which both axes are the same size. Note that motion of an object in an elliptical orbit is around the focus, which is where the Earth would be, not around the center of the ellipse.

The shape of the ellipse, i.e., how much is it elongated, is described by a number called the *eccentricity*, which ranges from zero to slightly less than one. See Fig. 10.1. If the eccentricity is exactly zero, we have the very special case of a circular orbit. The Clarke orbit is a circular orbit. (Orbits can be open, with no end to them, but these are not of interest to us here.)

If the orbit is elliptical, the object the satellite is orbiting is not at the center of the ellipse, but at one of two special locations called the focuses, or foci. For a satellite revolving about the earth, Earth is at one focus while the other is vacant. For circular orbits, the foci coincide in the center. A satellite moving in an elliptical orbit has two notable locations along the orbit: the outermost tip of the ellipse, where the satellite is farthest from Earth, called the *apogee*, and the closest point to Earth, called the *perigee*. The time it takes to complete one orbit, usually measured from perigee to perigee, is called the *period* of the orbit.

Note that while physicists would measure the distance of a satellite from the center of the earth, in the communications satellite business we are most concerned with the satellite's distance from the surface of the earth; this distance is called the *altitude*.

If the plane of the orbit is not the same as the equator, the angle the plane makes with the equator is called the *inclination* of the orbit, symbolized by i. The inclination can range from 0° (an equatorial orbit with the satellite moving west-to-east as the earth does) through 90° (a polar orbit) to 180° for a satellite moving backwards from the direction that Earth turns. Orbits with i less

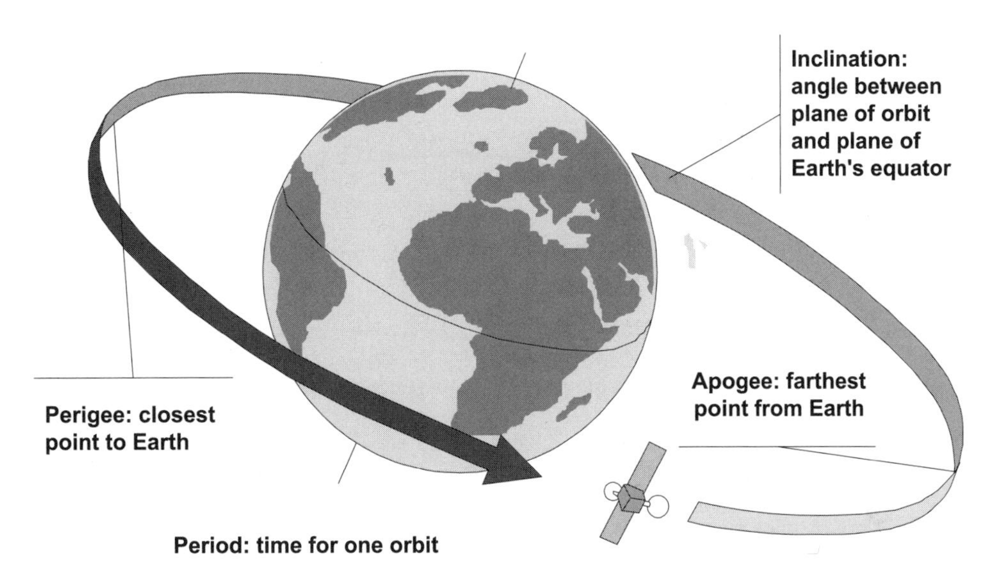

Figure 10.2 A closed orbit is defined by the size and shape of the ellipse, and by the orientation of that ellipse with respect to the sky and to Earth's equator. For our purposes, the most important parameters are the apogee and perigee distances (the same if a circular orbit), the part of the Earth they lie over, and the inclination of the orbit to the equator. Truly geostationary orbits are circular and have no inclination.

than 90° are called *direct* or *prograde* orbits; orbits with i greater than 90° are called *retrograde*. See Fig. 10.2.

For a truly geostationary satellite, the orbit must lie in the plane of the equator of the earth. The orbit will stay over the equator (unless pulled on by other forces, called perturbations, to be discussed later).

2. *A satellite will not move uniformly (at the same speed) if the orbit is elliptical, and will move fastest when close to Earth and slowest when far from Earth; if the orbit is circular, the motion is at a constant speed.*

   Thus a satellite in a noncircular orbit will spend more time hovering over the part of Earth that is under its apogee, and speed across regions of Earth around its perigee. An alternative way of stating this law is that a line from the earth to the satellite will sweep across equal areas in equal times, so this is sometimes called the "law of areas." See Fig. 10.3. This motion is somewhat like a pendulum: fast at the bottom and slow at the top. A couple of satellite systems—to be described in detail later—now make use of this property of orbits to concentrate service area by orienting the elliptical orbit so that the apogee is over areas with lots of potential customers.

3. *The period of the orbit depends only on the average distance of the satellite. For a circular orbit, this is the radius of the orbit.*

   To emphasize, realize that only the *altitude* of the satellite above the earth's surface determines the period, not the size or mass of the satellite. See Fig. 10.4.

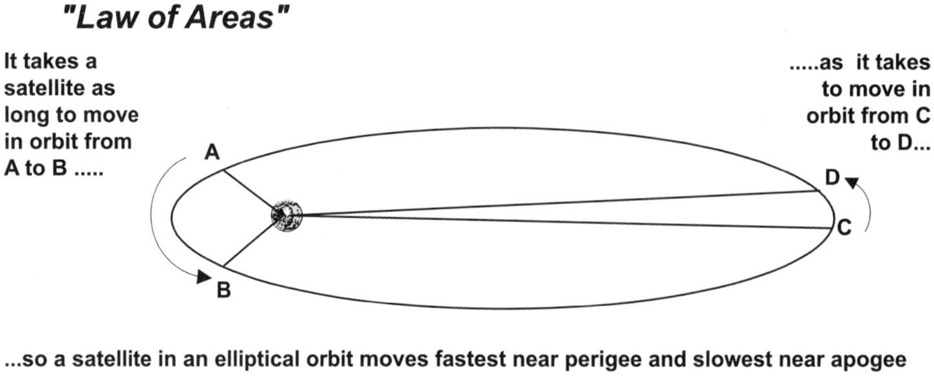

"Law of Areas"

It takes a satellite as long to move in orbit from A to B .....

.....as it takes to move in orbit from C to D...

A

B

C

D

...so a satellite in an elliptical orbit moves fastest near perigee and slowest near apogee

Figure 10.3 The "Law of Areas" says that the line from Earth to a satellite sweeps over equal areas in equal times. This means practically that a satellite in an elliptical orbit moves fastest when near Earth, and slowest when near apogee. This can be made to be an advantage for such satellite systems as Molniya and the Sirius SDARS satellites to allow the apogees to be over longitudes where service is needed most.

We are talking about satellites which are orbiting without any propulsive power. The launch vehicle lifts the satellite to its proper height, gives it momentum in the proper direction, then releases it. From then on, the satellite orbits by balancing its momentum against the pull of gravity. Since there is no friction (except for some very low orbits just within the outer layers of Earth's atmosphere), there is nothing to slow the satellite down, so it keeps on orbiting indefinitely, just as the moon has been orbiting Earth for eons without any propulsion!

For those wanting more mathematical background on Kepler's Laws, consult Appendix E.7. More details on all the so-called orbital elements can be found in Appendix E.8.

10.2 Geosynchronous and geostationary orbits

Since only the altitude of the satellite determines the period, if we wish to put a satellite into an orbit of the same period as that in which the Earth turns, there is one and only one distance from Earth for which this will happen. The Earth turns, from west to east, in one day, which we approximate by 24 hours on the clock, but it is actually 23 hours, 56 minutes, and 4.091 seconds. To have the period be exactly equal to this—to be truly *geosynchronous*—the satellite must lie 35,786 km above the surface of the earth. (Note that this is above the surface of Earth, not from its center. Using a radius of Earth of 6378 km, the distance of the geosynchronous orbit from the earth's center is 42,164 km.) If you prefer common units, the geosynchronous orbit is 22,236 miles above the surface and 26,199 miles from the center.

We often round these numbers for convenience, and so we loosely say that at an altitude of 36,000 km or 22,000 miles a satellite will orbit in 24 hours.

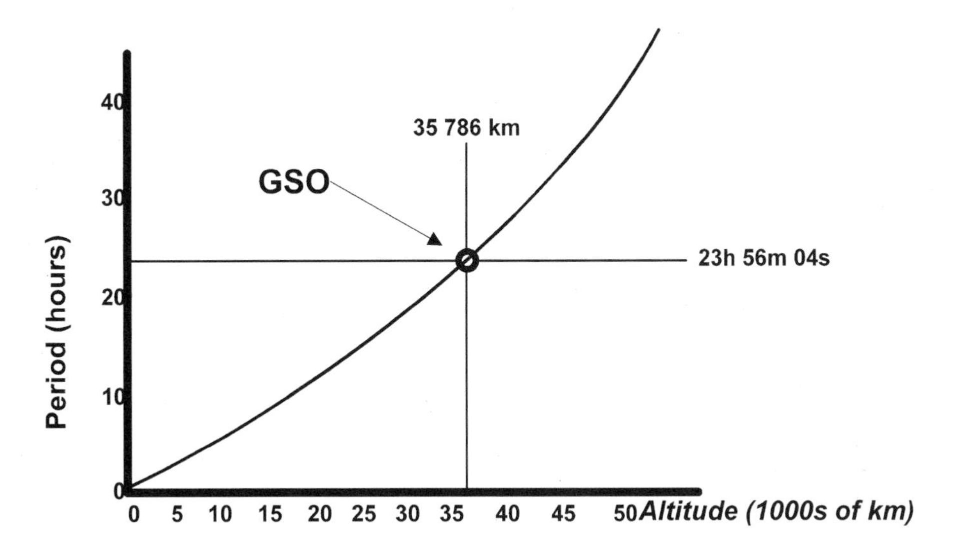

Figure 10.4 Satellite period and distance. The time it takes a satellite to make one full orbit depends only on its distance from Earth. Thus there is only one distance for which the satellite will orbit in 24 hours: about 36,000 km or 22,000 mi. The Clarke orbit at this distance above Earth's surface is circular and is directly over the equator.

Any satellite with an average distance above the surface of 36,000 km will orbit Earth in 24 hours. All such satellites are *geosynchronous*. You can see that more conditions are needed to make the satellite *geostationary*—stationary in the sky as seen from Earth's surface. Obviously, a satellite in a polar geosynchronous orbit would not be fixed in the sky as seen from Earth. To be geostationary, the orbit must satisfy three conditions:

- the orbit must be geosynchronous;
- the orbit must be circular; and
- the orbit must be over the equator, i.e., the inclination must be exactly zero.

To emphasize, a geostationary satellite is not stationary in space. In fact, the satellite is hurtling along through space at a speed of 11,069 km/hr or 6,878 mph. It is stationary only as seen from a point on the earth.

Thus, there can be *only one, unique geostationary Clarke orbit*. You cannot place a satellite a few hundred miles in, out, up, or down from that exact distance or vary the shape or orientation of the orbit and still have it be geostationary. A satellite in an orbit a bit "up" or "down" from the equator would have an inclined orbit, but would not be truly geostationary. (Such orbits are sometimes used to conserve the stationkeeping fuel of an aging satellite, as will be discussed later.) A satellite in an orbit a bit closer to Earth than 36,000 km will not be geostationary, either; it will be seen from Earth to move slowly eastwards through the sky. Conversely, a satellite slightly farther out than 36,000 km would be seen to move slowly westward in the sky.

For this reason, the Clarke orbit is considered to be, in the words of the lawyers, a "limited natural resource" that is the common property of everyone. It has a circumference of about 264,000 km, or about 165,000 statute miles. All geostationary satellites must lie along this line. Accordingly, an international body, the International Telecommunication Union, has been assigned the task of allocating positions along this orbit.

Other than the Clarke orbit, there are no other unique orbits around Earth of interest for communications. Nongeostationary orbits (*NGSO*) must be fully described by the size, period, eccentricity, inclination, and orientation of each orbit. Also, because there are infinitely many such orbits, they are not internationally regulated, and anyone can put anything into any NGSO orbit (subject to the laws of the country from which it is launched). However, since all NGSO satellites will cross the equatorial plane twice each orbit, regulations specify that they must not cause radio interference to GSO satellites.

As Arthur C. Clarke foresaw half a century ago, a satellite in a geostationary orbit (*GSO* or *GEO*) will have two major advantages for communication:

First, since the satellite is fixed in the sky as seen from an earthstation, the earthstation does not have to track the satellite across the sky. This makes the earthstation much less expensive. For comparison, consider the case of the early Telstar satellite, which was in a much lower orbit. To use it for communications, the earthstation antennas weighed hundreds of tons and had to be able to track the satellite across the sky to an accuracy of better than a degree of arc in something like 20 minutes. This made the earthstations cost millions of dollars. If all earthstations cost this much, there would be few earthstations and the communications satellite industry would not have grown very large.

(Indeed, this fact gives a good illustration of what you might call the fundamental economic principle of the industry: every dollar you spend on the satellite gets divided by the number of users, while every dollar required for a terminal gets multiplied by the number of users. One can see that this leads us to build more expensive satellites to keep the cost of user equipment low and build a large market. Today's satellites cost 10 to 20 times what Early Bird cost, but a terminal capable of linking to these satellites costs about a millionth of what the first earthstations cost. As electronics have gotten less expensive over the past couple of decades, it is increasingly easier and cheaper to communicate via satellite, and the basic principle still applies: keep user terminal costs low.)

Getting back to the advantages of geostationary satellites, and continuing our attempt to aggregate a market for communication services, the second advantage is that a GSO satellite, being 36,000 km above the earth, can see about 44% of the entire surface of the planet. (This number uses the common practical assumption that an earthstation should never look at a satellite that is closer than 5° to its horizon; if you allow viewing down to the horizon you get a couple more percent of the earth.) Thus, it would take only three GSO satellites, as in Clarke's early illustration, to cover (almost) all of the earth with much overlap.

As seen from a GSO satellite, our Earth is just a disk 17.3° in diameter. See Fig. 10.5. Since the earth is 6,378 km (4000 mi) in radius, and the Clarke orbit is 35,786 km (22,300 mi) in altitude, you can easily calculate that a GSO satellite is about 5½ Earth radii away from its subsatellite point. (For comparison, the moon is about 60 Earth radii distant.)

A further characteristic of GSO satellites is that all earthstations are approximately, but not exactly, the same distance from all other earthstations via satellite, no matter how far apart they are across the surface of Earth. Whereas the link between an earthstation located at the subsatellite point is 35,786 km, a station linking to a satellite just on its horizon (not a good idea, as we will see later) has a link length of 41,680 km. Thus, the time it takes for a radio signal to travel between the stations and the satellite will vary from 0.119 sec to 0.139 sec. That may not sound like a large change, but it can be a big effect for high-speed data protocols.

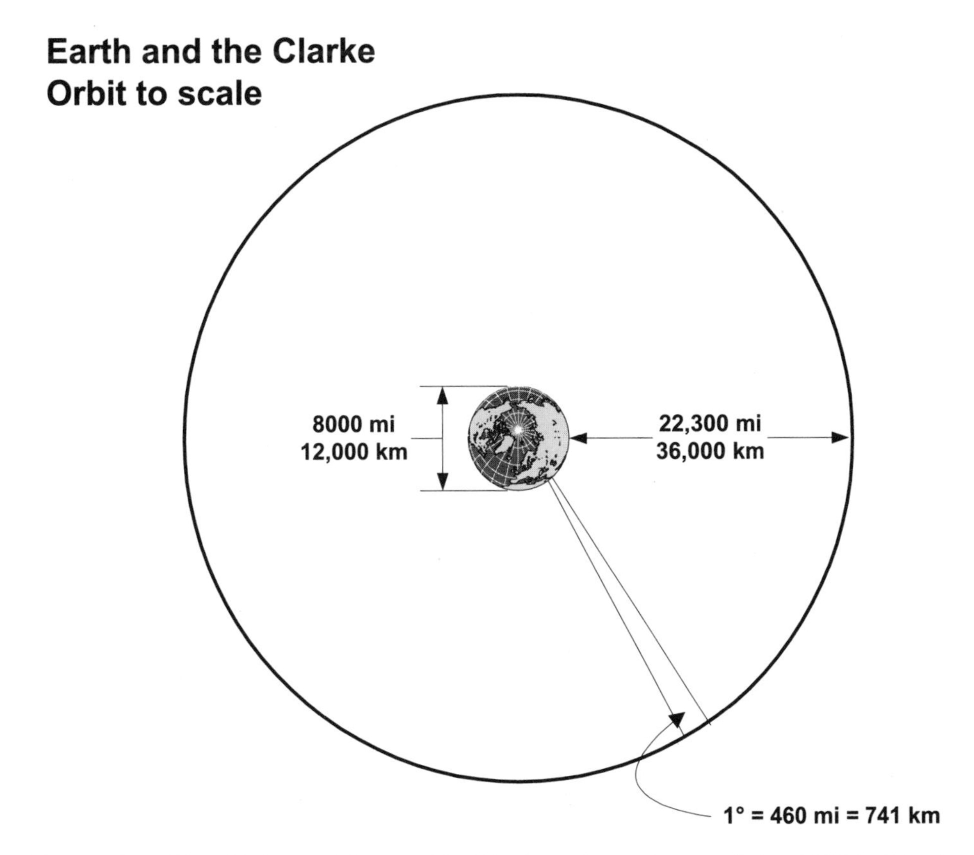

Figure 10.5 The Clarke orbit is about 5.5 Earth radii in radius, and about 265,000 km in circumference. A spacing of 1 degree along the orbit corresponds to a physical spacing of about 741 km, and conversely, a 1 degree spotbeam from a satellite toward its subsatellite point will have a diameter on the surface of Earth of about this size.

So what are the disadvantages of a GSO satellite?

First, the distance between the transmitter and receiver (satellite and earthstation) is always very large, so received signals are weak and thus require higher transmitter powers and wider bandwidths. Every earthstation is about 36,000 km from every satellite; so every earthstation is about 72,000 km from every other earthstation. The GSO satcomms industry has the basic property that the shortest distance between two points is 72,000 km, no matter how far apart they are on Earth!

A second disadvantage of GSO satellites is that they cannot be seen from polar regions. For our purposes this is at latitudes higher than 77°, both north and south. Fortunately, there is not a heavy telecommunications demand in these part of the earth.

10.2.1 Inclined geosynchronous orbits

Truly geostationary satellites orbit at the geosynchronous distance of 36,000 km in circular orbits lying exactly over the earth's equator. They have zero inclination. Since they are over the equator, earthstations at high latitudes have a problem seeing them, and they will be lower in the sky the higher the latitude of the earthstation, and the farther the satellite is east or west of the earthstation's longitude. With these low-look angles, terrestrial features such as mountains, buildings, and even tall trees may block a station's view of a satellite. This is particularly true for mobile users whose locations are constantly changing.

One suggestion to reduce this problem is to place several satellites in geosynchronous inclined orbits. The inclination can be chosen for the latitude of the desired service region. All of the satellites trace out a groundtrack that is a large figure 8. In such a system, the satellites—three minimum—circulate in and out of the region almost overhead of the service region. Thus, these are sometimes called *quasi-zenith* satellites or orbits, or *figure-eight* orbits.

In a three-satellite arrangement, one satellite handles the communication for the region for a third of a day, and as its motion carries it away from the zenith, traffic is switched off to the next satellite approaching the zenith. Such a configuration is especially good for mobile users because it reduces blockage by terrestrial obstacles. They can share the same frequencies as truly GSO satellites, and they can cover polar regions. The Japanese Space Agency is working on projects to use such orbits for mobile satellite communications systems and for a GPS-like navigation system.

10.3 Nongeostationary orbits

There is an infinite number of lower and higher orbits. The lowest practical orbit is set by the drag of the atmosphere, and is usually considered to be about 100 km up.

At that altitude, a satellite could probably make a single orbit before air friction pulled it down. Few satellites, maybe occasionally some spysats, orbit that low.

NGSO orbits are characterized by altitude, period, eccentricity, inclination, and orientation of the orbit. We often see the abbreviations LEO for low earth orbit, MEO for medium earth orbit, ICO for intermediate circular orbit, and HEO for highly elliptical orbit. There are no fixed definitions of the adjectives. "Low" is usually taken to mean approximately below 1600 km altitude; "medium" and "intermediate" roughly mean above that and below GSO; and "high" generally means at or above GSO altitude.

One major problem with NGSO satellites is that each one "sees" and thus serves less of the Earth than a GSO satellite. This means using more satellites. To provide more continuous coverage, one possibility is to launch several satellites into the same orbit, spacing them along the orbit so that they appear at a given location sequentially. Such a system is sometimes called a *string-of-pearls* orbit.

10.3.1 Low orbits

Low- and intermediate-orbit satellites have advantages and disadvantages compared to GSO. On the positive side, such satellites are usually smaller, lighter, and easier to launch. This makes the satellites, the launches, and the insurance for them less expensive. From a coverage point of view, with inclined orbits a system of LEO or MEO satellites can provide service to 100% of the earth's surface, including polar regions. If such orbits are elliptical, the satellites can dwell for a long time over areas beneath their apogees, and can concentrate a service area over places with high demand.

On the negative side, being closer to Earth, each satellite sees only a small fraction of Earth's surface, often only 2% to 15%. The radio beam to and from each LEO satellite sweeps across the surface as the satellite orbits. This means then that for wide or global coverage, there must be many satellites, partially offsetting the advantage that they are each typically smaller and cheaper. Such a group of satellites working together in a system is called a *constellation*. Depending on the design of the system, having many satellites may require many gateway earthstations or expensive satellite-to-satellite links to provide continuity across the globe. See Fig. 10.6.

This also means that a user on Earth, say someone making a telephone call via one of the LEO systems, will probably be using the services of more than one satellite during the course of a single few-minute call, requiring signal "hand-off" from one satellite to another, analogous to the way a terrestrial cellular telephone user moves from cell to cell. Indeed, the LEO telephone constellations can thought of as cellphone systems turned upside down: the user is fixed and the cell moves.

Because the satellite is moving, it is also impractical for the user's handheld telephone to have a directional antenna constantly requiring the caller to orient the telephone toward the satellite. Instead, the handset will have an omnidirectional or

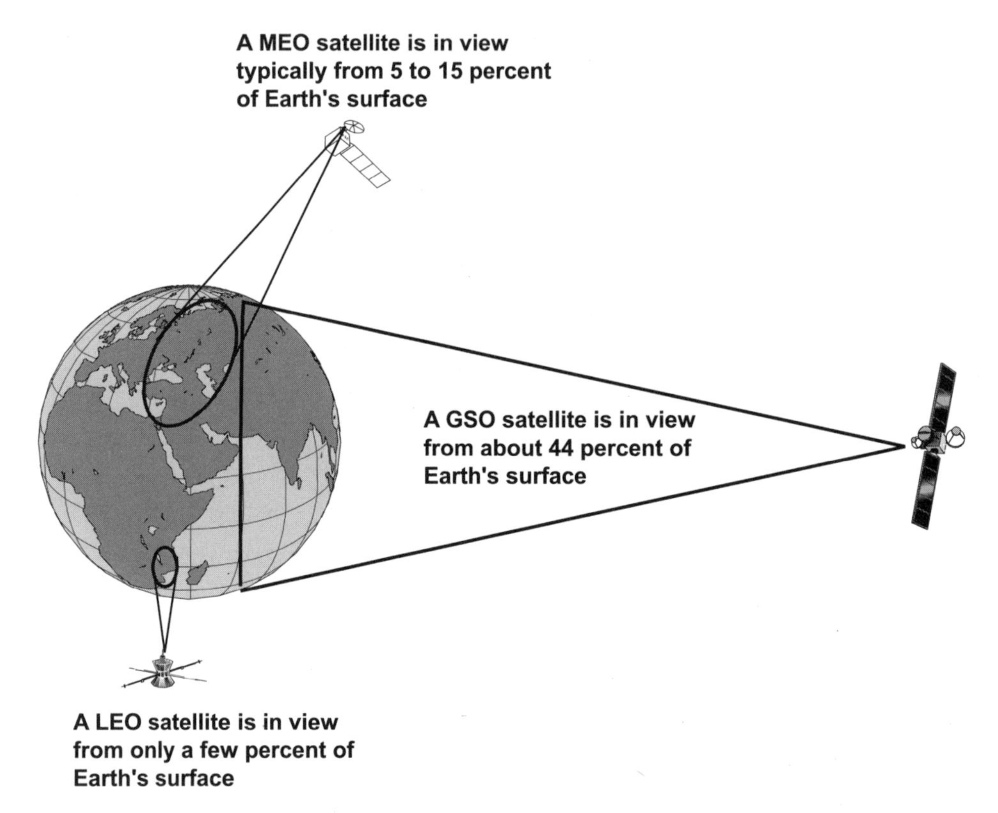

Figure 10.6 The closer to Earth a satellite orbits, the less of Earth's surface it sees. Thus, more low-orbit satellites are required for full Earth coverage.

hemispherically directional antenna, which is much less sensitive than a very narrowly beamed dish antenna, and partially offsets the fact that the satellites themselves are closer.

Lastly, LEO and MEO satellites are more subject to space debris collisions, atmospheric drag, and radiation—the amounts depending on altitude. Typically they will have shorter operational lives than GSO satellites.

10.3.2 "Little" and "Big" LEOs

The terms "Little LEO" and "Big LEO" are often mistakenly interpreted to imply something about the size or orbit of these satellites. Actually, the terms refer to the kinds of communications services they will supply. Further, the terms are often used to refer not only to LEO satellites, but also to those in medium orbits (and, sometimes even GSO), adding to the terminological confusion.

The so-called Little LEOs are designed to provide low-speed data services and operate in the VHF part of the radio spectrum. The official designation is "nonvoice, nongeostationary mobile satellite service." The OrbComm system is an example. They are used for such tasks as store-and-forward messaging (rather like e-mail), electrical meter reading, position reporting and fleet management, SCADA (supervisory control and data acquisition) for managing utilities, paging, and simple low-speed data relay.

The Big LEOs can provide these services, but what distinguishes them from the Little LEOs is that they can provide telephony as well. Examples are Iridium and GlobalStar. Another difference is the part of the spectrum they use for links to the users. Big LEOs operate in specially designated parts of the L- and S-band. (GSO satellites that provide telephony, such as Inmarsat and Msat, utilize mostly L-band frequencies. Most other GSO commsats operate in the C-, Ku-, and Ka-bands.)

As mentioned above, any NGSO satellite carries along with it a service area, defined by the region of Earth that it can see at any one instant (perhaps further reduced by using a narrow radio beam). Thus, no one satellite is permanently associated with any particular service region on Earth.

It is possible to take advantage of one of Kepler's Laws of orbits to place satellites in special orbits for special purposes. In particular, if a satellite is in an elliptical orbit, its motion will, as we have seen, be slower around its apogee. One of the SDARS satellite radio broadcast systems uses just such a scheme to maximize the time its satellite spends over its targeted service area. And there are other examples.

10.3.3 "Virtual GEO" orbits

One scheme developed and patented, but not yet used by any satellite system, is to use a multitude of nongeostationary satellites to mimic a geostationary system. In this scheme, called "virtual GEO" by its developers, there are 15 satellites placed in orbits with periods of 8 hours. Their apogees are about 17,000 to 27,000 km altitude, and the orbits are inclined about 63° to the equator. By choosing the proper orientation of all of the orbits, the system can ensure that at least one satellite is always in view. Because of the high inclination of the orbits, there are two advantages. First, they are well separated from GSO satellites, allowing them to use frequencies usually reserved for GSO satellites without interference; secondly, they will appear quite high in the sky as seen from medium latitudes, minimizing atmospheric problems and blockage by structures on the ground.

10.3.4 Molniya orbits

About 30 years ago, engineers in the Soviet Union invented a specialized communications satellite constellation to solve two problems they had using GSO satellites. First was the fact that they had a lot of very high latitude areas needing communications services. The second problem was that it is more difficult to launch from high-latitude launch bases to reach geostationary orbit—basically, the farther your launch pad is from the equator, the harder it is to place a satellite over the equator.

To deal with this situation, they devised the Molniya system. ("Molniya" means "lightning" in Russian.) These have proved so useful for certain applications that such systems are now generically called Molniya satellites. You can usually tell when an earthstation is looking at a Molniya-type satellite because it is aimed away from the equator, whereas all GSO satellites are over the equator.

Several satellites are placed in a highly inclined, highly elliptical orbit with a period of 12 hours. The inclination is about 63°. The perigee of each orbit is just a few hundred kilometers above the south polar region, while the apogee is high over the northern latitudes—actually farther from Earth than GSO satellites. From the laws of orbits, you know that around perigee the satellite is moving very fast, and is useless for communications. This is fine since there is little commercial demand around the South Pole. But satellites move slowly near apogee, and as it nears apogee, earthstations point to it and track it slowly over several hours. This makes for earthstations that are somewhat more expensive than truly stationary ones, but are a lot less expensive than stations that track from horizon to horizon in just a few minutes. Notice also that satellites will appear very high in the sky from earthstations in high northern latitudes. When a satellite being used begins to move too fast, another is timed to come into view so that the earthstations may shift their traffic to that one.

As seen from Earth, Molniya satellites appear to have a butterfly-shaped path (Fig. 10.7). Keep in mind that while the period of the satellite is 12 hours, the earth turns once in 24 hours, so in each day a given satellite makes one pass over the Eastern Hemisphere and the next over the Western Hemisphere.

The advantages of a Molniya system is that it gives excellent coverage of one polar region and can be more easily launched from high-latitude launch bases. Its disadvantage is that it does not serve one polar region, or even one hemisphere, at all well.

10.4 Geosynchronous transfer orbit

When a satellite is launched with an intended orbit along the Clarke Belt, it is often not launched directly into its final orbit, but first placed in an intermediate orbit called the *geosynchronous transfer orbit,* or GTO.

A GTO is a highly elongated orbit that has its perigee just a few hundred kilometers above the earth (Fig. 10.8) with a period of about 10.5 hours. For small- and medium-sized satellites, its apogee is tangent to the Clarke orbit; for very large satellites, the initial GTO may have its apogee well below the Clarke orbit. Once the satellite is in GTO, some of its unfolding and unfurling parts can be partially deployed, the electronics can be turned on, and the satellite's systems can be checked before it is finally inserted into its GSO orbital slot.

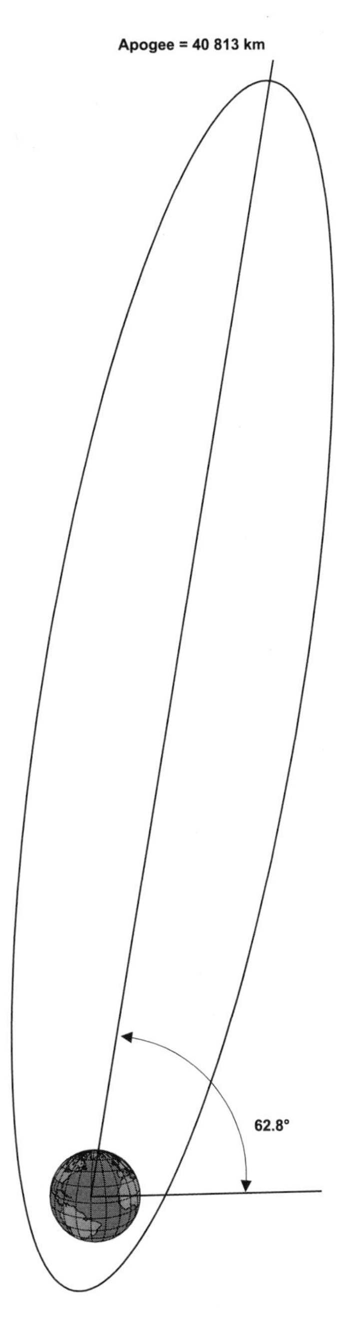

Figure 10.7 The Molniya satellites are in orbits with perigee only a few hundred kilometers above Earth's surface, but with apogees well beyond the Clarke orbit, with the orbit inclined about 63 degrees to the equator. The satellite has a period of 12 hours, and thus makes an apogee pass alternately above the Eastern Hemisphere and Western Hemisphere. Such an orbit also allows the satellite to be used only near apogee, but with very high elevation angles even from high latitude earthstations.

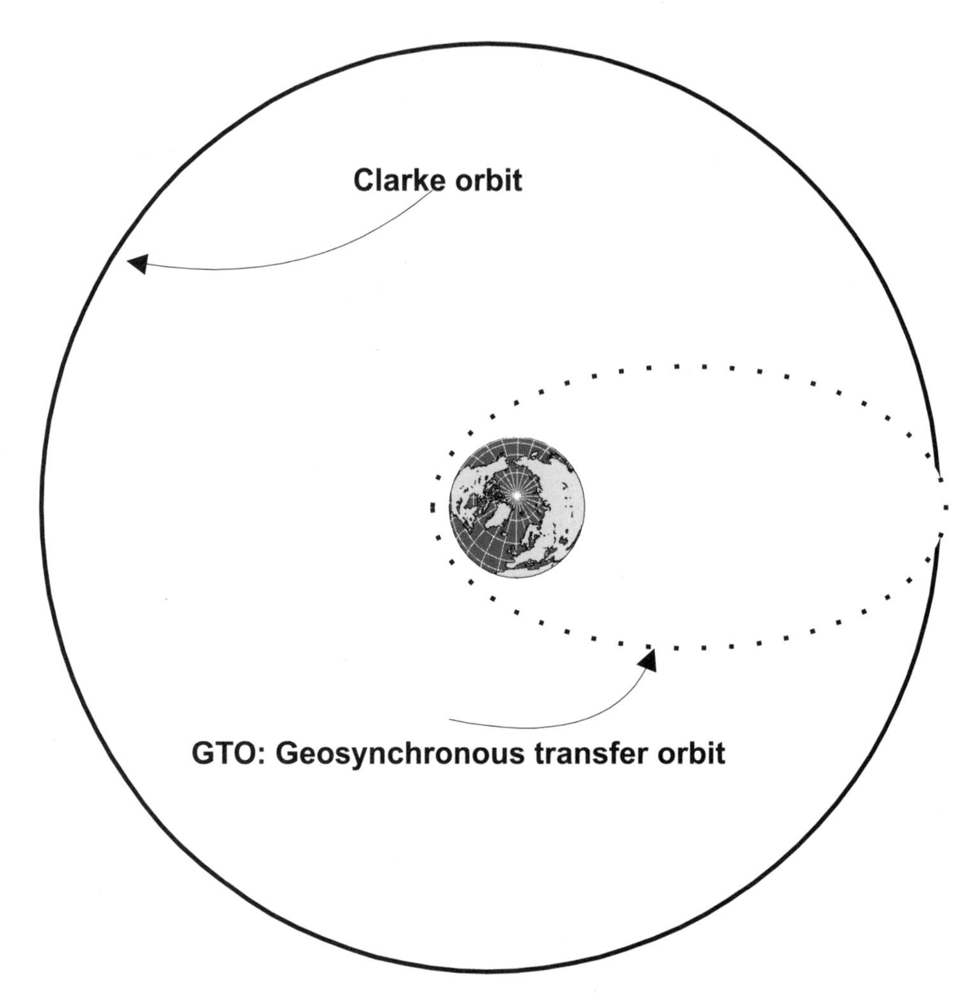

Figure 10.8 The geosynchronous transfer orbit is an elliptical orbit intermediate between a low orbit and the Clarke orbit. Most satellites are launched by having the launch vehicle place them in such a GTO, and later an apogee kick motor built into the satellite fires when at apogee to insert the satellite into the Clarke orbit. Some larger satellites are launched into a lower GTO which does not reach geosynchronous altitude, and then several thruster firings gradually raise the orbit to the Clarke orbit.

The satellite may be allowed to swing around its GTO for hours to weeks, depending on the satellite. After the satellite is checked out and the spacecraft reaches its apogee, a command is sent to fire a solid-fueled motor built into the satellite, called the *Apogee Kick Motor, AKM*. This rocket motor changes the satellite's motion by between 1500 and 2000 meters per second. This "burn," which is the final major propulsive event in the satellite's life, eliminates the inclination of the GTO and raises the perigee, converting the satellite's orbit from GTO to GSO. This procedure is called *orbit insertion* or *circularization* or *apogee firing*. Very large satellites may have several apogee firings, each of which gradually raises the perigee of the orbit until the satellite is in GSO.

When you attend a launch or launch party (this is highly recommended—launches are exciting and the food is usually great!), a half-hour or so after launch, the launch controller will (we hope!) make an announcement that the satellite has achieved "nominal orbit." That orbit is the GTO, not the satellite's final place in the Clarke orbit.

In the next chapter, we will see how a satellite's orbit determines the areas of the earth it can serve.

Chapter 11

Orbital Slots, Frequencies, Footprints, and Coverage

Since all GSO satellites are in an orbit lying over the earth's equator, the way to refer to such a satellite's orbital position is by the longitude over which it remains stationary. The location on Earth's surface directly below any satellite is called the *subsatellite point*. For GSO satellites, these are always on the equator. The position in orbit, denoted by its longitude, is called the *satellite slot*. See Fig. 11.1.

11.1 Satellite longitude and spacing

Confusingly, there are two common ways of measuring a satellite's longitude. The preferred method is to begin counting at the Greenwich Meridian of 0° longitude,

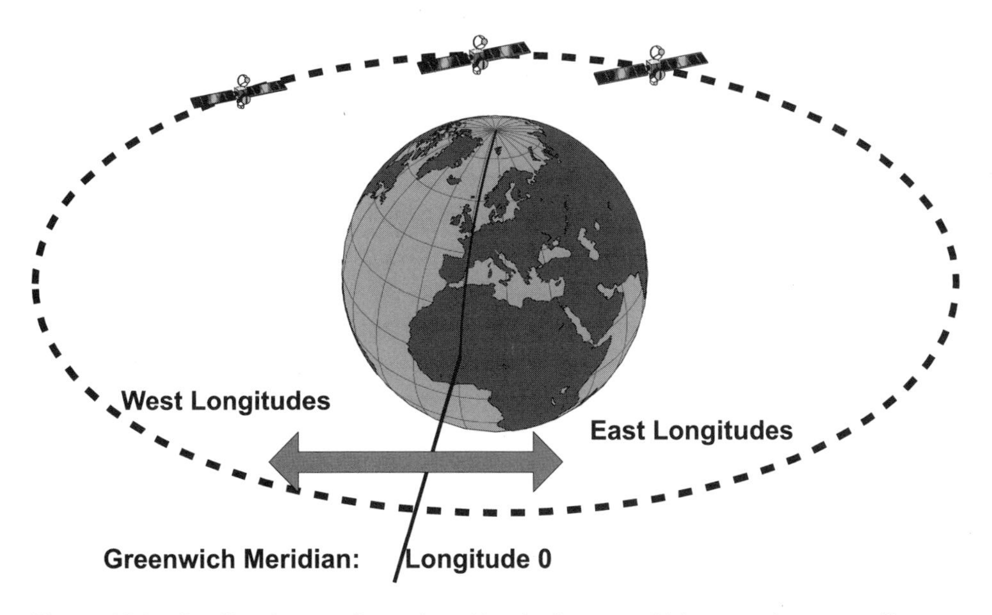

Figure 11.1 Satellite slots are the assigned longitudes over which geostationary satellites are placed. The longitudes may be measured from 0° to 360° to the east (the preferred method) or to the west, or from 0° to 180° east and west, measured from the Greenwich Meridian of 0° longitude.

and increase toward the east, all the way around to 360°. This is the method always used by the International Telecommunication Union, and by European and Asian nations. When it is not clear that such an east-going longitude is being used, the satellite's longitude is followed by the letter E. Thus the satellite PAS-4 is at longitude 64.5°E, over the Indian Ocean.

Many people in the Americas, however, prefer to measure from the Greenwich Meridian westward, and the satellite's longitude is followed by the letter W. To convert an east longitude to a west longitude, or vice versa, subtract from 360°. For example, Orion-1 is at 37.5°W, which is the same as 322.5°E, over the Atlantic Ocean.

Sometimes people measure only 180° both east and west. If you are compiling a list of GSO satellites visible from some earthstation, you have to be careful when using more than one source: you can erroneously end up with twice as many satellites as there really are.

Sometimes only an approximate location for a GSO satellite is all that is required to specify its area of service. One common group of abbreviations is to name the ocean over which the satellite orbits: *AOR* for Atlantic Ocean Region, *IOR* for Indian Ocean Region, and *POR* for Pacific Ocean Region. Sometimes these regions may be further subdivided, such as AOR-East and AOR-West.

Like frequencies, orbital slots are allocated by the ITU, and then assigned by individual nations. Because there is only a single Clarke orbit, it is considered to be a limited natural resource and the "common heritage of mankind." There are no rights to ownership of orbital slots.

Each kind of satellite service—such as fixed services, mobile services, and broadcast services—is assigned specific ranges of frequencies to use. For instance, there are C-band fixed services, Ku-band for both fixed services and a different part of the Ku-band for broadcast services, and the newer Ka-band for fixed and mobile services. It would be a waste of valuable spectrum to allow a single satellite to exclusively use a specific frequency band, so all satellites operating in a given service and frequency band use the same range of frequencies.

This naturally means that satellites could potentially interfere with one another if an earthstation cannot distinguish between the signals from two satellites using the same frequencies, or if the uplink beam from an earthstation is so broad that it hits more than one satellite. It is important to remember that, with minor exceptions, *all satellites operating in a specific service are sharing the exact same frequencies.*

Some satellites operate in two or more frequency bands; these are often called *hybrid* satellites. For example, some mobile communications satellites connect to gateway stations in the C-band, but connect with the mobile users in L-band or S-band. Some fixed-service satellites have a mix of C-, Ku-, X-, L-, and/or Ka-band transponders for versatility. This adds to the difficulty of minimizing interference between satellites.

An analogous situation on Earth is that of broadcast radio and television stations. For example, all FM radio stations in the United States broadcast in the 88–108 MHz band of frequencies, each station's assigned frequency is centered on an odd decimal place (e.g., 101.1, 101.3, 101.5, etc.), and each has the same 0.2 MHz bandwidth as every other FM station. However, they may differ in transmitter power from one another. Television channels are always the same frequency, no matter what city the station is in. For example, every U.S. television channel 13 uses the frequency range of 210–216 MHz.

So—how to keep the stations from interfering with one another?

11.1.1 Orbital spacing

For both terrestrial stations and satellites sharing identical frequencies, the twofold answer is the same: limit the power transmitted, and keep a minimum physical separation between them. Thus, you can have a television station on Channel 2 in Miami and another on the same channel in Baltimore because their power is limited and they are well separated. Thus, you can have a C-band satellite located at orbital slot 221°E and at 223°E.

In orbit, the physical separation is termed *orbital spacing* (Fig. 11.2). The actual spacing between similar satellites varies greatly, and the minimum that can be tolerated without interference varies with the frequency, satellite power, satellite beam patterns, earthstation dish size, satellite polarization and orientation, and other

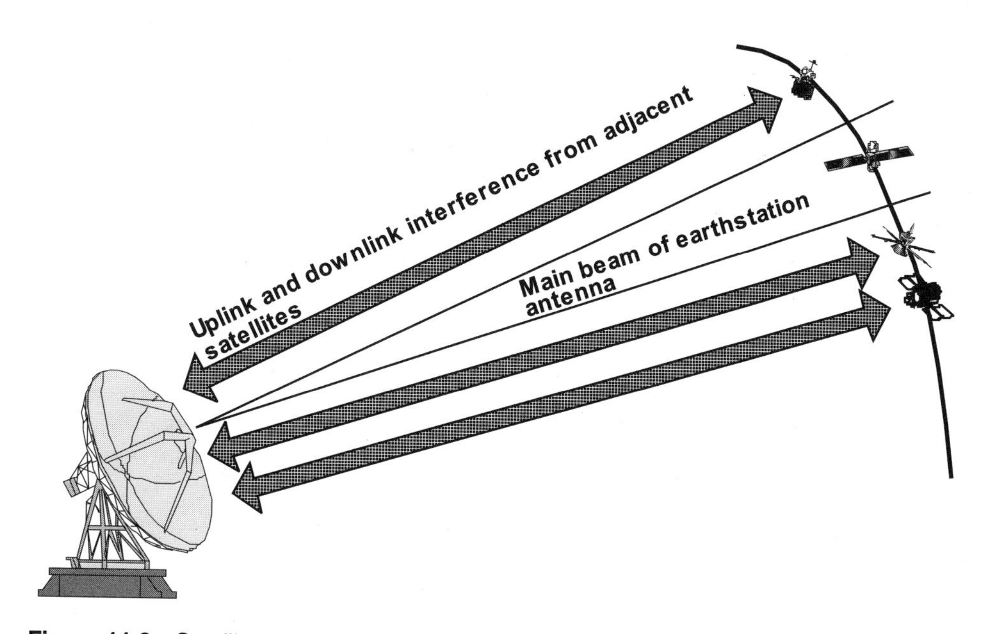

Figure 11.2 Satellite spacing. An antenna beam from an earthstation may be several degrees wide. If satellites are spaced too closely together in orbit, the beam may cover not only the intended satellite, but adjacent ones, creating interference. Note that the problem is one of electronic collision, not physical collision. The ITU suggests a spacing of 2°. Because beamwidth varies with frequency, satellites using the lower frequency bands need to be spaced farther apart than do ones using the higher-frequency bands.

factors. The complex and contentious topic of minimizing interference is called *frequency coordination.* (The problem of coordinating—minimizing interference—between GSO systems is bad enough; it is a much more complex task to coordinate problems between non-GSO systems and between non-GSO and GSO systems.)

The goal is to maximize the use of the Clarke orbit though control of the spacing of the satellites and their transmission characteristics. Proper choice of modulation, multiplexing, and error control affect how closely satellites may be spaced, and how strong their signals may be. For example, for analog signals, frequency modulation with a wide swing of frequencies around the center frequency (technically called a high-modulation index) promotes spectral efficiency in the GSO. For digital signals, using four- or eight-phase shift-keying is most efficient. Other techniques such as compression also affect how much total traffic a satellite can carry.

The problem is partially caused by the fact—as we will see in more detail in Chapter 16—that when an earthstation dish is pointed at a satellite, the beam or range of directions into which it sends a transmitted uplink radio signal, or from which it is sensitive to receive downlink radio signals, is not a perfectly cylindrical shape. Instead, it is actually a cone of directions that may be several degrees wide. This is called the antenna's *beamwidth.* If other satellites lie in this cone, the uplinked signal will be seen by the intended satellite and also by the nearby ones as well, causing them *uplink interference*. Similarly, the downlink signals from these nearby satellites will also be received by the earthstation along with the signal from the intended satellite, causing the earthstation *downlink interference*.

As an analogy, think of shining a flashlight on a wall. The lightbeam will be brightest in the center and drop off in intensity away from the center. If the flashlight was an earthstation, any satellite within the beam, not just at the center, would be getting some signal. The same is true in reverse, if the flashlight was a telescope receiving signals. All real-world transmitters and receivers, except for some lasers, have a range of angles into which they send and from which they can receive.

Another analogy may help explain this situation. You may have had the experience of driving a car at night on a long, straight, flat highway. Occasionally you see a light approaching you. If the other vehicle is very far away, you don't know if it is a motorcycle with a single headlamp or a car with two headlamps spaced a couple of meters apart. That is because there is a "beamwidth" for your eye, and if two objects are less than this angle apart, you cannot distinguish them as distinct objects. Only when the object gets closer do you discover if it was one light or two. To continue this analogy, assume it is a car with two headlamps, and now pretend that for some reason each light is sending you a message blinked in Morse code. Of course, when that car is close enough, you can focus on one or the other headlight and get the message; but when the car is too far away to distinguish between the two "signals," you get a mix of both and thus, your received signal is so noisy that you cannot tell one message from the other. The same thing happens if the receiving beam of an earthstation is so broad that two (or more) satellites are within the beam.

11.2 Once around the Clarke orbit

If we do a simple calculation to see just how much space we have available along the Clarke orbit, we have a circumference of about 265,000 km. Dividing by the 360° in a circle, we find that each degree along the orbit is about 740 km, or 460 miles. Since typical earthstation beamwidths are often a degree or more, we see that satellites at the same frequency must be typically spaced a few degrees apart along the orbit. If satellites are spaced a few degrees apart as seen from Earth, that means that their real physical separation in space is hundreds, even thousands, of kilometers. It is important to realize, then, that the need for satellite spacing comes about because of potential electronic interference between them, not because of possible collision of the satellites.

Another technique can be used to allow satellites to be spaced close together without interference. If two closely spaced satellites have same-frequency antenna beam patterns pointed at different regions of the earth, then they will not interfere, because an earthstation in one region will not be able to see the satellite beam pointed to the other region. This is called *frequency re-use*.

The problem with interference between satellites operating in the same band is that they all operate on exactly the same frequencies. However, satellites operating in different frequency bands, say C-band and Ku-band, will not interfere. Thus, they could be placed at approximately the same orbital slot. This is called *co-location.* It is conceptually equivalent to having the antennas of several radio stations attached to one tower; they don't interfere because each is on a separate frequency. By a combination of different frequencies and different beams, one orbital location can host several satellites.

The ITU suggests that ideally, FSS satellites should be able to be spaced 2° apart to maximize the number of satellites in the GSO. (National regulatory authorities, especially the FCC in the U.S., have also adopted their own spacing regulations for satellites.) This is not always possible for many reasons. But if it could be adhered to, this would mean that the maximum possible number of such satellites at a given frequency is 180. (The typical spacing is actually about 3°.)

This simplistic answer is totally pointless because we have not taken one paramount item into consideration: to be useful, a satellite has to be visible from an earthstation. Thus, a satellite over the Pacific Ocean is hardly useful for European communications. Therefore, there is a complex interrelationship between orbital slot, frequency, satellite spacing, and coverage of the earth's surface.

Obviously, there will be more satellites in those parts of the Clarke orbit that serve the most people. The most crowded portions of the arc are the longitude ranges 1°W to 35°W (serving trans-Atlantic communications), 87°W to 135°W (serving mostly North America), and 49°E to 90°E (covering the Indian Ocean and connecting Europe with Asia).

11.3 Satellite coverage

From the Clarke orbit, each satellite sees about 44% of the earth's surface, centered on its subsatellite point. If the antenna on the satellite covers this full area, this is said to be a *global beam* pattern. Sometimes it is desired to serve a smaller region of Earth with a *regional beam* or *spot beam*, which will limit the part of the earth served. This is done by limiting the range of directions to which the satellite's downlink antenna points.

With a global beam, a satellite can see a region of Earth's surface that extends along the equator to longitudes 77° on either side of the subsatellite point, and also 77° north and south of the equator. Because of the geometry of the cone-shaped beams from the satellite intersecting the sphere of the earth, the service region when drawn on the usual Mercator map projection looks like a big square with rounded corners, like that in Fig.11.3. This is somewhat misleading, however. Note that because the meridian lines of longitude on Earth converge toward the

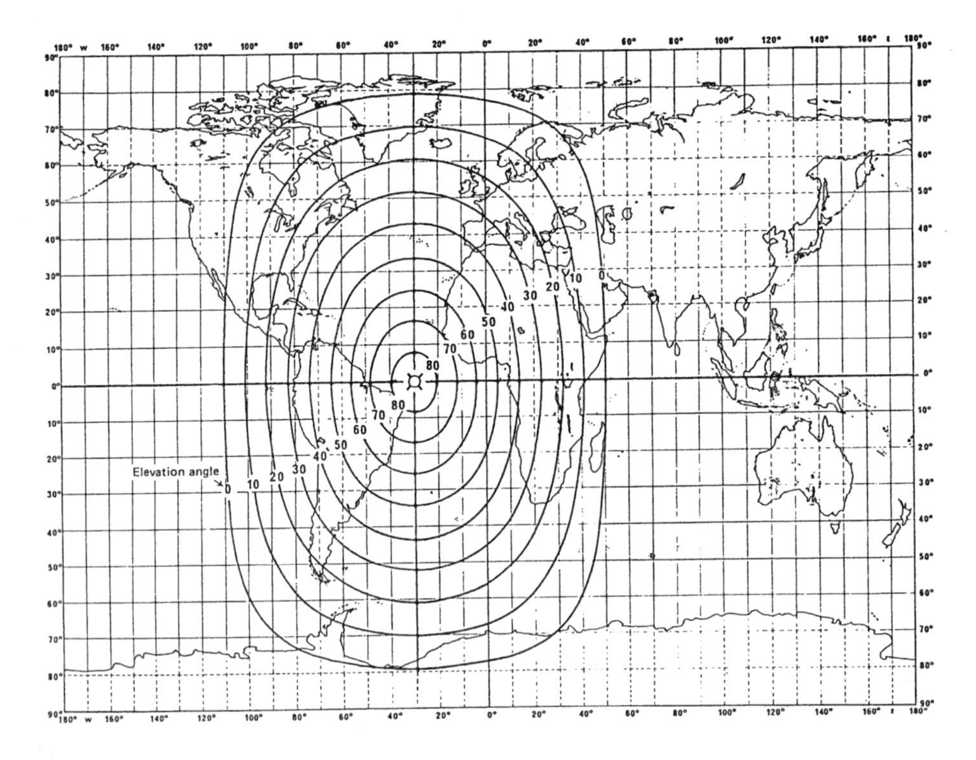

Figure 11.3 A single GSO satellite can see only about 44% of the surface of the Earth, and cannot see any areas with a latitude more than 77° north or south. The farther an earthstation is from the equator, the smaller the range of longitudes it can see along the Clarke orbit. An earthstation on the equator can see satellites about 77° to the east and west of its longitude, but an earthstation at latitude 77° can see on a satellite at its own longitude. (Guidelines suggest that an earthstation should not try to connect with a satellite that is less than 5° above the horizon.) (Source: ITU.)

poles, a satellite sees a narrower and narrower strip of geography as the service region is farther and farther from the equator.

To give a numerical example, any earthstation along the equator within 77° of the subsatellite point can see and use the satellite. As you try to serve points off the equator, this maximum range narrows, slowly at first. But for an earthstation at latitude 77°, the earthstation must be at the same longitude as the satellite to see it. In summary, earthstations at high (northern or southern) latitudes have a narrower arc of the Clarke orbit that is useful for interconnecting them with other earthstations. This puts countries and regions distant from the equator at a disadvantage.

For example, to serve all of the 48 contiguous United States, a region often denoted by the acronym CONUS, any satellite between west longitudes 54° and 143° could provide a link between any two places in this area. But if we try to provide coverage of all 50 states, we have to add the far-west state of Hawaii, and, worse yet, the high-latitude state of Alaska. So, to serve the entire U.S., we are now limited to satellites between orbital slots 119° and 143°, a much smaller part of the arc.

Canada, farther to the north, has it much worse. Even ruling out the Arctic regions, to see the majority of occupied Canadian provinces, a satellite can only be in the range of orbital longitudes between about 90° and 114°. This, of course, limits the number of satellites that can be used because of minimum spacing requirements, and thus limits the total amount of satellite communication.

This problem with high latitude service is one of the two major reasons the Soviet Union invented the Molniya system of satellites. In fact, if you look at a globe, you will realize that Russia is so vast in an east-west dimension, and so much of it is at high latitudes, that there is no single GSO satellite that can see all parts of the country.

NGSO satellites see less of the earth than GSO satellites (except for Molniya satellites near their apogee). The closer they are to Earth, the less of the earth's surface is visible. Whereas GSO satellites see about 44% of the earth's area, the LEO satellites, only a few hundred kilometers in altitude, see only a few percent.

This, of course, means that for global coverage, an NGSO satellite system must have more satellites. A GSO system would require only three satellites for global coverage (excepting the polar regions). The Big LEO system Iridium needs 66 for complete coverage. MEO systems need an intermediate number.

Thus, the trade-off of satellite altitude and number significantly affects the total cost of the system. GSO systems require few satellites, but they and their launches are expensive. LEO systems require many dozen satellites, but each is less expensive, as is their launch. MEO global systems often have the lowest overall cost as they require a moderate number of satellites placed at a medium altitude.

Sometimes several satellites of the same or different operators are placed near one another along a small range of orbital longitudes along the Clarke orbit. This makes it easy for earthstations to switch from one satellite to another as demand dictates. Such a grouping of satellites is sometimes referred to as a *satellite neighborhood.*

11.3.1 Orbits and groundtracks

GSO satellites are, by definition, stationary over one subsatellite point located along the equator. All NGSO satellites move relative to the surface of the earth, so their subsatellite point is moving. To send to and receive from such satellites, the earthstation must either track the satellite across the sky or have an antenna which can handle signals coming from anywhere overhead (called omnidirectional or hemispherically directional antennas). The closer they are to Earth, the faster the satellites move as seen from a location on the planet, and the smaller the service area of each satellite.

Choice of orbital altitude and orbital inclination control the coverage and the elevation angles of the satellites as seen by terrestrial users. (Looking ahead to the next section, they also control how much time the satellite is in sunlight, and thus the availability of electrical power.) The Molniya system takes advantage of a high inclination and high eccentricity by using the satellites only when they are near apogee, so their motion across the sky is very slow. For NGSO systems, the choice of a number of satellites and the service areas is also influenced by the choice of the orbit inclinations. A quasi-zenith system is designed to keep at least one satellite high over a desired service area by placing several satellites in a highly inclined orbit. The two "Big LEO" systems, Iridium and Globalstar, made very different decisions about orbital altitudes and inclinations while trying to provide the same kinds of services: Iridium went with 66 highly inclined low-orbit satellites, whereas Globalstar decided on 48 MEO satellites. If you are interested in the details, you can find them on their webpages.

LEO and MEO systems (and satellites on their way to Clarke orbit) pass across the earth along what is called their *groundtrack*. Depending on the nature of the orbit, this may be a simple line around the earth, or a complex set of loops. As the satellite orbits, its antenna beam patterns sweep across Earth as well. Depending on the altitude of the satellite, this motion can be very fast. A LEO satellite that orbits the earth in two hours is moving at almost 18,000 km/hr over the surface.

11.4 Satellite orbits and the Sun

Geosynchronous satellites are 35,786 km above the surface of the earth. To better appreciate the scale of this orbit, given that the earth's radius is 6,378 km, we see that these satellites orbit 5.6 Earth radii, or 2.8 Earth diameters, from the surface. (Add 1 to the radius, or 0.5 to the diameters numbers to get the corresponding distances from Earth's center.)

From the Clarke orbit distance, the earth subtends an angle of 17.3°. Thus, to provide a global beam, an antenna on the satellite must have a beamwidth that wide. A narrower beam would not cover the full visible portion of the earth; a wider beam would waste power into the void of space.

GSO satellites orbit over Earth's equator. As you may remember from high school science courses, the earth itself orbits the Sun once a year, and the earth's poles—and consequently its equator—are tilted from the vertical by about 23.5°. This tilt produces the seasons. Combined with the distance of the Clarke orbit above the earth, this tilt also means that GSO satellites are almost always in full sunlight.

This is important, since 100% of the electrical energy needed to relay signals and run the satellite's systems comes from sunlight converted into electricity by solar cells. Depending on the type of cells and how long the satellite has been in space (recall the degradation due to radiation), about 10–20% of the sunlight falling on them is turned into electricity, which is then stored in batteries aboard the satellite. (A very few military satellites, mostly for radar reconnaissance, not for communication, have used nuclear reactors for power. This power source is too heavy, too dangerous, and much too expensive to be cost-effective for commercial communication satellites.)

The closer a satellite orbits Earth, the more often it spends some of its orbital period in the earth's shadow. This is called an *eclipse*, and is exactly like the lunar eclipses you may have watched in the sky when the moon passes into the earth's shadow. For the LEO satellites, almost all of them—with the exception of a few with carefully oriented near-polar orbits—experience an eclipse every orbit. The eclipse may last up to about half of the period of the orbit. Look at Fig. 11.4 and you can see the geometry of the situation.

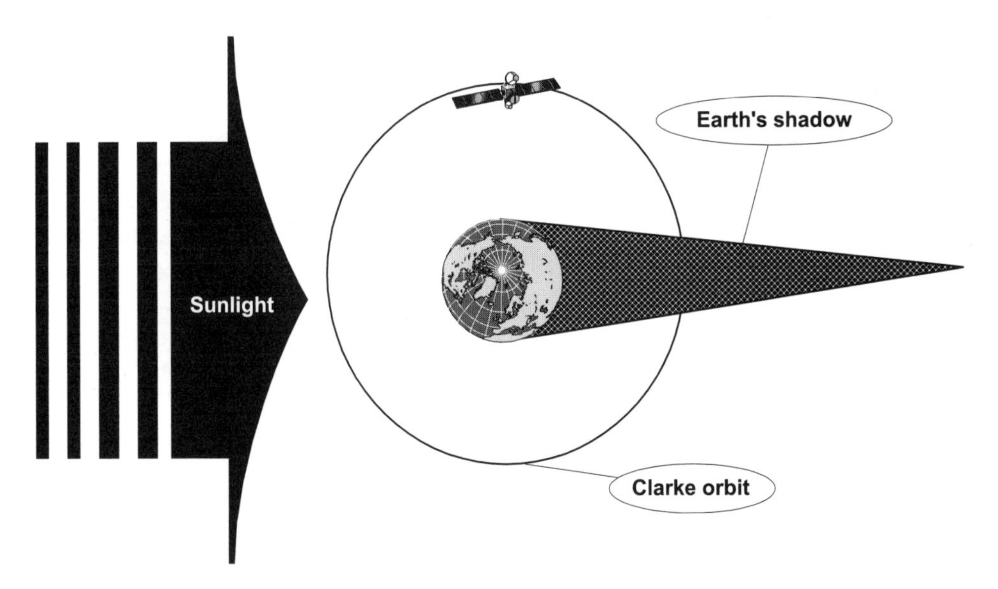

Figure 11.4 Satellite eclipses occur when the satellite passes into Earth's shadow. Most of the year, the satellite orbits above or below the shadow, remaining in sunlight. Around the times of the vernal equinox (about March 21) and the autumnal equinox (about September 22) the shadow falls near the Clarke orbit, and so GSO satellites spend anywhere from a few minutes to a bit over an hour in darkness. During these predictable "eclipse seasons," the satellite must run on batteries, and keep critical systems, such as stationkeeping fuel, from freezing by turning on heaters.

11.4.1 Eclipses in the Clarke orbit

GSO satellites, being farther from Earth and over the equator, only experience eclipses around the two times a year that the sun also appears near Earth's equator. These predictable events occur every night for about six weeks, twice a year, centered at the time of the vernal (spring) equinox, about March 21, and on the autumnal (fall) equinox, about September 22. These periods are called the satellite *eclipse seasons*. Thus, the eclipse seasons are approximately from the beginning of March to mid-April, and from the beginning of September to mid-October.

As you can see from Fig. 11.5, the earth's shadow, stretching from our planet of course away from the sun, has a completely dark inner part called the *umbra*, and a lighter outer part, the *penumbra*, where the sun is still partially visible. When a satellite is within the penumbra, its power from the sun is reduced; when within the umbra, no sunlight at all falls on the solar cells and the satellite is without an external power supply.

Each eclipse season lasts 44 days if you consider only the dates when the satellite passes into the umbra: it begins about 22 days before each equinox, and ends 22 days after the equinox. This gives each satellite in GSO 277 days each year during which it is always in full sunlight. (If you include moving through the penumbra, each eclipse season is 72 days long, leaving 221 eclipse-free days. The dates and duration can vary by a day or so because of leap years, etc.)

If you could look out from the dark side of Earth and see its shadow projected onto the sky at the GSO distance, you would see a big dark circle. As the equinox

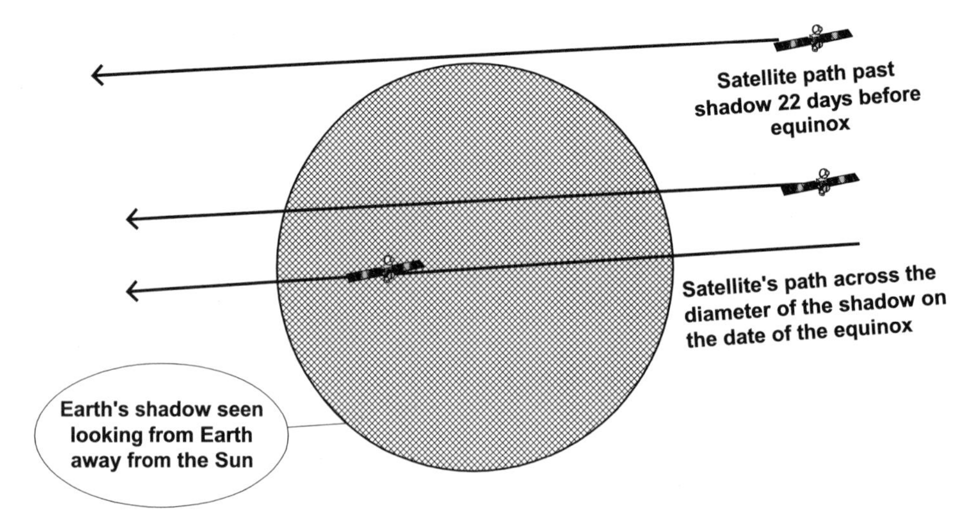

Figure 11.5 If you could look away from the Sun down the Earth's shadow and see it, it would appear as a dark circle. (You can see this during lunar eclipses.) As a GSO satellite orbits, it may pass into the shadow. About three weeks before an equinox it just grazes the shadow, then each night it moves deeper into the shadow. On the night of the equinox, it transits the diameter of the shadow, taking somewhat over an hour. After that date, each night for the next three weeks, the period of eclipse is shorter and shorter. After the three weeks, the satellite will not experience eclipse for about five months.

approaches, each night around midnight the satellite would seem to move nearer and nearer the shadow. About 22 days before each equinox, the satellite would just skim by the shadow; each night after that, for the next 3 weeks, it would dip farther and farther into the shadow, spending a longer time each night in total darkness. On the night of the equinox, it would take the satellite about 72 minutes to traverse the shadow. For the 3 weeks after the equinox night, each night it would spend less and less time in the shadow, until about 22 days after the equinox when it misses the shadow and does not have another eclipse for almost 5 months. Figure 11.6 is a graph showing the duration of the eclipse by date.

From the satellite's point of view, the midpoint of each eclipse comes at its midnight, when it is opposite the sun. What is more important for linking to the satellite, however, is when the eclipse will occur as seen from your earthstation. If your earthstation is at the same longitude as the satellite's orbital slot, the eclipse will occur at or near your local midnight (corrected for any time zone shifts). If the satellite is east of your longitude, the eclipse will occur before your midnight, i.e., during prime-time hours. If the satellite is west of you, the eclipse will happen after your local midnight in the pre-dawn hours. See Fig. 11.7.

Since all of a satellite's power comes from the sun, spacecraft controllers must make sure that the satellite's batteries are fully charged before the eclipse. Immediately after each eclipse, the batteries must be recharged for the next night. The satellite must have enough batteries onboard to keep the basic housekeeping functions of the satellite going, such as running the computers, stationkeeping, telemetry and command systems, etc. In addition, there must be enough battery power to run the heaters that are used only during eclipse. These keep the fuel aboard from freezing, the bearings and batteries from getting too cold, and sometimes perform other temperature-sensitive functions.

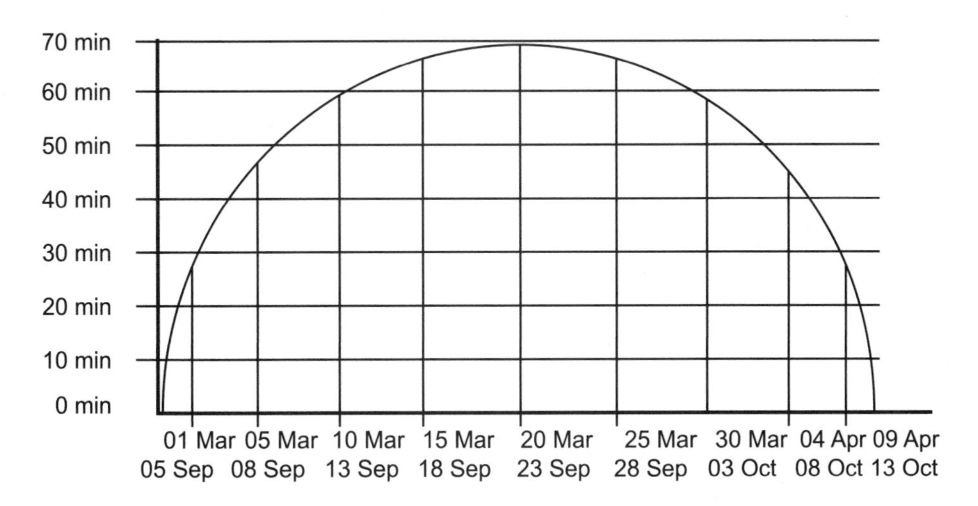

Figure 11.6 The duration of eclipse as a function of date.

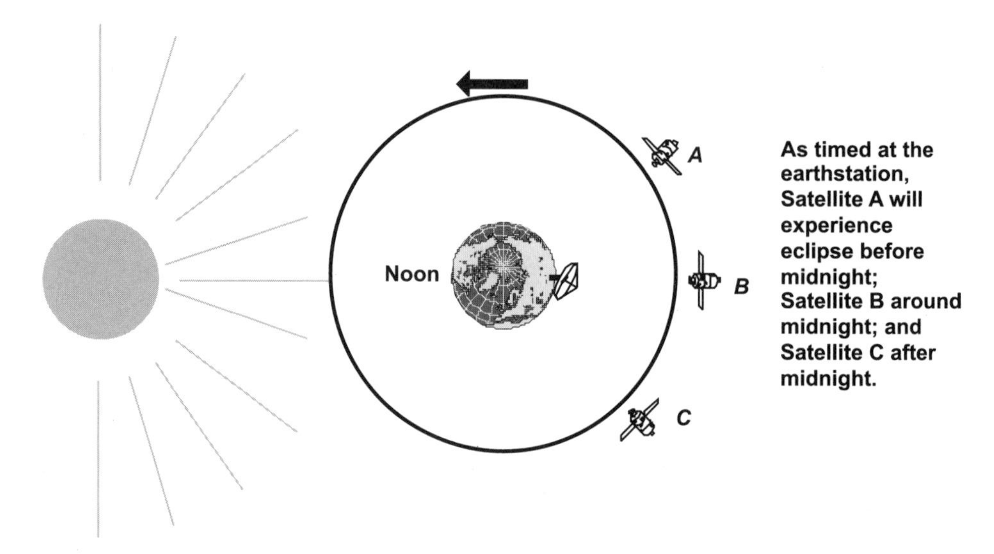

Figure 11.7 Local time of an eclipse of a GSO satellite depends on the relative longitudes of the earthstation and the satellite. A satellite east of an earthstation will be in eclipse before the earthstation's local midnight; a satellite at the earthstation's longitude will be in eclipse at local midnight; a satellite west of an earthstation will be in eclipse after the earthstation's local midnight.

11.4.2 Eclipse protection

There is no law that says you have to keep the communications going through your satellite during eclipses. That choice is a business decision that you need to make when you design the satellite. You should, however, keep the satellite's systems up and running, and the tracking, telemetry, and control functions operational. That requires some battery power. If a satellite has enough battery storage capacity aboard to keep its communications going throughout an eclipse, it is called *eclipse protected.*

The main reason why you may choose not to put enough batteries aboard is mass. Most satellites are designed to some maximum mass that a launch vehicle can carry, so the design process is a "zero-sum game" in which every additional kilogram of one component means one less kilogram of something else. That "something else" is usually fuel—the fuel that keeps the satellite on station and oriented properly. There is thus sometimes a trade-off between battery power and satellite lifetime.

In the old days (10–20 years ago) batteries were less efficient, and this trade-off was often necessary. Full eclipse protection could require a mass of batteries that would replace a year or two of stationkeeping fuel, thus shortening the revenue life of the satellite. As the watt-hours-per-kilogram efficiency figure for batteries has been improved, today most (but not all) GSO communication satellites are eclipse-protected. This means that while the satellite operators must go through the procedures to charge batteries, etc., during each eclipse, the user of the communications through the satellite never notices the eclipse at all.

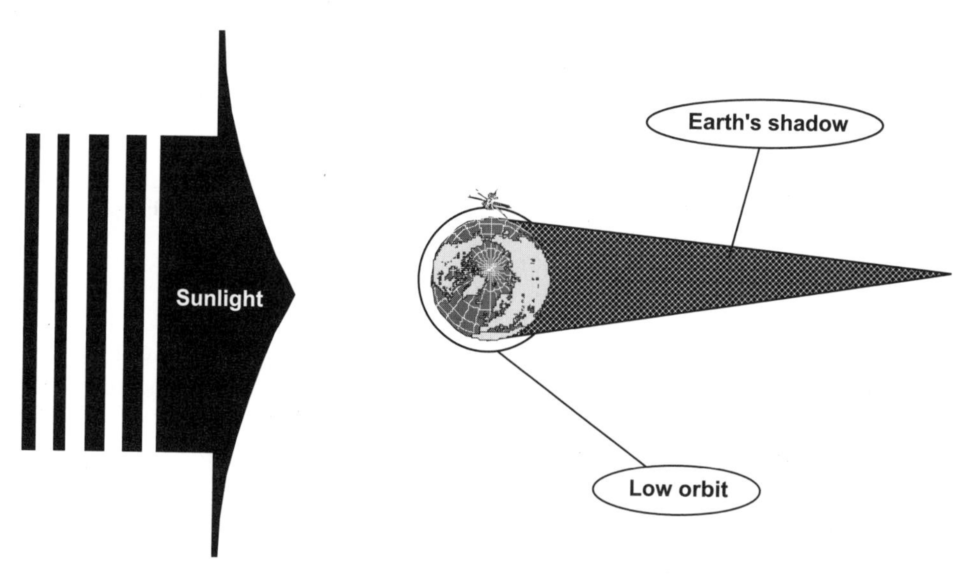

Figure 11.8 Satellites orbiting lower than the Clarke orbit will have more frequent eclipses. Those in the lowest orbits may experience an eclipse every orbit, depending upon the orbit's orientation, and the eclipse may last half of the orbit period, greatly reducing the ability of the satellite's power system to recharge its batteries during sunlight to use the power during eclipse.

11.4.3 NGSO eclipses

Satellites closer to Earth experience more eclipses. The details of how often and for how long depend on a complicated mixture of orbital altitude, shape, and orientation, but in general—the closer to Earth, the more common the eclipses. See Fig. 11.8.

In particular, the LEO satellites, just a few hundred kilometers up, typically experience an eclipse every orbit, and it may last up to half of the orbital period, around a couple of hours. Some of the planned LEO systems, for which the satellites are much smaller and less massive than typical GSO satellites, simply cannot put enough batteries or solar cells onboard to collect enough electrical power during the sunlit portion of their orbit to provide full eclipse protection during their dark times. This means that some of the LEO satellites are not operational during their entire orbit (and means that even more satellites must be added to the constellation if the operator wants to provide continuous coverage).

11.4.4 Solar outages

Something else affects the operation of a GSO satellite around the equinoxes. Since these are the times of the year during which the sun is near the equator. From a given earthstation the sun may appear to pass directly behind the satellite the dish is linking to. The sun is about 1/2 degree across and is a strong radio transmitter too! Since a typical transponder output is the equivalent of a few thousand to a few million watts, and that of the sun is hugely greater than that (about 380 million billion billion watts!), most communication services suffer an outage of a few minutes during which the satellite is useless because of the signal competition with the sun.

In such cases, there is nothing wrong with the satellite: it is in sunlight and functioning normally. If you stood behind your earthstation dish, what you would observe is the sun moving into the beam of the dish. What you would detect on your receiver is an increase of background noise. The background noise temperature will go from a few hundred degrees to several tens of thousands of degrees, thus greatly decreasing the *C/N* of your signal. For a while, you are simply "off the air" due to interference. This is called a *solar outage, sun outage* or *sun transit*. When the sun leaves the antenna's beam after a few minutes, the noise level decreases, and the *C/N* returns to normal again. (A side note: if your dish is a highly polished reflector of visible light, when the sun is in the beam it can become a solar cooker, literally melting the electronics located at the focus of the dish! This is why most dishes are painted a dull color.)

The amount of the effect on your radio signal depends on dish size and frequency. It is worse for large dishes at lower frequencies and for small dishes at higher frequencies. Below 1 GHz, the effect is minimal because the sun is not a good transmitter at those frequencies.

Like eclipses, sun outages occur in seasons, about the same dates as eclipses but for fewer days. The exact dates, times, and duration, as observed at a particular earthstation, depend on both your earthstation's longitude and latitude and on the satellite's longitude. Obviously, if the satellite is east of you, the outage will occur between dawn and noon; if it is at your longitude, the outage is around noon; and if it is west of you, it is in the afternoon hours. Solar outages occur typically for 5 or 6 days, within 21 days of the equinox. Thus, duration is from a few seconds to a maximum of about 15 minutes. If your earthstation is in the northern hemisphere, the outages will occur in the three weeks before the vernal equinox (March 21) and after the autumnal equinox (September 22); if the earthstation is in the southern hemisphere, the outages will happen in the three weeks after the vernal equinox and before autumnal equinox.

You can find tables and calculators on the web pages of satellite system operators that will allow you to find out when any particular earthstation-to-satellite link will be affected.

Most earthstation-satellite links are simply "off-the-air" during the outage. You can either wait it out, or switch traffic to another satellite. By using some frequency-hogging spread spectrum techniques, you can continue to operate during these outages, but most operators do not.

NGSO satellites are much less concerned with sun outages. They are moving so fast in orbit that they cross the sun's disk rapidly as seen by a user on the ground. Further, the fact that the terminals used to contact the satellites have omnidirectional antennas dilutes the noise contribution from the sun into the antenna. (One of their problems, depending on frequency, is passing between the beam from a GSO satellite and an earthstation.)

Now that we have discussed how and why satellites orbit, and how their orbit properties influence their operation, we need to briefly turn to the subject of just how we get them up there in the first place.

Chapter 12

Out To Launch

What the public calls a "rocket," the industry people usually call a *launch vehicle*. From the standpoint of a company wishing to send a communications satellite into space, the most important technical characteristics of a launch vehicle are its capacity; the number of kilograms it can place into various orbits; the physical size of the satellite it can accommodate, typically described by the maximum diameter and length; and its reliability. The satellite is referred to as the *payload* for the launcher.

12.1 The launcher's job

The basic task of the launch vehicle is to release the satellite at the correct altitude and with the proper velocity (speed and direction). At that point, the launcher's job is over. As you would expect, it takes more energy—basically more fuel—to achieve a higher orbit than it does to place a satellite in a lower orbit. Of course, it also takes more energy to launch a massive satellite than a smaller one. Recall that once on-orbit, the satellite continues to orbit by its momentum; at that point, it is not being pushed around. See Fig. 12.1.

The weight-lifting capacity of a launch vehicle is determined by a complex interrelationship of fuel capacity and type, rocket engine thrust, efficiency, launch pad latitude, and the altitude, shape, and inclination of the desired orbit. Launch of a GSO communications satellite will cost roughly between $50 million and $200 million, with the cost to orbit about $10,000–$30,000 per kilogram.

The timing of a launch is controlled by the desired orientation of the orbit and launch pad location. The time period during which a satellite can be launched to reach its intended orbit is called the *launch window*. To reach equatorial orbit, the launch direction, or *launch azimuth*, will always be toward the equator and toward the east, which is the way the earth is turning. Thus, launches from the northern hemisphere aim toward the southeast; those from the southern hemisphere go northeast. Other orbital inclinations can be achieved by changing the launch azimuth or by maneuvering the satellite once it is in space. Most launch sites have a range of launch azimuths that are safe for descending parts of the launcher to fall back to Earth harmlessly, usually over water (Fig. 12.2).

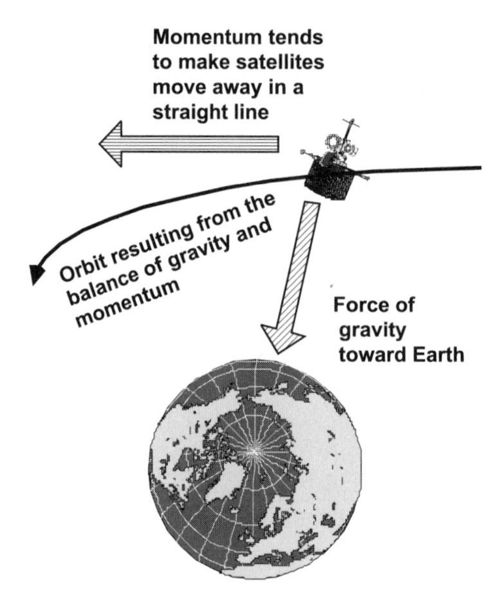

Figure 12.1 A closed orbit results when the momentum of the satellite, which would tend to keep it moving, is balanced by the downward force of gravity. Once in a stable orbit, no propulsion is needed to keep it moving, and the orbit will remain the same as long as no other forces affect the satellite.

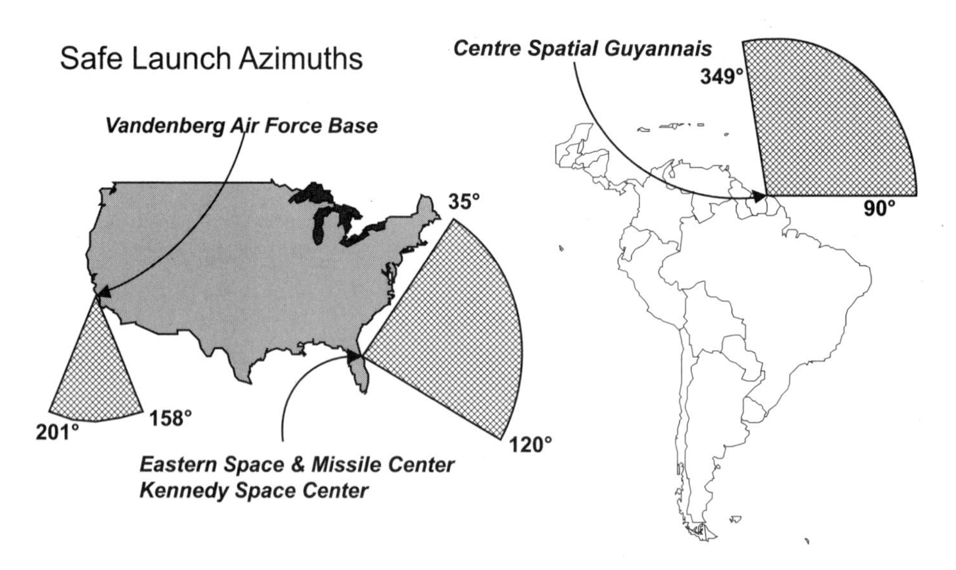

Figure 12.2 Safe launch azimuths are the range of directions in which discarded parts of a launcher may safely fall back to Earth without causing damage. This figure shows the safe launch azimuths for the two major launch sites in the United States, and for the Guyana Space Center.

Satellites launched into LEO and MEO orbits are usually injected directly into these orbits by the launch vehicle. For satellites intended for GSO, the launch vehicle typically injects the satellite into the highly elliptical GTO, which has its perigee usually a few hundred kilometers up and an apogee at or near the intended GSO. After the satellite has orbited in the GTO for a while (hours to weeks, depending on how the satellite checks out), a smaller rocket engine built into the satellite's body, called an *apogee kick motor,* AKM, is fired on command to *circularize* the orbit into the GSO. This AKM firing occurs at the apogee of the orbit, and is the last major propulsive event in the satellite's life. After that, the flimsier parts of the satellite can be deployed, such as extendable solar panels and unfurlable antennas. Once on-station, very small thrusters are used to maintain its location and orientation.

Thus, if you look at the launch capacity specifications for launchers, they will typically specify that they can place so many kilograms into such-and-such a typical orbit. The orbits typically mentioned are LEO equatorial, LEO polar, and GTO. For example, the Ariane 5, which has a take-off mass of 713,000 kg, can put 18,000 kg into low-inclination LEO and 6,800 kg into GTO.

12.1.1 A fuel and its rocket are soon parted

Rocket fuel works by producing huge amounts of hot gas that expands rapidly and rushes out the nozzle of the rocket motors, causing a reaction that thrusts the vehicle in the opposite direction. The efficiency of a rocket fuel is measured by a number called the *specific impulse*, measured in seconds. Basically, it is the number of seconds that a unit mass of fuel will produce a unit amount of force. Most rockets fire for only a few minutes, then are discarded. Launchers may have some stages using solids, and some using liquids.

Major rocket fuels come in three types: solid, liquid, and hybrid. This does not count ion thrusters (which work electrically) or pressurized gas squirted out of jets, both of which are used only for very low thrust applications such as station-keeping and orientation.

12.1.2 Where the rubber meets the road to space

Solid rocket fuel is much like the rubber eraser on a pencil; in fact, its major component is a rubber-like substance. Added to this are oxidizer chemicals and sometimes other ingredients such as aluminum powder. The advantage of solid fuel is that it is very stable and safe: outside a rocket motor you can hit it, burn it, and otherwise mistreat it, and it will not explode. Another advantage is that, once cast into a rocket motor, it is storable for long periods. The big disadvantage is that you need to know how much you will need before you start. Solid fuel is mixed as a liquid, and a carefully calculated number of kilograms of it is poured into the rocket casing. See Fig. 12.3.

Figure 12.3 A payload assist module. The bulbous object with the nozzle at the bottom of the photograph is a solid-fuel rocket motor that will boost this spinner satellite from low Earth orbit into a geosynchronous transfer orbit. (Source: NASA.)

The calculation is based on the mass of the satellite to be propelled and the change in speed that you want to provide the satellite. (Rocket scientists refer to such a velocity change as *delta-V*, often symbolized by the uppercase Greek letter delta plus the letter *V* for velocity: ΔV; it is measured in meters per second.) Once a solid fuel is ignited, it burns to completion (as long as combustion conditions in the rocket are adequate). It cannot be started and stopped as can liquid fuels.

Such solid-fueled motors are used for many purposes. Many launch vehicles may have a few small solid-rocket motors, *SRMs*, attached to the base to get them going quickly; others use one or two huge SRMs as part of the first stage; many launchers use solid-fueled upper stages; and the AKM is almost always a solid-fueled motor.

12.1.3 How liquids move your assets

Many launch vehicles use liquid fuels for power. Any combination of chemicals that when combined produce large amounts of gas can be used. Some of these combinations "burn" in the familiar sense, such as the popular combination of kerosene and liquid oxygen. Others, such as nitric acid and hydrazine, combine chemically to produce the same kinds of results. Such self-combusting combinations are referred to as *hypergolic*. The most powerful combination in use today is liquid hydrogen burned with liquid oxygen. Since both of these gases must be cooled to very, very low temperatures in order to liquify them, such propellants are often referred to as *cryogenic*. Liquid fuels in general give more thrust than solid fuels, but are much more difficult and dangerous to handle and store.

Cryogenic fuels, in particular, must be loaded onto the launch vehicle no more than a few hours before launch. The launcher stages (which are mostly huge fuel tanks) that contain cryogenic fuels must be heavily insulated. (Often these insulating panels intentionally fall off at launch, giving uninitiated observers the impression that the vehicle is coming apart!) If the launch is postponed ("scrubbed") the fuels must be drained from the launcher's tanks.

Some experimental rockets use a *hybrid* system of a solid burned with a liquid or gas. Rubber and liquid oxygen is an example of such a combination. Such vehicles can be throttled, stopped, and restarted, unlike totally solid fuel stages. So far, none of the major launch vehicles uses such a system.

12.1.4 Small thrusters

Onboard a typical satellite, liquid fuels are most commonly used for the small stationkeeping maneuvers necessary to keep the satellite in place and pointed correctly. We will look at these more later. Whereas the launcher's stages may produce hundreds of thousands of newtons of thrust, these small onboard devices produce tiny thrusts, typically from a few tenths to a few newtons. (A newton is the metric unit of thrust, equal to 0.22 pounds-force of thrust.)

The most common stationkeeping fuel used is hydrazine. Such a single-fuel system is called a *monopropellant*. Hydrazine is a nasty, poisonous liquid, but it gives a fair amount of "bang for the buck." A typical medium-sized satellite may have several tanks containing a total of a hundred kilograms of hydrazine. When it is to be used, valves open and allow the liquid to expand into a gas (often with the help of small heaters), which rushes out of small nozzles, providing tiny amounts of thrust. There are no huge blasts of flame and smoke as there are during launch.

Larger satellites may use two chemicals to provide the greater thrust needed for their greater mass. These are called *bipropellant* systems. Typically hydrazine is again one of the chemicals used.

A very new development is the *ion-thruster*, or electric thruster. These use a single gaseous fuel that is ionized by electricity from the solar cells and accelerated out a nozzle to produce a very efficient but very tiny amount of thrust. Typical thrust is on the order of a few thousandths of a newton. Because the thrust is so low, instead of being operated periodically upon command from ground controllers, as for chemical thrusters, ion thrusters often work continuously.

12.2 The launch vehicle and launch program

A rocket typically consists of several *stages* stacked on top of one another. Figure 12.4 shows a cutaway diagram of an Atlas launcher. This staging is done for efficiency, since the launcher not only has to launch the payload, but must launch itself as well. Most of the mass of a launch vehicle is fuel, and most of the rest is the fuel

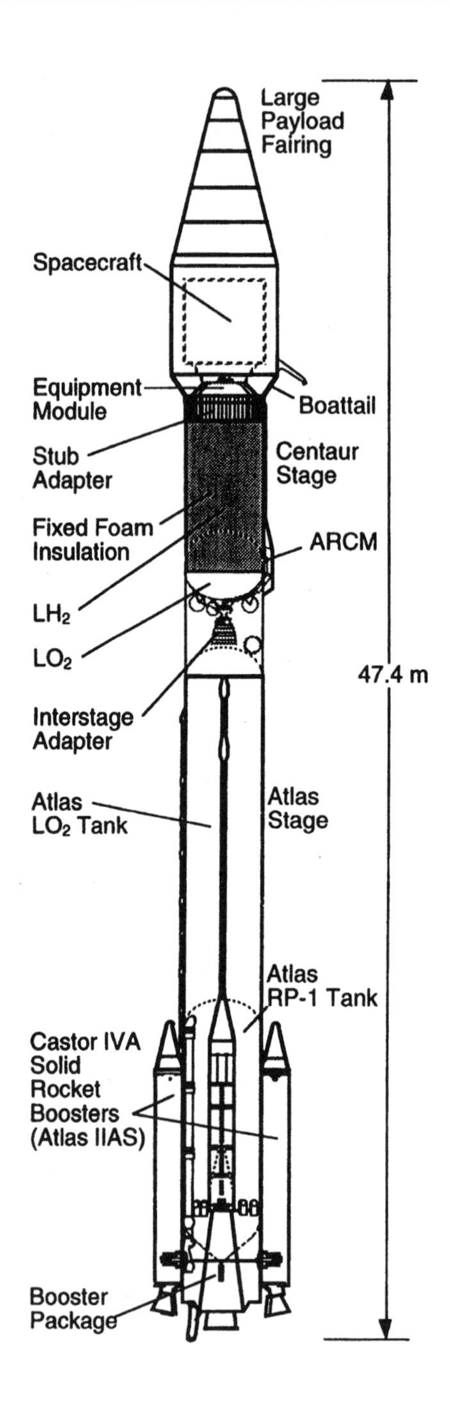

Figure 12.4 A cutaway diagram of an Atlas 2 launch vehicle. The solid-fuel booster and the first stage get the vehicle going, then the boosters fall off. When the vehicle has reached maximum velocity, the first stage is empty of fuel, detaches, and falls back to Earth. The upper stages continue to increase the altitude and speed of the payload until it is injected into a GTO. (Image courtesy of International Launch Services.)

tanks. Typically only about 2.5% of the total mass launched makes it to a final low-Earth orbit, and only about 1% to GSO! (There have been proposed *single-stage-to-orbit* launchers, SSTO, but none is operational so far.)

To make the launcher more efficient, it throws away parts no longer needed, meaning fuel tanks. Such launchers are called *expendable launch vehicles*, or ELVs, in the trade. (In contrast, the Space Shuttle recovers and reuses its solid-fuel boosters, and the orbiter part returns to Earth for reuse. A couple of the proposed future launchers will also be reusable.)

Figure 12.5 shows a typical launch flight profile, in this example, for an Atlas launcher. The first stage is the most powerful and contains the most fuel. It may be augmented by several small solid-fueled motors to help it get going, and then it must boost the entire vehicle out of the atmosphere and to high speed. When it has reached maximum velocity and its fuel tanks are nearly empty, the first stage separates from the rest of the vehicle to save weight and eventually falls back to Earth.

A second stage then ignites and increases the speed. Since most launches are not from sites on the equator, there may be a coasting period during which the motors are shut off and the vehicle coasts to the equator. The altitude of the vehicle will decrease somewhat during coasting. Then the motors reignite, increasing the speed even more.

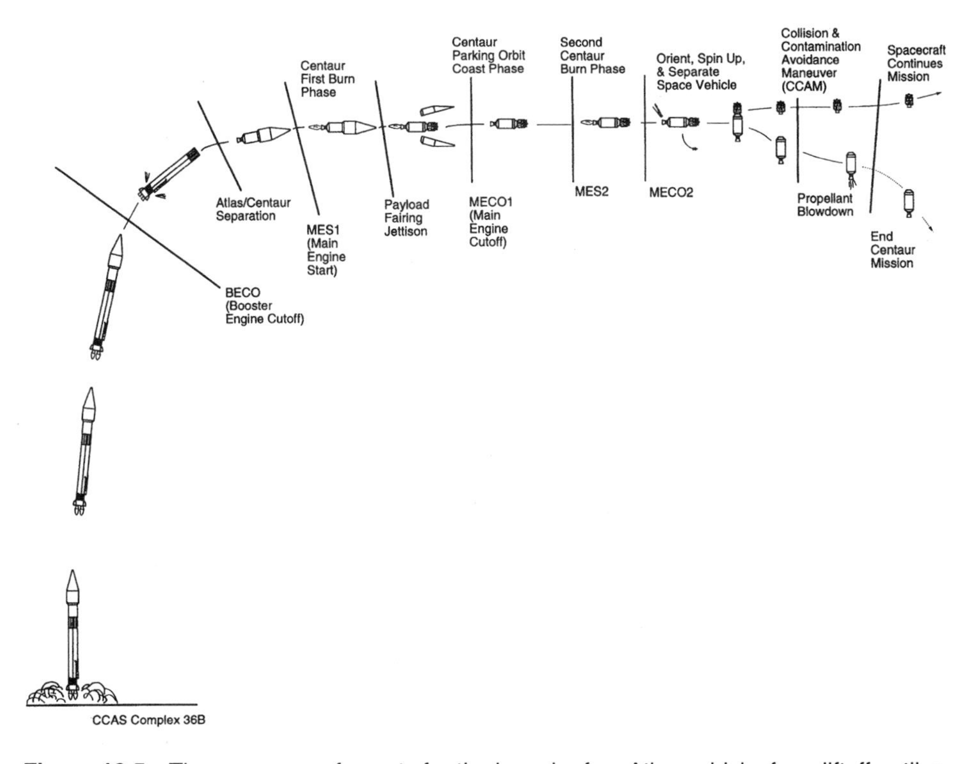

Figure 12.5 The sequence of events for the launch of an Atlas vehicle, from liftoff until release of the satellite into GTO. This sequence takes less than a half-hour. (Image courtesy of International Launch Services.)

Typically, during the first-stage burn or the second-stage burn, the vehicle has reached an altitude where air resistance is no longer a problem. This means that streamlining is no longer necessary, and so, to save still more weight, the *fairing* (nose cone) of the launcher typically splits open like a clamshell and falls back to Earth, exposing the payload.

If there is a third stage, it ignites when the second stage has reached maximum speed and separates. Sometimes there is a fourth stage, or a payload assist module. However many stages there are, the uppermost stage is responsible for injection into orbit. It must orient the satellite into the proper direction, achieve the desired orbital speed, and then eject the satellite. For safety reasons, the upper stage often then maneuvers so as not to hit its own payload (this has happened in the past!). The upper stage has achieved orbit too, and sometimes the operators purposely deorbit it to reduce space junk. Depending on the orbit, upper stages that are not purposefully deorbited may remain in space for days to centuries. To avoid the upper stages from later exploding from residual fuel, these upper stage tanks may be vented into space to empty them.

If there are several satellite payloads, the upper stage will orient itself and adjust speed for the first one, and eject it; then it will reorient itself and eject the second payload, and so on. As we mentioned, for LEO and MEO, the orbit achieved by the upper stage is (at least close to) the final intended orbit. For satellites heading for Clarke orbit, the upper stage places the payloads into GTO.

12.2.1 The launch campaign

From the time you order a launch vehicle, it typically takes from half a year to three years for the vehicle to be constructed, made ready, mated with your payload, and launched. The vehicle, or its parts, are shipped from the factory to the launch site for final assembly in vehicle and payload processing buildings. The same holds for the satellite payload. Once it is near the launch site, the satellite is fueled with its stationkeeping fuel, the apogee kick motor is installed, and the satellite is attached to the launcher in a process called *payload integration*. Depending on the specific launcher, this may occur on the pad once the vehicle has been erected or in a launch processing building, after which the entire launch stack is rolled out to the pad itself. These final steps typically take a few weeks to a couple of months.

The countdown itself typically lasts a few days. These include final checks and tests and fueling of liquid-fuel tanks if used. Cryogenic fuels are loaded only within the final few hours since they are very volatile. In fact, hoses keep the launcher's tanks topped off until a few seconds before launch. Other services, such as power to the vehicle's guidance systems, power to the satellite, and even air conditioning for the payload are supplied up to the last few seconds through umbilical cables which connect the launcher to the pad.

Usually the final few minutes of the countdown are completely computer-controlled, as humans would be too slow to accomplish the enormous number of tasks occurring. The status to the payload and launcher, and also that of support services, such as down-range tracking, must be in perfect order. Often the launcher's rocket motors are ignited a few seconds before lift-off. Finally, at T-0, the rocket leaves the pad, heading for space. The vehicle's successive stages burn and drop away, with the final stage placing the payload into orbit. The payload is usually in orbit within 20 to 30 minutes (or, if something went wrong, it is in pieces somewhere by then!).

The launch process, from ignition to orbit, typically takes less than half an hour.

12.3 Launch vehicles

While there are more than a dozen launch vehicles available to the communications satellite industry, the field is dominated by just a few aerospace firms that have most of the market. Further, not all launchers are suitable for all satellites. The following is a brief description of what is now available and those which will probably enter the launch market within the next few years.

Antrix is the commercial spin-off of the Indian Department of Space. It has recently begun offering the Polar Satellite Launch Vehicle, PSLV, (Fig. 12.6) and the

Figure 12.6 The PSLV, a launch vehicle built by the Indian Space Research Organization and launched from Sriharikota, India. (Photograph courtesy of ISRO.)

Geosynchronous Satellite Launch Vehicle, GSLV. These launchers were designed and built by the Indian government. So far, there have been few commercial users. They are launched only from the Satish Dhawan space center (SHAR) at Sriharikota on the southeastern coast of India.

Arianespace is a European-based consortium that markets the Ariane family of launchers. The current model is the Ariane 5 (Fig. 12.7). These were designed by the European Space Agency. They are launched only from the Centre Spatial Guyanais (CSG) at Kourou, French Guiana, on the northeast coast of South America. Ariane has about 40% of the total commercial launch market. Arianes are used for both LEO and GSO satellites. Arianespace is planning to offer launches of two other launch vehicles. One, the Soyuz (Fig. 12.8) is from Russia, and will gain increased launch weight capacity by lifting off from the low-latitude CSG pads. The second vehicle is a new one intended for small and medium-sized payloads, called the Vega (Fig. 12.9), planned for first launch around 2006.

Boeing purchased the MacDonnell Douglas Company and with it the Delta launcher, which began life in the 1950s as the Thor military missile. There are launch pads for Deltas at the Cape Canaveral Air Force Station (CCAFS) at Cape Canaveral, Florida, and at Vandenberg Air Force Base (VAFB) near Santa Barbara, California. Deltas can send satellites to both LEO and GSO.

Figure 12.7 An Ariane 5 launcher on its mobile launch platform, which carries it from the vehicle assembly building (where the parts are assembled and the payload installed) out to the launch pad itself. (Photograph courtesy of Arianespace.)

Figure 12.8 A Soyuz launcher, built in Russia and marketed by Arianespace. Hampered in payload capacity by the high-latitude launch bases in Russia, the payload is increased by launching from the low-latitude Guiana Space Center. (Photograph courtesy of Arianespace.)

Figure 12.9 The Vega launch vehicle being developed by Arianespace for the future. (Photograph courtesy of Arianespace.)

The *Brazilian Space Agency* has begun to market the VLS, a small launcher most useful for small payloads to LEO. It is launched from the new Alcantara Launch Base in northern Brazil. They have had discussions with other launch vehicle manufacturers about the possibility of building customized launch pads so these other launchers could take advantage of this low-latitude base.

China Great Wall Industrial Company (CGWIC) offers the Long March (Fig. 12.10) series of launch vehicles. These are capable of placing satellites in LEO or GSO, and are launched only from Chinese military bases, usually from the Xi Chang base near the Chinese deserts. Since CGWIC is a government-owned company, commercial firms have complained that it can offer below-cost launches, thus creating unfair competition in the launch market.

Eurockot is a joint venture between German and Russian aerospace firms to market the Russian-designed and built Rockot (Fig. 12.11) vehicle. It is launched from Russia.

International Launch Services, ILS, is a joint venture between Lockheed-Martin (LM) and ex-Soviet firms. LM earlier purchased General Dynamics, which had developed the Atlas missile, which became the Atlas launch vehicle (Fig. 12.12). Russian firms developed the Proton launcher (Fig. 12.13). ILS markets both of these, and

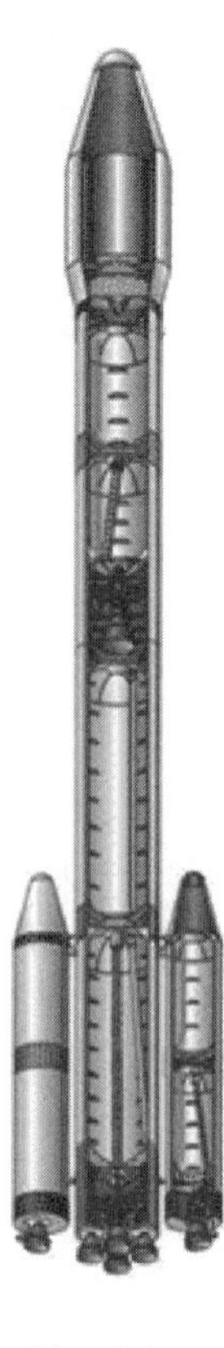

Figure 12.10 An example of the Long March launchers, built in China. (Photograph courtesy of CGWIC.)

Figure 12.11 A Rockot launcher lifts off carrying a payload into orbit. (Photograph courtesy of DASA.)

Figure 12.12 An Atlas 5 launch vehicle, latest in the Atlas family, sits on the launch pad. (Photograph courtesy of International Launch Services.)

Figure 12.13 A Russian-made Proton launcher just after launch. Note the shockwaves in the engines' exhaust plumes. (Photograph courtesy of International Launch Services.)

both are capable of LEO or GSO missions. The Atlas can be launched from either CCAFS or VAFB; the Proton is launched from Baikonur, which is a Russian-built and Russian-run base located in Kazakhstan. Both have a decades-long track record and a substantial share of the market.

Orbital Sciences Corporation (OSC) has developed two small launch vehicles, mostly intended for small satellites to LEO. The Pegasus launchers (Fig. 12.14) are unusual in that they do not take off from a launch pad. Instead, they are mounted under a large aircraft and air-launched from 40,000 feet. The aircraft with its rocket payload can thus use any airport in the world with a runway long enough, and can fly down to a launch site on the equator to achieve maximum efficiency. The other vehicle is the Taurus, which is launched vertically like most other rockets (Fig. 12.15). They also offer the Minotaur (Fig. 12.16), which is a converted ICBM.

Rocket Science Corporation (RSC) is a commercial firm set up by the Japanese space agency to commercialize the H-2 launcher (Fig. 12.17). While the Japanese have long had a launch capability with an enviable success record, the rocket technology they used was partially licensed from other nations with the restriction that they could not launch third-party payloads. This limitation is removed with the 100-percent-Japanese H-2. The goal of RSC over the next few years is to greatly reduce the cost while increasing the capacity, and to bring an upgraded H-2A to market. Currently, all Japanese rockets are launched only from spaceports at Tanegashima and Kagoshima, in Japan.

Figure 12.14 The Pegasus small launcher is seen mounted underneath its Lockheed 1011 carrier aircraft. The plane flies the missile to about 40,000 feet altitude, at which point the Pegasus falls away and ignites its engine, carrying its payload into space. Pegasus is primarily intended for low orbit payloads. (Photograph courtesy of Orbital Sciences Corporation.)

Figure 12.15 The Taurus launcher, intended for smaller payloads to low or geostationary orbits. (Photograph courtesy of Orbital Sciences Corporation.)

Sea Launch is a consortium with a unique idea. The Norwegian firm Kvaerner, the world's largest builder of ships and floating oil-drilling platforms, converted a drilling platform into a launch base and built a control ship. A Ukranian company supplies the Zenit launch vehicle (Fig. 12.18). Satellite payloads are assembled in a plant in Long Beach, California, and loaded onto the control ship. The ship and the self-propelled launch platform then sail down to a spot on the equator southeast of Hawaii in international waters. The launcher is transferred to the platform from which the launch takes place. See Fig. 12.19.

Figure 12.16 A Minotaur launch vehicle. These are converted ballistic missiles. (Photograph courtesy of Orbital Sciences Corporation.)

Starsem is a joint French-Russian venture to commercialize the Russian Soyuz launcher using French aerospace capabilities.

Several other companies and nations may emerge as competitors within the next few years, not including upgrades and additions from existing launch vehicles. A number of very experimental firms in Canada, the U.S., and Australia are working on launchers or new launch bases. Israel has the Shavit launcher, which could be commercialized. Another firm is trying to commercialize the air-launched Polyot launcher (Fig. 12.20).

12.4 Launch bases

With the exception of air launches, a rocket without a launch pad is useless. Each vehicle is unique, and each pad must be designed for it specifically. This limits the launch rate, because whereas the launch industry is commercialized—that is, you

Figure 12.17 The Japanese H2A launcher. (Photograph courtesy of JAXA.)

Figure 12.18 A Russian-built Zenith launch vehicle, seen here being used in its Sea-Launch configuration. (Photograph courtesy of Sea Launch.)

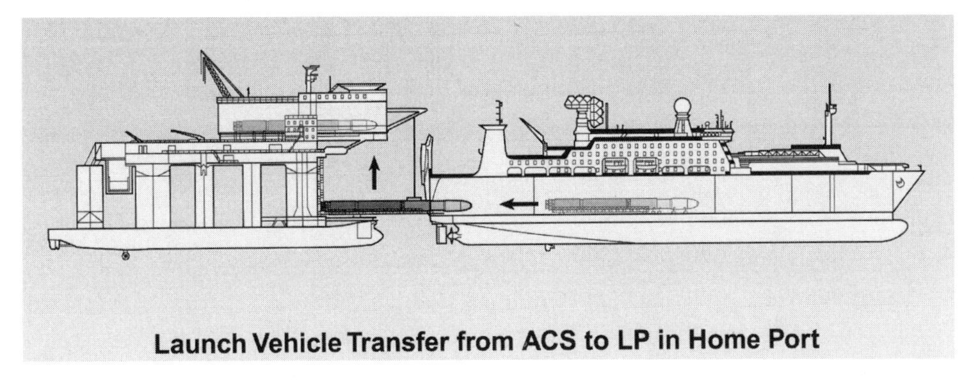

Figure 12.19 Cutaway diagram of the Sea Launch command ship, on the right, transferring a Zenith launcher into the Odyssey launch platform. This operation is performed at the home port and then the ship and self-propelled platform sail to a launch location on the equator. (Photograph courtesy of Sea Launch.)

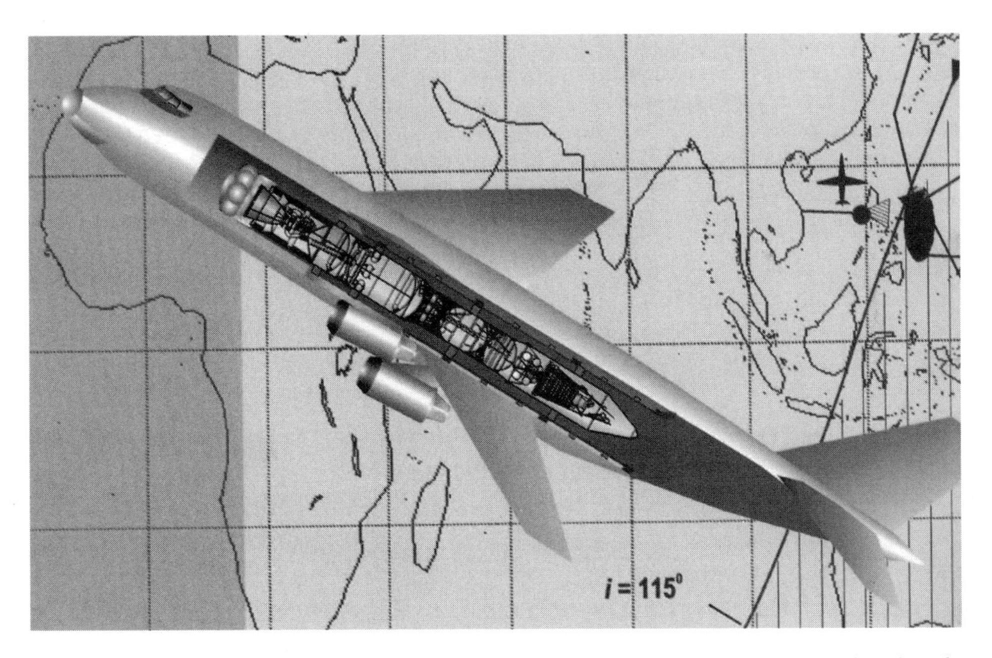

Figure 12.20 The planned Polyot launcher, air-launched from an Antonov carrier aircraft. (Photograph courtesy of Polyot.)

go to the launcher company, not NASA or some other space agency to get a rocket—almost all of the launch pads are owned by governments, usually by the military. Figure 12.21 show the locations of major launch sites.

In the U.S., many people mistakenly think that NASA has the launch pads. Actually, the famous Kennedy Space Center (KSC) has pads only for the Space Shuttle. It is from the next-door CCAFS that even NASA launches its expendable rockets if an east coast site is needed, and from VAFB on the west coast. Similarly, the French Space Agency owns the CSG and rents its facilities to Arianespace.

Figure 12.21 Some of the world's major launch sites. Russia, Kazakhstan, and China all have launch bases well inland; the other nations' bases are all near an ocean. Smaller, research-oriented launch bases are omitted. The air-launched Pegasus and Polyot rockets can be launched from almost anywhere within range of a large airport. The SeaLaunch liftoffs take place southeast of Hawaii near the equator.

For this reason, several entrepreneurs are attempting to establish nongovernmental launch bases that would be free of many of the restrictions. Since it takes a few days to a month from the time a launch vehicle is first erected on its launchpad until launch, the number of pads really determines the number of launches possible per year. New launch bases will help with this problem.

In the U.S., private spaceports are being planned in Florida, Virginia, California, Alaska, and Hawaii. In Australia, several sites are being explored, including the venerable Woomera base in the desert, and a spot on the low-latitude Cape York Peninsula. Other spaceports have been proposed for other nations.

Two major concerns influence the location of launchpads. One is safety: since the first couple of stages and the payload fairing fall back to Earth during the launch, they should not fall on populated areas. For most nations, this has meant locating spaceports on coasts, however both Russia (and ex-Soviet Kazakhstan) and China have only inland bases. They also have much empty land area for parts to fall in.

The Japanese have a unique launch safety requirement that currently restricts their launches to only a few weeks a year in times around the end of January and in August. This results from the lucrative fishing area that lies downrange from Tanegashima and Kagoshima. By treaty with the Japanese fisherman's union, launches can only take place during the period that the fishing fleets return to port on holidays! If RSC intends to be a commercially viable launching company, either this restriction has to be eased, or another spaceport must be found.

The second concern is launch efficiency. To place a satellite into an equatorial orbit, a launchpad on or close to the equator is best. It can make a significant improvement in launch capacity. For example, if you took two identical launch vehicles and launched both to GTO intended for GSO from CCAFS, at latitude 28.5°N, and from Kourou, at 5.2°N, the one from French Guiana could carry a payload about 15% greater. For a launch site at a latitude of 45°, the launcher can carry a satellite to GSO that is only 65% of what it could loft from an equatorial base. That's a lot of lost capacity!

In general, if you launch due east from any launchpad, the satellite will go into an orbit that has an inclination equal to the launchpad's latitude. Launching in any direction other than east will result in a higher inclination. To achieve an equatorial orbit from a nonequatorial launchpad, you have to build into the launch trajectory an orbital *plane-change maneuver* (sometimes called a *dogleg maneuver*); this is the single most energy-consuming thing you can ask a rocket or satellite to do. The greater the latitude of the launchpad, the greater the plane change required, and the greater the amount of rocket fuel that goes into this maneuver. This is one of the reasons why the Soviet Union invented the Molniya satellite system back in the early days of launch vehicles.

Having looked at orbital properties of our satellites and how to get them into orbit, we must now turn our attention to how the satellites themselves are built so as to satisfy the communications needs of the operator.

Chapter 13

Satellite Systems and Construction

Satellites are complex electromechanical devices composed of many interconnected systems designed and constructed to support the specific functions for which the satellite is intended. Because satellites are very expensive and are inaccessible once in operation, satellite systems must be designed and constructed with the highest-quality components. At each stage of manufacture and deployment, punctilious quality control must be in place, and redundant systems are used where possible for backup purposes. Small errors in components, manufacture, assembly, and even shipping can result in multi-hundred-million-dollar losses. All of these factors contribute to the high cost of the typical satellite.

(There is said to be a standing joke in the naval architecture business that when designing a ship, you are not ready to "cut metal" until the weight of the blueprints equals the projected weight of the ship. In the business of satellite construction, the weight of the blueprints and documents exceeds the satellite's final weight many times over!)

13.1 Satellite manufacturers

While there are thousands of companies worldwide supplying components for satellites, there are only a few large firms that specialize in being the general contractor, putting it all together to make the satellite itself. Some make a wide range of satellites; others specialize in military satellites, small satellites, or some other niche market. Some have mostly constructed satellites for their own nation's space program, while others have a global customer base. These major manufacturers and integrators of commercial communications satellites for LEO or GSO include

- Alcatel Espace
- Alenia Spazio
- Astrium
- Boeing Space Systems
- EADS
- Lockheed Martin Commercial Space Systems
- Mitsubishi Electric

- NEC Toshiba
- Orbital Sciences
- Space Systems/Loral

It is not uncommon for satellite manufacturers to partner with other firms to respond to a request for proposals for a large system.

During construction, all components of the satellite are rigorously tested for proper operation, then retested after they are installed in the satellite. The completed satellite is tested repeatedly as a whole. Among the severe tests are a thermal vacuum test, in which the satellite is put into a vacuum chamber and heated to the conditions close to those it will encounter in space. Another test vibrates the satellite to make certain it will withstand the launch forces and vibration. Electrical and communications tests ensure the correct operation of the communications payload.

Once the satellite arrives at the launch site, typically the apogee kick motor is installed, and the tanks are filled with fuel(s) that will be used for stationkeeping. The satellite is then integrated onto the top stage of the launch vehicle. After this *payload integration*, both the satellite and launcher receive power and other housekeeping needs through umbilical cables and tubes that are connected to the launch pad tower. These disconnect at launch.

Until recently, it typically took about three years from contract to launch for a GSO communications satellite. Recently, operators have wanted to deploy systems faster, and so manufacturers have tried to cut the process down to 18 to 24 months. The challenge has not only been to build faster, but to maintain quality while doing so. Manufacturers also like to use space-tested parts, but meeting the increasing power and versatility demands from customers means sometimes using new components, with the consequent possibility of component failure.

Manufacturers of the LEO satellites have smaller systems to build, and so can build faster. They also build many more satellites to fill out LEO and MEO constellations, and plan for continual replacement of the satellites once the system is operational. Some of the small satellite builders have managed to cut the time "from loading dock in to loading dock out" to less than two weeks.

13.2 Major satellite subsystems

Just as a vehicle will have a series of common systems, regardless of whether it is a battle tank, a passenger car, or a motorcycle, so, too, do satellites have basic systems of hardware and software that enable their intended capabilities. The major subsystems of a communications satellite are

- structure
- power
- propulsion
- thermal control

- monitoring and control
- communications.

Each system must be highly reliable, and this contributes to the complexity and high cost of satellites. Figure 13.1 shows the basic components of a typical commercial communications satellite.

These basic structures and capabilities of a satellite are often called the satellite *bus*. Each satellite manufacturer has one or more standardized buses available to carry a variety of communications payloads. Thus, you will see mentioned, for example, the Lockheed Martin A2100, or Astrium's Eurostar 3000 bus, or the Orbital Sciences bus (Fig. 13.2). Specifying the bus for a satellite is rather like specifying the model of a vehicle, such as Jaguar XJ6 or BMW 318 or Mac truck. This tells you the basic size, mass, power supply capability, fuel capacity, and other basic features. A satellite buyer then builds on this bus to provide the specific

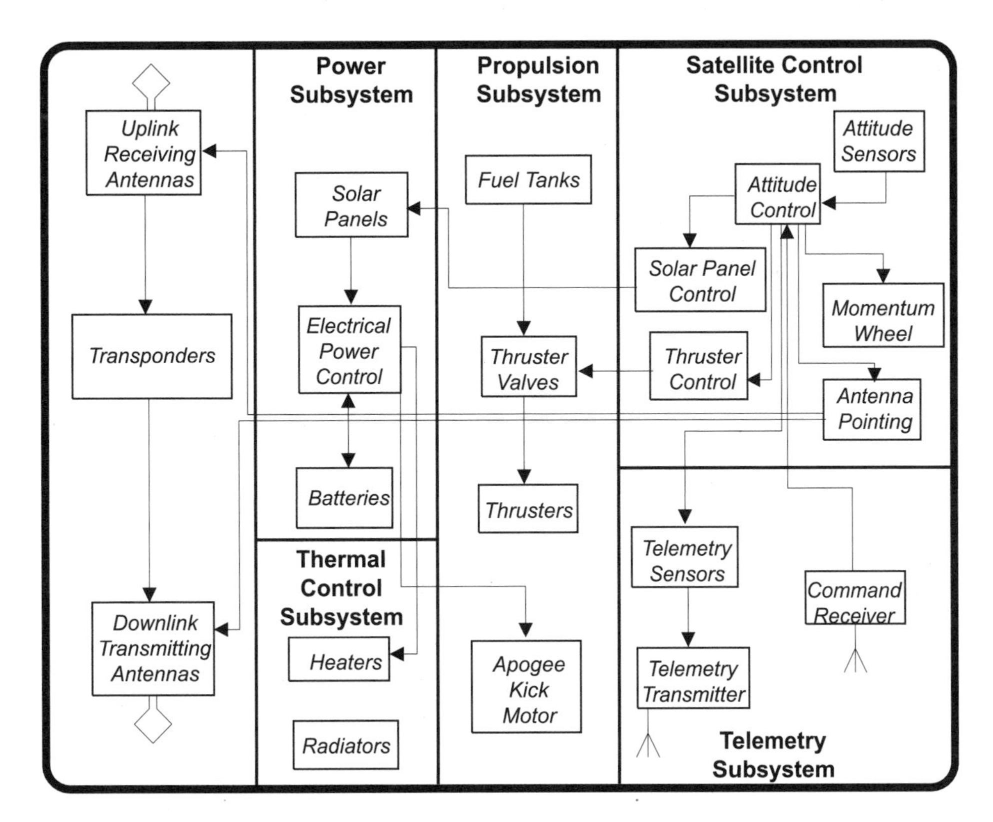

Figure 13.1 A schematic diagram of generic major satellite systems. The basic subsystems, shown in the boxes occupying the right-hand three-quarters of the diagram, are the systems that comprise the satellite bus. They would be present whatever the payload of the satellite, for they provide the "housekeeping" functions and supply the power to the payload systems. In the left-hand box are the components of the communications payload, the hardware that provides the intended functions of the satellite. If the satellite is to have onboard processing capability, this section will be much more complex.

Figure 13.2 An example of a three-axis-stabilized satellite bus, the medium-sized STAR bus from Orbital Sciences Corporation. Different buyers might wish somewhat different details of the solar panels, antennas, etc., to carry out their particular mission objective. (Photograph courtesy Orbital Sciences Corporation.)

equipment to provide the services desired, just as you would specify details you want when buying a car, such as automatic transmission, color, sports suspension, and antilock brakes.

A satellite system operator first determines just what communications capacity is needed, based on a knowledge of its markets and customers. This dictates where the satellites are to be located in orbit and other details of the communications subsystem, also called the *communications payload*. These include specification of such details as

- Antennas, their beam patterns and polarizations
- Center frequency, bandwidth, and power of each transponder
- Level of redundancy for various systems
- Battery power for eclipse protection
- Desired operational lifetime
- Any other details specific to the mission of the satellite.

These requirements, in turn, then dictate how much the components will weigh, how much power they need to operate, how many cubic meters of volume they will occupy within the satellite bus, and other demands on the basic structure. Once all of this has been specified, these determine the cost of the satellite as well as its final mass and size. These in turn determine which launch vehicles could place the satellite in orbit. Adding together the cost of the satellite, the cost of the launch, and insurance (both to cover the satellite owner for losses and for liability and other items) determines the total cost to bring the satellite into operation on-orbit. (After deployment in orbit, there are ongoing operational expenses as well, which we will consider later.)

13.2.1 Structural subsystem

These are the basic mechanical structure and components of the satellite. These must be designed and constructed to withstand the space environment as well as the rigors of the launch process, which include high acceleration forces, vibration, heat, and radiation. They must be highly tested, reliable, and where possible, provided with backup systems because repair is usually impossible. Structural components must be strong, lightweight, and resistant to the space environment. Further, the satellite must be designed to fit within the physical dimensions of the payload fairing of the intended launch vehicle(s). Most GSO satellites have some mechanical components that expand or deploy once on orbit—most commonly antennas and solar panels. The deployment mechanisms must be reliable and surefire. (There is at least one case of a communications satellite rendered completely useless simply due to the fact that someone forgot to remove a small plastic shipping clip from the solar panels after the satellite was installed on the launcher!)

There are two generic structures for GSO satellites; many NGSO satellites are variants of these, or may be entirely different. There are relative advantages to each, depending on the mission of the satellite.

One basic structure looks like a tin can, and is called a *spinner* (Fig. 13.3), for the reason that some of its orientational stability comes from the fact that most of the satellite is spinning around a longitudinal axis at typically 50–100 revolutions per minute. An antenna structure and the communications electronics are part of what is termed the *despun platform*. This is necessary, of course, because the antennas must be pointed at the earth. Between the spinning part and the despun platform are very critical components called *BAPTA*s, *(bearing and power transfer assemblies)*. These are among the most crucial mechanical components of this type of satellite.

On the outer skin of the spinning structure lie the solar cells, which feed the raw electricity to the power supplies and thence to the batteries and other electronics inside the structure. Also inside are the fuel tanks for stationkeeping, the apogee kick motor (empty once the satellite is on-orbit), and other mechanical components. The spinning provides some gyroscopic stability to the orientation of the satellite, which is one of the advantages of this type of satellite structure. The other major advantage is that because of the spinning, thermal buildup is minimized on most of the structure. (Such spinning is jocularly termed "barbeque mode.") The big disadvantage of this kind of satellite is that as you try to build larger and larger satellites, a higher percentage of the total mass must be taken up

Figure 13.3 An artist's rendition of a typical spinner-type communications satellite, in this case the B393. Note that the solar cells are on the outside of the "tin can" structure, which rotates at 60–100 rpm, while the antenna stays pointed down toward Earth as part of the despun platform. While this spinner satellite has only one main communications antenna, others may have several. (Photograph courtesy of Panamsat.)

by the basic structure, rather than going into the communications payload for which you put the satellite up in the first place. There may be a single antenna on one end of the can, rather like a tin can with its lid open, or a more complicated set of antennas attached to an *antenna mast*.

The other basic structure is called the *body-stabilized*, or *three-axis-stabilized* structure. These satellites look boxy, and usually have one or two large *solar panels*, like wings, sticking out to the sides. Antennas are permanently mounted on the sides of the satellite. Such a satellite contains one or three gyroscopes that maintain its orientation such that one side of the satellite is always facing the earth. The antennas are mounted on that face or on the adjacent sides. (If there is a pair of opposite dish-shaped antennas sticking out the sides, this is sometimes joking called a "mickey mouse" configuration.) Figure 13.4 is a diagram of the TDRS, a NASA body-stabilized satellite.

The electronics of a body-stabilized satellite are usually on the interior and sometimes exterior surfaces of the box. The AKM, batteries, fuel tanks, etc., are within the box. The solar cells are on one or two panels that always face the sun,

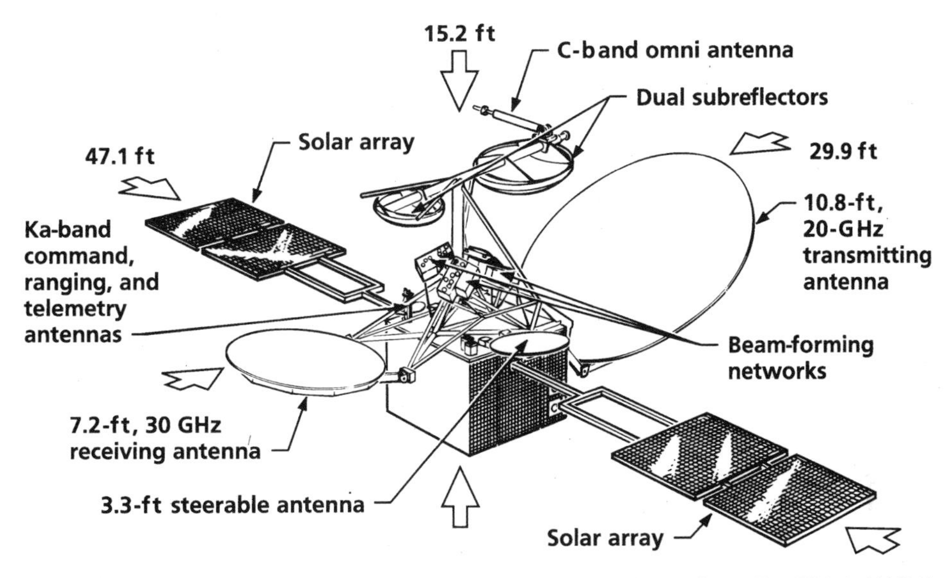

Figure 13.4 A diagram of large three-axis-stabilized (body-stabilized) satellite. This is NASA's Tracking and Data Relay Satellite. Note the two winged solar panels which remain always pointed at the sun. Multiple antennas allow for multiple broad and spot beams. This was one of the first satellites to experiment with the use of the Ka-band. (Source: NASA.)

thus operating at full efficiency. These panels must rotate once a day around small BAPTAs so that as the satellite faces the earth, the panels continually face the sun. The satellite is oriented with the panels in the "up and down" direction, i.e., perpendicular to the orbital plane. (The panels are usually folded up against sides of the satellite during launch, and spread out, or deployed, after the satellite reaches orbit. They are too flimsy to withstand major propulsive maneuvers while spread out.)

The relative disadvantages of such satellites are that they require more control over their orientation, and that the sun shines for many hours on the same parts of the structure, causing thermal buildup. The major advantage is that these structures get more efficient as you make bigger satellites, allowing a higher percentage of the total satellite mass to go into the communications payload.

The basic structures for LEO and other small satellites may be similar or quite different from those typical for GSO satellites. Figure 13.5 shows an example of one of the OrbComm "Little LEO" satellites, and Fig. 13.6 shows an Iridium "Big LEO" satellite.

Another mechanical component of a satellite is the structure that allows the satellite to be carried by the chosen launch vehicle. This is sometimes called the *payload adapter.* Each satellite type and each launch vehicle has a unique design. Occasionally a satellite operator wants to provide for the possibility of launch on more than one launch vehicle, and so has the manufacturer produce a payload adapter for each possible launcher. This, of course, increases the cost of construction. After the launch vehicle to be used is selected, the unused adapters are discarded (or possibly sold if they could be useful to some other satellite).

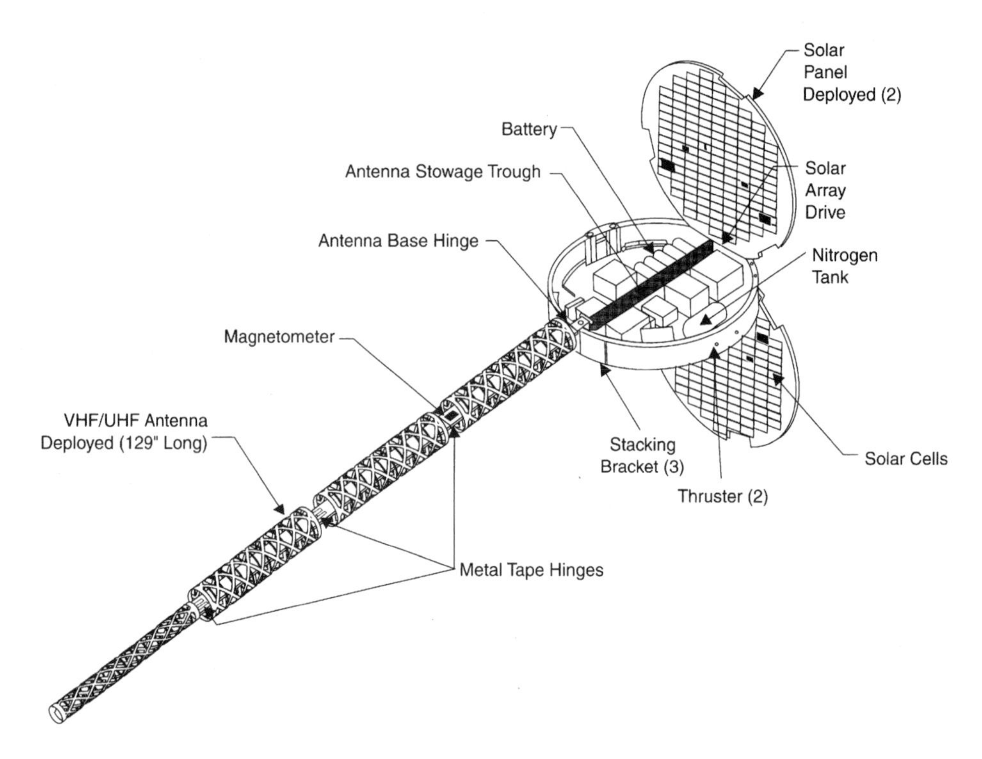

Figure 13.5 An example of a "Little LEO" satellite, this one from OrbComm. The antenna is the long Earth-pointing structure, while the flat panels hold the solar cells. The disk-like main structure is about a meter in diameter and about 17 cm thickness. (Photograph courtesy of Orbital Sciences Corporation.)

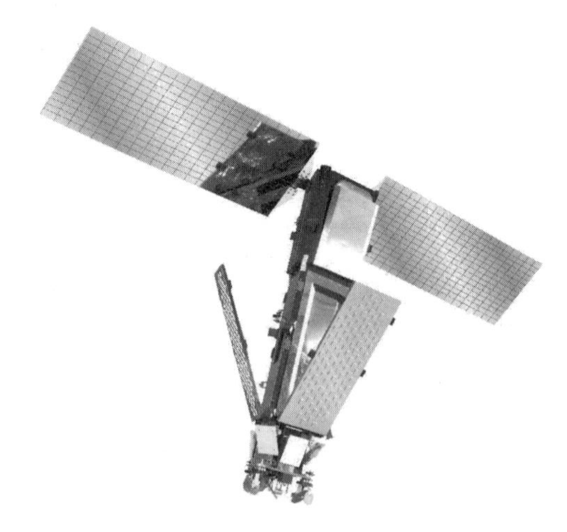

Figure 13.6 An example of a "Big LEO" satellite, part of the Iridium constellation. The large flat panels angling from the sides are the main communications antennas. (Photograph courtesy of Iridium.)

Special provisions also must be made to safely contain the satellite during its transportation to the launch site. This is often done by airplane, and occasionally by truck or ship.

13.2.2 Antenna system

The communications antenna(s) on the satellite are part of the communications subsystem, and an important component of the satellite. They may be small or large, depending on the satellite.

Smaller satellites may have only a single dish antenna, typically 2–3 meters in diameter. Larger satellites may have several antennas (Fig. 13.7). If a satellite has

Figure 13.7 Satellites often have multiple antennas either to produce stronger beams or multiple beams. This photograph during construction shows the two main antennas of the Echostar 103 satellite. Note the solar panels folded down along the sides in their launch configuration. (Photograph courtesy of EchoStar Communications Corporation.)

a global beam, covering all of the earth visible from the satellite, it may use a simple conical horn antenna for this. Dish antennas, with beamwidths often from 1° to 10°, are used for smaller beams. Since the dish structure is only a reflector, a dish may be used simultaneously for receiving the uplink from Earth and transmitting the downlink to Earth. This is done by using receiving and transmitting feeds, or beam-forming networks of feeds, located side by side at the focus of the dish. If a satellite uses more than one frequency band, say C-band and Ku-band, typically it has antennas dedicated to each band.

Often the dish antennas are folded down or along the side of the satellite during launch, and then deployed to working position once on orbit. Some of the newer, larger satellites have antennas up to about 10 meters in diameter, which is much too large to fit within the payload fairing of the launch vehicle. These are folded or rolled up, often made of a metal mesh (Fig. 13.8), and then unfolded or unrolled once the satellite is on orbit. Some satellites have the ability to tilt or move the antennas in order to aim the areas of beam coverage.

For specialized purposes, such as providing a global beam that covers the entire area of Earth visible to the satellite, a simple type of antenna, called a *horn*, may be used. It is simply a cone-shaped tube, with the open, larger end pointed down to the earth.

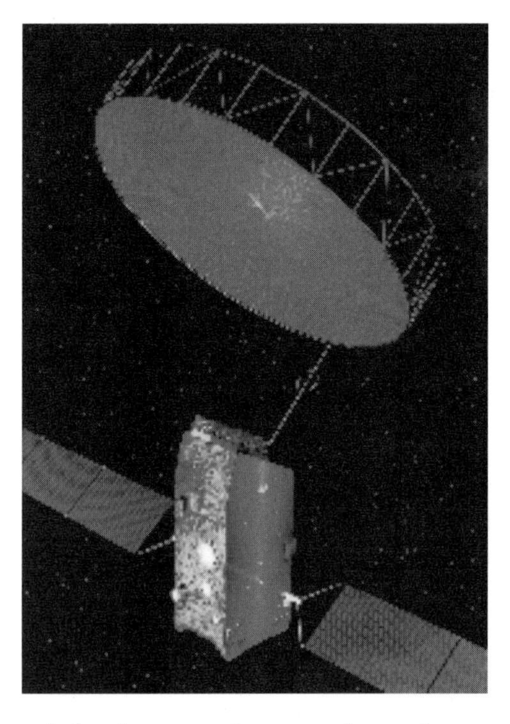

Figure 13.8 As demands for stronger and narrower beams increase, some satellites are being built with unfurlable mesh antennas that unfurl to diameters otherwise too large to fit into the nose cones of launch vehicles. This drawing shows the design of the Inmarsat-4 satellite. (Photograph courtesy of EADS Astrium.)

A few satellites have a *phased-array antenna*. These are composed of many smaller antennas, which may be coils of wire, or a flat plate with slots in it, or some other configuration. They work by using multiple feeds to jointly send the radio waves. The direction of the antenna's beam and the shape of the beam is controlled by altering the intensity and phase of the waves coming from each feed.

The communications antennas handle the radio beams to and from Earth. A very few satellites (Iridium is the notable example in the commercial sector) have *crosslink* or *intersatellite link* antennas as well to communicate with other satellites in the same constellation. Such intersatellite links use frequencies in the Ka-band.

13.2.3 Power generating, storage, and conditioning subsystem

This system provides the electricity to power the satellite functions. All communications satellites derive their power from conversion of sunlight into electricity by solar cells. This raw electricity is gathered and regulated to the proper voltages by a power supply system, supplied to the electronic components of the satellite such as heaters, thrusters, computers, amplifiers, etc., and also stored in special space-rated batteries for use during eclipses.

The most common are silicon-based cells, but newer satellites may use the more expensive gallium arsenide cells which are more resistant to radiation degradation. As we saw in Chapter 9, about 1420 W fall on each square meter of the satellite's surface facing the sun. Of the sunlight energy falling on the solar cells, only about 12% is converted into electricity. A typical small solar cell produces only about half a volt, producing about 50 mW of power. A satellite requires about 45 kg and 15 m^2 of solar cells for every 1000 W of (end-of-life) power it requires. Smaller GSO satellites may consume only a kilowatt or so of power; some of the larger satellites are planned to need 25 kW.

Of the total power supplied by the solar panels and stored in the batteries, about 75% is consumed by the communications subsystem; about 10% by the thermal control system; about 3% for tracking, telemetry, and control systems; about 4% for stationkeeping operations; and about 8% for battery charging.

Remember that during eclipses the satellite is in shadow, generating no electricity. Further, the electronics, gyroscopes, servomotors, and heaters that keep components from getting too cold require power. Battery storage capacity must be sufficient to carry the satellite through the eclipse. (As noted earlier, some of the smaller LEO satellites simply cannot hold enough batteries to do this, and so are nonoperational for part of each orbit.)

A fully eclipse-protected satellite must have enough battery capacity to provide up to about 72 minutes of power; actually, the capacity must be significantly greater, as there is a rule that you should never drain a battery of all of its charge. The capacity of the solar cells and batteries must be designed to provide the satellite's systems with enough electricity at the end of its lifetime—after years of degradation from radiation—which means that at the beginning of its life it has much more than it needs.

Batteries weigh a lot, and since the total weight of the satellite is limited by the capabilities of the bus and launcher, there is sometimes a trade-off between total battery capacity and some other massive items on board, particularly stationkeeping fuel. Fortunately, over the past several decades, the efficiency of the storage batteries has improved, and this has reduced the need to make this trade-off between operational capability and satellite lifetime (and hence revenue).

Smaller communications satellites may consume a few hundred watts of electrical power; typical GSO satellites may use 1 to 5 kW; some of the newer proposed satellite buses plan to use 20 kW and more.

13.2.4 Stationkeeping and orientation subsystems

Keeping the satellite at its assigned orbital slot and its antennas pointed in the proper direction is critical to providing satellite services. Maintenance of correct orbital position is called *stationkeeping*. Maintenance of the pointing is called *orientation* or *attitude control.*

Sensor systems on board the satellite measure whether the satellite is in its proper orbital location and whether the antenna is in the correct orientation with respect to Earth. Their results are part of the telemetry stream of data sent back to satellite control stations.

Often the satellite has a special tracking radio transmitter, called a *beacon*, that sends a signal that earthstations receive to accurately determine the satellite's position. The satellite may also track a beacon from control stations on the ground. Gyroscopes and other sensors on the satellite also help determine its orientation. Some satellites also have small telescopes and sensors to track the position of the sun, earth, and stars to maintain orientation.

Lower-orbit satellites do not have orbital slots, and so do not have to maintain a fixed position. They may or may not need a system for orientation. Some smaller satellites use what is called *gravity-gradient* stabilization, which uses the unequal force of gravity on a long boom or antenna to approximately maintain attitude.

Inevitably, due to slight perturbations, a GSO satellite moves out of its assigned orbital slot, and its antennas may point slightly off from the regions of Earth that they are supposed to cover. Motors and mechanical flywheels, called *momentum wheels*, can partially correct for deviations in orientation. There is even a small perturbation to the satellite's position and orientation due to the very tiny pressure of sunlight on its solar panels.

When the satellite gets too far out of position, periodically ground-based controls send commands to the satellite to fire small compensating thruster systems. For attitude control, when the orientation of the satellite shifts, it can be partially corrected by speeding up or slowing down the gyroscopes inside the satellite. However, there comes a limit where the gyros are spinning either too fast or too slow to work well, so commands must be sent to the satellite to slow them down or speed them up, respectively. Unfortunately, this would cause the satellite to spin out of proper orientation as well, so when these gyro adjustments are made, the thrusters are fired to counteract the gyros. These are sometimes called *momentum dumps*, for they effectively remove momentum from the gyros and throw it into space as the thruster fuel escapes. Typically, position corrections are required every few weeks. These operations are explained more fully in the next chapter.

Sometimes a satellite must be moved from its original orbital slot longitude to another position. While such maneuvers are not very common, they are sometimes necessary. Most commonly, these maneuvers are intended to reassign a satellite to another slot to cover for one which failed, or to move a satellite when it is sold to another operator. Such relocation maneuvers also consume fuel. To accomplish such a relocation, controllers send commands to the satellite to fire its thrusters to either speed up or slow down the spacecraft. Over a period of days or weeks, it then drifts slowly along the Clarke belt (hopefully its transponders are turned off during this to avoid interfering with the nearby satellites it is passing!). When it reaches its intended new slot, the thrusters are again fired to stop the drift. Such a maneuver can consume months to years of stationkeeping fuel.

The system for providing these small thrusts is called the *reaction control system*, or RCS, and typically consists of fuel tanks, usually containing hydrazine or (for large satellites) two propellants connected through a series of valves to small jet orifices which can squirt the gas into space. Use of hydrazine alone is called a *monopropellant* system; use of two fuels is called *bipropellant*. Hydrazine is a nasty, toxic, carcinogenic liquid that must be handled with care. It is used because it is the best propellant for such systems to date. (Companies are working on less toxic propellants.) These produce only a few tenths or a few newtons (the metric measure) of thrust. Obviously, more massive satellites require more fuel. A typical commercial communications satellite using a monopropellant may have about 2.5% of its total on-orbit mass in the fuel; a bipropellant system may require a fuel capacity of about 1.8% of the spacecraft mass.

A newer technology, so far used on only a few satellites, uses what are generically called *ion thrusters*, or sometimes *electric propulsion*. These work by ionizing a gas with some of the electrical capacity of the satellite, and squirting this high-speed gas outward like miniature jets. While the thrust provided by such a system is very small—typically thousandths of a newton—it makes very efficient use of the stationkeeping fuel. Unlike the periodic thruster firings for stationkeeping with conventional thrusters, ion RCS systems fire almost continually, so the stationkeeping operations for the satellite are different.

These brief thrusts of power result in a change of velocity of the satellite, or ΔV as the engineers call it, of a few meters per second. As we will see in the next chapter, there are forces altering both the north-south position, and the east-west position (longitude) of the satellite.

13.2.5 Thermal control subsystem

A lot of heat is absorbed by satellites from sunlight and more is generated by the internal electronics and electromechanical components. Internally generated heat must be moved to the surface for it to be radiated away. The larger the satellite, the bigger the thermal control problem.

It is necessary to keep the components of the satellite's systems within the proper operational temperature ranges. Thermal control measures include such so-called passive systems as insulating thermal blankets, gold-flashed Mylar shielding, and reflective paint, as well as such active measures as moveable louvers. One common device, called a *heat-pipe,* is used to transfer heat from the interior to the outside. These are closed pipes containing a volatile liquid that act as closed convection cycles. Unwanted internal heat is absorbed by evaporation of the liquid in the pipe's hot end then carried to the other end by the gas inside, which condenses, releasing the heat, which is then radiated away into space.

While most of the time the satellite must rid itself of excess heat, during eclipses the temperature of the satellite drops hundreds of degrees, so heaters are turned on to keep fuel tanks and other systems from getting too cold. Such heaters require electricity stored in the satellite's batteries.

13.2.6 Telemetry and command subsystem

Satellite controllers on Earth must know and be able to control the functions of the satellite. Control computers aboard the satellite monitor and control all of the satellite's operations. This includes recording basic housekeeping information such as battery charge, voltages, currents, orientation, fuel levels, and the configuration of the transponders. All of this data is reported to the control station on Earth by a telemetry system.

Other onboard systems receive commands from the controllers to maintain the satellite, switch communications functions, and perform orbital maneuvers. Some of the systems are automatic, and some require commands from the ground. There is at least one backup control computer system as well.

While operating properly on-station, telemetry and commands may use a small part of the communications bandwidth of the satellite. Systems must also be designed to operate not only when the satellite is properly functioning and oriented toward Earth, but also during launch, orbital maneuvers, and emergencies during which the satellite may lose proper orientation. For this reason, most satellites have dedicated omnidirectional tracking, telemetry, and control (TT&C) antennas in addition to the antennas being used for the communications functions through the satellite.

13.3 Communications payload subsystem

These are the components that provide the communications services of the satellite and include antennas, filters, receivers, switches, amplifiers, transmitters, and other devices. The most failure-prone systems, such as receivers and amplifiers, typically have redundant components which can be switched into operation—either automatically or on command—in case the primary component fails.

The electronics aboard the satellite that provide the communications functions are called *transponders,* sometimes abbreviated as *xpndr.* Sometimes they are called *repeaters.* The basic functions of a transponder are

- receive an uplinked signal from an earthstation
- change it to the proper downlink frequency
- amplify the power of the radio waves
- transmit the radio signal down to another earthstation.

When an earthstation uplinks to a satellite, there is a minimum level of radio power that must reach the receiver, called the *threshold* of the receiver. If the received signal strength is below this, the satellite will not recognize the signal or be able to handle it properly. On the other hand, since the electronics will malfunction if overpowered, the receivers usually also have a *saturation level* that is the maximum power they can receive without overloading. To prevent this, receivers have *limiters* that allow no higher than some predetermined power level to be received to avoid overdriving the circuitry. In specifications for a satellite, you will commonly see numbers for a transponder's *threshold flux level* and *saturation flux level.*

Satellites have many transponders, typically several dozen. Some high-capacity satellites may have over 100 transponders, each with similar or different characteristics. Satellites that utilize two different frequency bands are called *hybrid* satellites. Each frequency band in use (such as C-band, etc.) will usually use the full legally authorized bandwidth for the service that the satellite is providing. For example, FSS satellites are typically permitted a total bandwidth of 500 MHz. (We will see in the next chapter how transponders are used, and in Chapter 20 how frequency reuse allows us more effective bandwidth by using spot beams.)

A transponder's basic components are a receiver section and a transmitter section (Fig. 13.9). The receiver section consists of the receiving antenna, filters to make sure the signal is in the correct frequency range, possibly a limiter to keep the receiver from being overloaded by a too-strong signal from the ground, a preamplifier, and a *frequency converter.* This last device simply converts the received uplink frequency to the desired downlink frequency (which is always different to avoid interference). The transmitter section consists of a *high-power amplifier*, or HPA, another filter to ensure the downlinked signal is at the proper frequency, and the transmitting antenna.

The signal level coming into the uplink antenna on a satellite is very small, and so the total amount of amplification that takes place in the transponder chain has to be very large to produce enough of a signal to get back to the earth. In the

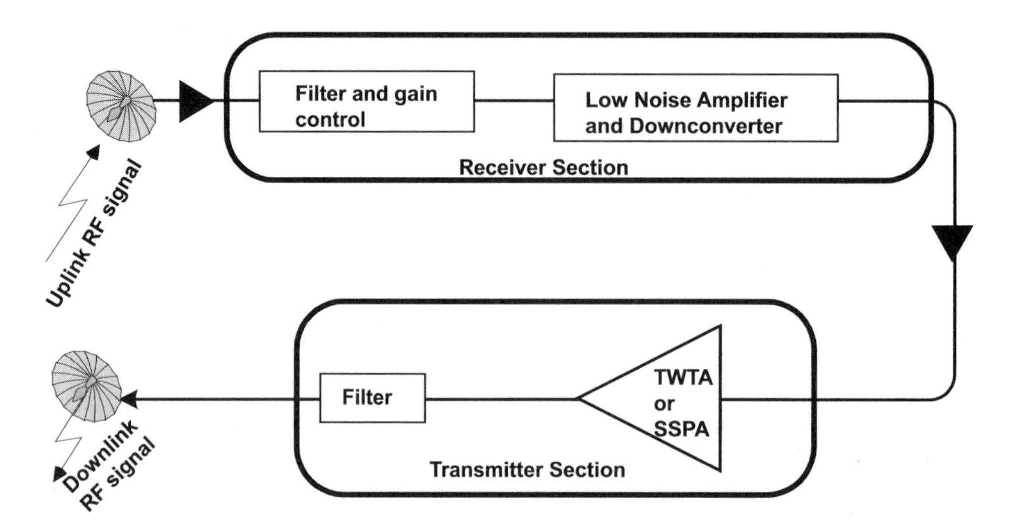

Figure 13.9 A very simplified schematic diagram of a "bent-pipe" transponder. The receiver section picks up the uplink from Earth, amplifies it, and converts it to an intermediate frequency (IF). The transmitter section converts the IF to the downlink frequency, amplifies it still more, and directs it to the downlink antennas for relay to Earth.

receiver section of the transponder, the amplification may be on the order of 60 dB, or a millionfold increase in the small uplink signal, with another 50–60 dB of amplification in the transmitter section. Subtracting 10–20 dB of loss to the signal that occurs in the other components of the transponder (filters, couplers, multiplexers and demultiplexers, etc.), the total net gain in signal strength through the transponder is around 100 dB.

13.3.1 Transponder amplifiers

There are two generic electronic technologies used for HPAs: traveling wave tubes and solid state amplifiers. Some satellites use one or the other; some use both.

Traveling-wave tube amplifiers (TWT, or TWTA; sometimes pronounced "twit") are rather like the vacuum tubes that were once used in all consumer electronics. For the highest powers and frequencies, they are still in use. Figure 13.10 shows a diagram of a typical TWT; Figure 13.11 shows a photograph of one. A TWT gets its name because weak radio waves enter the tube at one end and travel down its length, becoming amplified in the process. TWTs are older technology. They are heavy and expensive, and only a very few companies have a track record of manufacturing space-qualified models. They cost several hundred thousand dollars each. However, they are the technology of choice for the highest powers and highest frequencies because solid-state electronics gets trickier as powers and frequencies increase. TWTs also require more complicated—hence more massive—

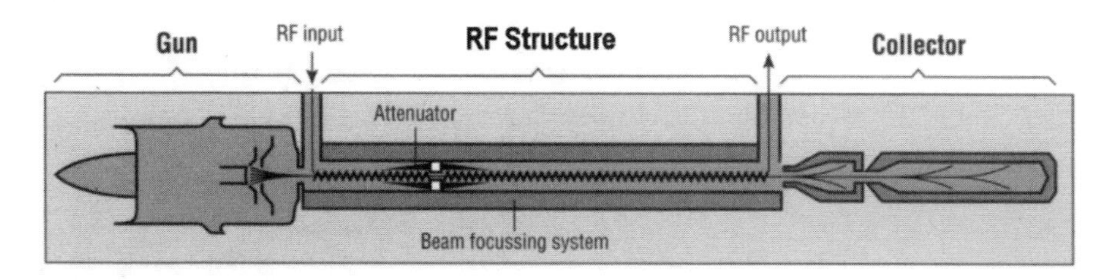

Figure 13.10 Simplified diagram of a travelling-wave tube amplifier. (Image courtesy of Thales.)

Figure 13.11 Photograph of a space-qualified TWT. (Photograph courtesy of Thales.)

power supplies on the satellite, and are less linear; that is, they produce more intermodulation distortion.

Solid state power amplifiers (SSPA) are more linear, lower in weight, easier to manufacture, and require only a single low voltage, making the satellite's power supply lighter and simpler. Once checked out, they are more reliable. But to date, they are used only for lower power and frequency applications. Manufacturers are striving to produce SSPAs with higher power and frequency performance.

13.3.2 Redundancy

HPAs are typically the least reliable components on the satellite. It is common, especially in GSO satellites with their typically longer lifetimes, to provide more amplifiers than transponder circuits. If the amplifier in a transponder fails, another amplifier can be switched into use, either automatically or on command. This is

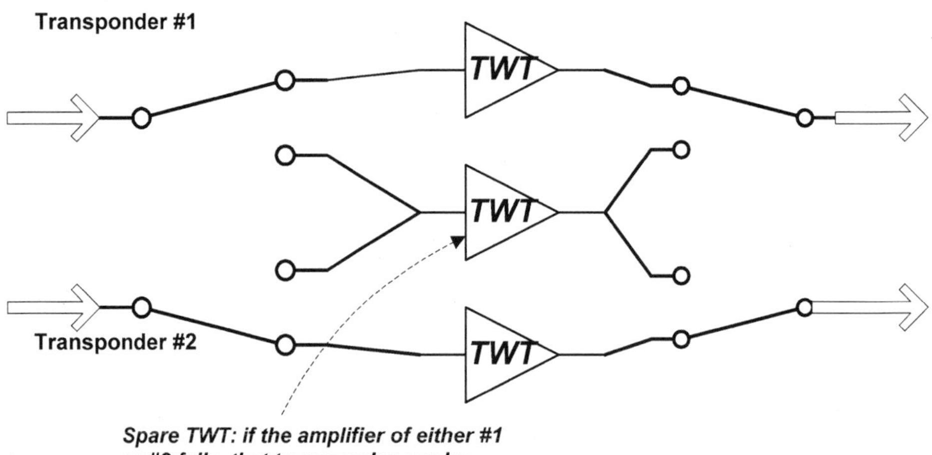

Figure 13.12 Simplified schematic of one way to provide redundant amplifiers for a transponder. Both Transponder #1 and Transponder #2 have their own TWT. Should either TWT go bad, either of these transponders can switch to using the spare TWT shared between them. This would be a "3-for-2" redundancy. In this particular configuration, if one of the transponders has switched to using the spare, then the spare is no longer available for the other transponder should its TWT fail.

called *transponder redundancy.* Typical satellites may have half-again as many amplifiers, or even twice as many, as there are transponder circuits. The former would be called a "three-for-two" redundancy, the latter "two-for-one." See Fig. 13.12.

Receivers, being lower-power components, are somewhat less failure prone. A typical redundancy level for these might be a 5-for-4 arrangement.

A satellite's command and control computers and other related components can also fail. Thus, almost all commercial satellites have a backup control computer that functions in parallel with the main computer, and which can take over if the primary one fails.

13.3.3 Transponder characteristics and uses

The number of transponders on a satellite varies greatly, depending on the purpose of the satellite. Twenty to thirty transponders is common, but some satellites have many more. There may be transponders of different frequency bands, of different powers, and different bandwidths on the same satellite.

Operationally, the best way to think of a transponder is not as a single piece of equipment, but rather as a signal path through the satellite. A transponder is made up of a chain of components, some of which can be switched in and out of the circuit, or altered in their effect of the signal. (As an analogy, think of your home hi-fi system. There is no one component—FM tuner, CD player, equalizer, etc.—to which you can point to and say that that is the system; different things can be switched in and out to make the system do what you want, just as with transponders.)

A transponder is characterized by four major things:

- Center frequency
- Bandwidth
- Sensitivity to uplinked signals, and their polarization
- Power of the downlink beams, and their polarization.

In addition, in some satellites a transponder may be identified with some specific beam patterns providing services to a particular region of the Earth.

As we will see later in Part 6, charges for the communications services of the satellite are based primarily on duration, bandwidth, and power. It is the transponders that provide these technical capabilities.

The center frequency and bandwidth are measured in megahertz or gigahertz. The sensitivity to incoming signals is measured by a number of decibels called the *figure of merit*, symbolized by *G/T*. The downlink power is measured in decibels and called the *effective isotropically radiated power*, abbreviated *EIRP*. (Both G/T and EIRP will be explained more fully in Chapter 17, where we talk about the properties of antennas.) The polarization may be either unpolarized (which is uncommon), vertical or horizontal linear polarization, or right-handed or left-handed circular polarization.

The way in which the allowed bandwidth of the satellite is divided up among the various transponders and antenna beams is known as the *transponder frequency plan*. For commercial satellites, these are published by the satellite operators so users of the satellite know how to tune their earthstations. For fixed services and some mobile services, frequency plans are determined by the satellite operator before manufacture, depending the operator's estimate of traffic demand by its expected customers. DBS satellites are predetermined by international agreements, so DBS operators have no flexibility in how their transponders are assigned. Figure 13.13 is a diagram of a fairly simple transponder frequency plan; Figure 13.14 shows a plan for a more complex satellite.

13.3.4 "Bent-pipe" satellites

Most communication satellites are called so-called "bent-pipes"; they are simply radio relays. The transponder is controlled, and if necessary partitioned, by the earthstations it links to. The transponder itself doesn't "know" where the signals are coming from or going to, whether they are analog or digital, whether they are telephone calls or datastreams, or anything else. As long as the uplinked signal is of some minimum power level and proper frequency so that it can be detected by the receiver part of the transponder, the transponder will relay the signal back to the earth according to its design.

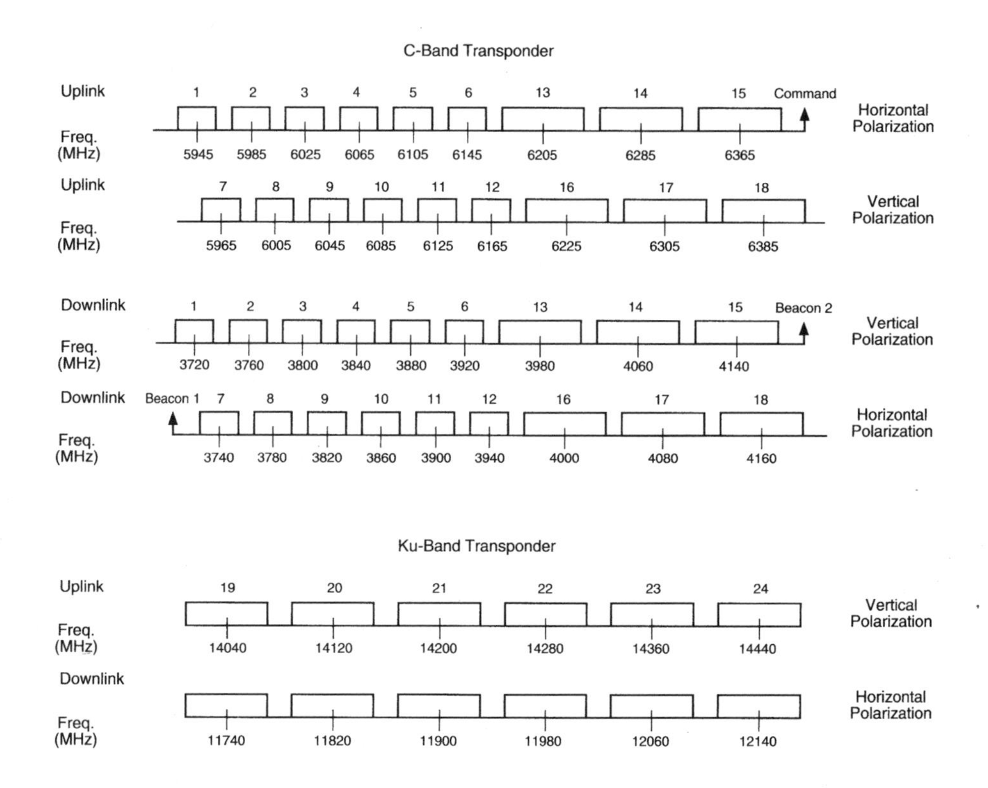

Figure 13.13 The frequency plan of the Spacenet-4 satellite. It is fairly simple, but with transponders of several powers, frequencies, and bandwidths. (Diagram courtesy of Gilat Satellite Networks.)

Basically, a transponder receives a signal from Earth, converts it to the downlink frequency, amplifies it, and beams it down to Earth. All of the uplinked frequencies spanning the entire frequency range of the band in use (e.g., 3.7–4.2 GHz) are received by the receiving antenna and feed on the satellite. This range of frequencies is filtered to be within the band limits. This very weak combined signal, typically billionths of a watt, is amplified by a wideband amplifier with a gain of perhaps 60 dB (a millionfold) and converted into a convenient intermediate frequency of the same bandwidth. This wideband signal is then put through a demultiplexer to separate each transponder's signal from the others. These individual signals, one for each transponder, are then amplified by narrowband amplifiers by about another 50 or 60 dB and converted from IF to the frequencies and bandwidths of the various transponders. All of these now relatively high-power signals (whatever the wattages of the transponders are) are combined by a multiplexer and sent to the transmitting antenna feed for transmission on the downlink to earthstations.

In some satellites, there may be a few extra functions, such as cross-strapping from one transponder to another, assignment of a particular transponder to a particular antenna beam covering some region of Earth, or other functions. None of these manipulations affects the nature of the signal itself.

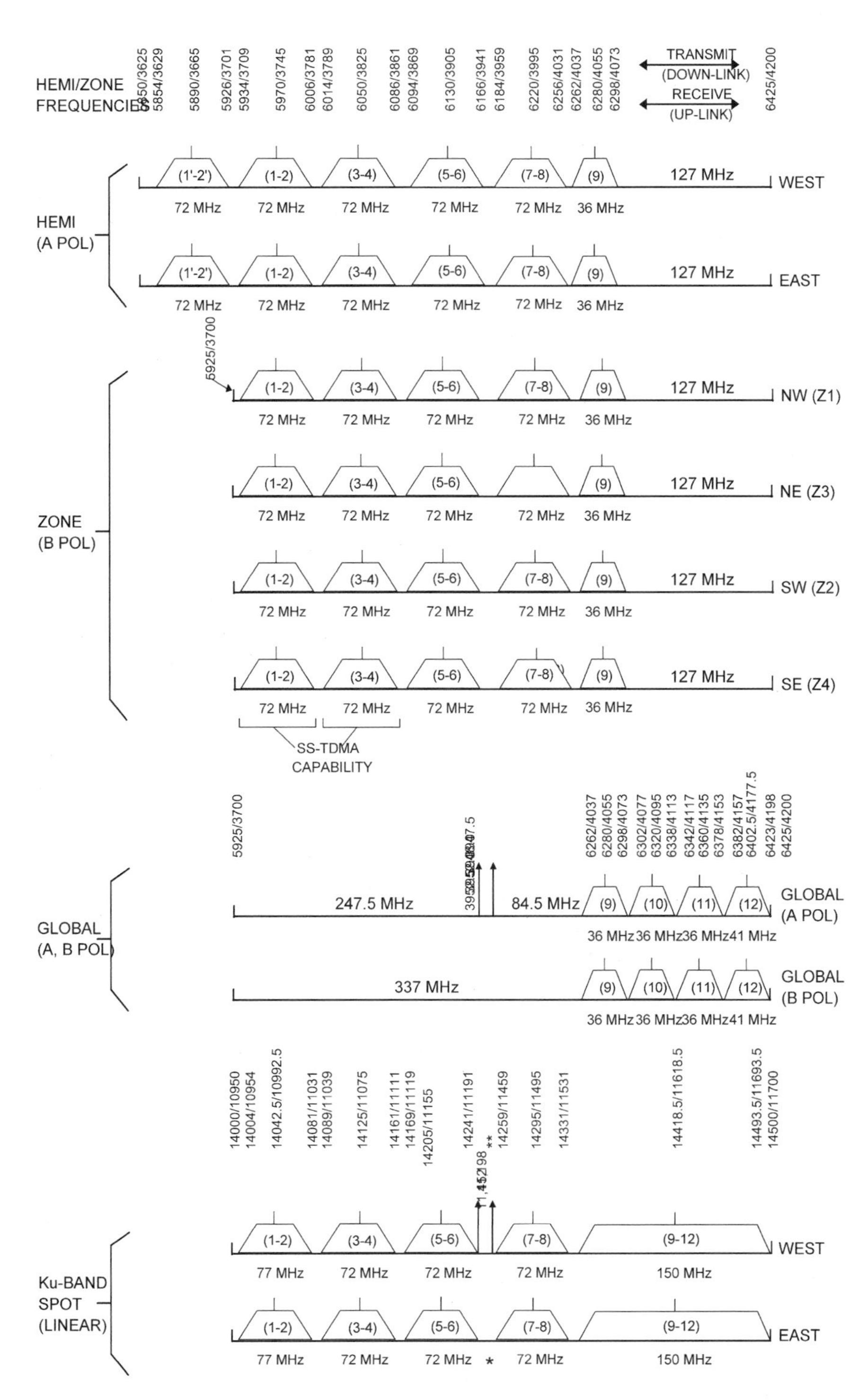

Figure 13.14 A more complex frequency plan, this one for an Intelsat 10 satellite, showing many transponders with differing frequencies, powers, and bandwidths. (Image courtesy of Intelsat Global Service Corporation.)

13.3.5 Onboard processing satellites

Satellites with onboard processing have considerably more complicated communications systems because they manipulate the signal going through the transponder. Some satellites, both in GSO and NGSO, do more than simply rebroadcast a radio signal. They may provide for demodulation and remodulation, signal routing and switching, and other functions. These are generically known as *onboard processing, OBP,* or *onboard switching*. One of the major uses of OBP is in efficiently implementing a mesh network. Another, detailed below, is onboard multiplexing of video signals to allow smaller, less powerful, and geographically dispersed uplinks to a single transponder.

For example, the Intelsat satellites have OBP that increases their flexibility in routing telephone calls. As another example, the Iridium Big LEO satellites have switching capabilities. This ability to process, manipulate, and route signals is sometimes called *switchboard-in-the-sky*. The planned high-speed data satellites designed for Internet traffic will have complex data handling and switching abilities, sometimes called *Internet in the sky*.

There can be several types of OBP. One involves RF or IF switching, common in the Intelsat satellites for telecomms traffic. Another kind of OBP provides baseband processing and routing ability aboard the satellites. The processing may be handled completely by the satellite, or under control from some master earthstation. Other OBP functions may include the hand-off of signals from one satellite to another for Big-LEO telephony satellites, antenna pointing calculations for aiming spot beams, or communications protocol conversions, for instance.

There are a number of advantages to using OBP. Among them, and without getting into too much technical detail, signals can be treated individually. Noise is lessened, and power requirements at both earthstations and on board are reduced. Bitrates of digital signals can be changed, a process called *bitrate conversion*, enabling each channel to be optimized to the traffic it carries. Further, the modulation scheme of the downlink need not be the same as that of the uplink. OBP increases the throughput of a satellite of a given mass, or, alternatively, could allow a lower-mass satellite to carry the same amount of information as a more massive one.

As we saw earlier, in a bent-pipe transponder, the RF signal from the uplinker is received, converted to an IF, then converted to the downlink frequency, amplified, and sent back to Earth. Signals arriving at a sophisticated OBP satellite may go through many more steps, such as

- RF received by receiver antenna and electronics of transponder
- error correction
- decoding routing information for a particular data packet
- buffering (briefly storing on board) the data
- multiplexing datastreams
- switching the datastreams to the appropriate transponder or downlink antenna
- converting the datastream to the downlink frequency, amplifying, and transmitting.

13.3.5.1 Onboard multiplexing

Often, particularly with broadcast television services, there is a need to combine video feeds from various locations for purposes of performing data compression and statistical multiplexing to conserve bandwidth and increase the number of channels offered to customers. To do this, the various programming providers must send the various video feeds from their origin points to a common uplinking earthstation where the compressing and multiplexing is performed.

An alternative, pioneered by Eutelsat, is to build statistical multiplexing equipment into the satellite transponder. Then, each programmer can independently uplink the video to the satellite where the signals are combined and broadcast to homes on Earth. This saves the costs of the extra links to a central earthstation. This patented system is known by the trademarked term Skyplex. It is one type of onboard processing. The Worldcom satellites use a similar system to allow geographically separated programmers to feed their signals to the satellite separately to be combined on board.

While OBP makes a satellite very versatile, it can greatly increase the complexity and cost of the system.

13.3.5.2 Intersatellite links

The Iridium satellites, and so far no other commercial satellites, have one other capability, that of sending signals from one satellite in the constellation to another without going through an earthstation. Such signals are called *intersatellite links, ISL,* sometimes called *satellite-to-satellite links, SSL.* These links take place in a specially allocated frequency band around 24–25 GHz. Like OBP, ISLs can greatly increase the complexity and cost of a system. (For historical completeness, two Orion satellites did have ISLs, but chose not to use them.) Many military and NASA satellites have ISLs.

NASA and the military make use of ISL, but these are not commercial satellites. The European Artemis satellite has a laser communications crosslink designed to work with remote-sensing satellites and other low-orbit satellites. These may become more common in future satellites, depending on their application.

ISL greatly increases the complexity and pointing requirements of a satellite, but provides several advantages. The most obvious is expanded connectivity, since the same satellite need not be in view by both earthstations of a point-to-point link, since the satellites link to each other across the orbit. For example, while it is possible to establish a direct link through a single satellite connecting earthstations in New York City and in Moscow, it is not possible to establish such a link directly between San Francisco and Moscow because they are too distant. A possible solution is to use two satellites with ISLs some tens of degrees apart in the GSO, one visible from San Francisco, the other visible from Moscow. Then, a channel would go from Moscow to the easternmost of the two satellites, then via the ISL to the westernmost satellite, and then down into San Francisco. The alternative—not using an ISL—would be to do a "double-hop" link in which the signal is uplinked from Moscow to a satellite it can see, then downlinked to an earthstation, say in Mexico, which then "turns around" the signal to yet another satellite that is also visible from San Francisco. This would cost the user the link charges on each end, the turnaround service in Mexico, and two space segment charges. It would also, of course, double the delay in the transmission.

So far, except for low-orbit Iridium (and some government satellites), ISLs have not been an economic way of doing business; but some will eventually be used.

The next chapter will explain how the satellites are operated and used to provide communications services.

Chapter 14

Satellite Operations: Housekeeping and Communications

There are two kinds of communication going through a typical communications satellite: the user traffic for which the satellite was placed into operation, and the continual everyday communications that maintain the satellite and its systems in operational status. The former are the communications operations (which make the money), and the latter are the satellite operations (which cost money). Ideally, the user of the satellite's communications services should never have to be aware of the details of keeping the satellite on-station, charging batteries, eclipses, switching in backup amplifiers, and the myriad other items that keep the satellite working and producing revenue.

14.1 Satellite operations

The functions of *tracking, telemetry, and control, TT&C*, also called *tracking, telemetry, control and monitoring, TTC&M*, or sometimes *telecommand*, are usually performed from earthstations different from the ones sending the communications links through the satellite, and sometimes at frequencies different from those used for communications.

Typically, a satellite is controlled from a *Spacecraft Control Center, SCC*, which receives and sends TT&C information to dedicated TT&C earthstations that communicate with the satellite. Such operations deal with the "health" of the satellite itself. Often there is a separate *Communications Control Center, CCC*, or *Network Operation Center, NOC,* to control the operations of the transponders handling telecommunications traffic through the satellite. Figure 14.1 shows the inside of a typical SCC.

14.1.1 Tracking

To *track* a satellite is to know where it is. This task is very different for GSO satellites than for NGSO satellites, because those in Clarke orbit remain relatively fixed over the earth, whereas NGSO satellites are in continual motion relative to

Figure 14.1 An overview of the Intelsat Satellite Control Center, where the global fleet of satellites is monitored and controlled. (Photograph courtesy of Intelsat Global Service Corporation.)

the surface. Tracking can be accomplished by a variety of means, both from the ground and from the satellite. Radar from the ground, and radio tracking signals received on the ground, can tell satellite operators where the satellite is located and how it is moving. The satellite itself can determine its position by triangulation on the earth, sun, and stars, or by monitoring a beacon at an earthstation.

Some satellites can even use other satellites to find out their positions. The *Global Positioning System, GPS*, using the *Navstar* satellites, is a U.S. military navigational system of a couple dozen satellites in nearly semisynchronous orbits.

14.1.2 Telemetry

Telemetry is the information on satellite systems status sent back to Earth for controllers to monitor. Such operational parameters as electrical bus voltages and currents, battery charge level and temperature, fuel tank pressure and temperatures, momentum wheel speed, amplifiers switched on or off, orientation angles, antenna configurations, and a host of other items are of interest to controllers. Onboard sensors monitor the various operational parameters and relay the data to the TT&C stations, where the readings are monitored and often displayed on computer monitors in the control room (Fig. 14.2). If some parameter gets out of line, warning signals alert the engineers to the possible trouble.

Figure 14.2 A close-up of one of the controller positions in the Satellite Control Center at Intelsat. Telemetry from the satellites is displayed on video monitors, and operators can perform routine and emergency maneuvers on each satellite. (Photograph courtesy of Intelsat Global Service Corporation.)

14.1.3 Satellite control

To *control* the satellite is to command the satellite to do what you want it to. This can include such everyday routine items as cross-strapping transponders, aiming antenna beams, and adjusting transponder powers to occasional-but-expected events such as battery charging before eclipses, periodic stationkeeping maneuvers, or handling unexpected emergencies such as having the onboard computer blasted out of operation by a solar storm.

In the old days (more than about 10 years ago), many of the TT&C links to commercial satellites were, surprisingly, not electronically secure. Today, the information coming from and going to the satellites is encrypted, and other security features are also used to make sure some would-be satellite terrorists or "crackers" do not disrupt operations.

When the satellite is on-station, the TT&C signals are often sent to and from the satellite in small frequency bands alongside the main communications frequencies, using the main communications antenna on the spacecraft. This allows for high-speed data transfer. However, while the satellite is going into orbit, or if it alters its orientation toward Earth, there is a backup TT&C system that uses very robust communication techniques to ensure that controllers can always monitor and control the satellite. These techniques include using VHF frequencies, which are largely unaffected by the atmosphere, and equipping every satellite with an omnidirectional TT&C antenna that does not have to be pointed at Earth.

Most satellite system operators have a *spacecraft control center* (SCC), a network of control earthstations, and often a backup control center to monitor and control their satellite fleets. If such a system is global, obviously a single earthstation cannot see all of the satellites in the fleet, so geographically dispersed control earthstations are needed. The master SCC and the earthstations are in communication with each other via small bandwidth voice channels over the satellites called *orderwires,* and they usually have backup terrestrial interconnections as well. Most satellite systems operate their own control facilities, but systems with smaller numbers of satellites may hire out the job to some large system operator.

Increasingly, satellites have onboard control mechanisms that perform many of the housekeeping operations without direct control from the ground. They also have built in "fail-soft" procedures to their control computers in case contact is lost with the controllers.

14.1.4 Satellite stationkeeping and orientation operations

One of the most important TT&C functions is keeping the satellite at its assigned slot in orbit (for GSO satellites) and oriented properly toward the earth so its antennas can provide communications services.

Satellites are typically kept within 0.05° to 0.1° of their assigned position. Satellites are therefore controlled to keep them within about 37–74 km of their nominal slot longitude. Figure 14.3 shows a conceptual view of the imaginary box that defines a satellite's orbital slot.

If the earth were a perfectly symmetrical sphere and there were no other massive bodies in the solar system, a satellite would stay at its assigned GSO orbital slot. But this is not the case, and so GSO satellites slowly drift away from their slots. The forces that cause this drift are called *perturbations*, and there are two kinds: those which move the satellite east or west of its slot along its orbit (thus changing its longitude), and those which move it out of a true equatorial orbit (thereby increasing its inclination). As you would expect, these are referred to as *east-west perturbations* and *north-south perturbations*. Their effect is to slightly change the velocity of the satellite, and thus its position. There is also a rather small third effect of the pressure of sunlight on the satellite; its major effect is to slightly alter the eccentricity of the orbit.

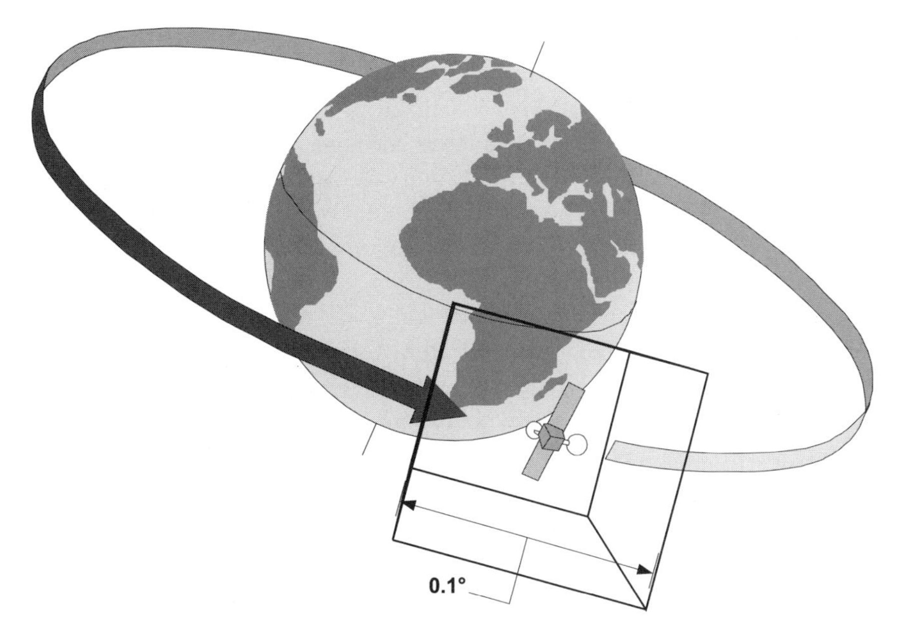

Figure 14.3 Every operational satellite is kept within an imaginary box centered on its nominal orbital slot. Terrestrial, solar, and lunar perturbations pull the satellite away from its assigned position, and controllers must maneuver the satellite so that it is not more than typically 0.05° from its nominal position. Inclined-orbit satellites are allowed to have larger variations in their north-south position, but must still be kept close to their assigned longitudes to minimize interference with adjacent satellites.

The amount by which the satellite's motion is altered is measured in meters per second and called "delta-*V*," symbolized by ΔV (the Δ is technical shorthand for change).

The east-west forces arise from the fact that our planet is not perfectly spherical and the material inside it is not quite evenly distributed. This lumpiness of the earth pulls each satellite along the Clarke orbit very slightly; the strength of the pull depends on the longitude slot of the satellite. It averages to a drift of about 0.01° per day. There are four spots along the GSO where all of the forces balance out and there is no east-west drift. These longitudes are called *equilibrium points*. Two of these are stable, in the sense that a satellite there, if pulled slightly away, will tend to move back to its original spot; two others are unstable, meaning that although at these longitudes the forces are balanced out, if the satellite moves slightly, it will tend to keep moving. The stable equilibrium points are at eastern longitudes 75°E and 252°E. Satellites with slots near the equilibrium points have to make few east-west correction maneuvers in order to remain at their assigned positions.

The east-west perturbations are important, because they cause the satellite to move out of the orbital slot assigned to it, thus potentially causing electronic interference with adjacent satellites. For this reason, the longitude of the satellite is carefully controlled. East-west correction maneuvers consist of firing the RCS thrusters for typically 6–60 seconds about every 2–3 weeks. The ΔV required for these corrections is about 2 m/s each year of operational lifetime to keep a satellite within about ±0.1° of its nominal longitude.

The north-south perturbations are caused by the sun and moon, neither of which orbits in the plane of Earth's equator, and thus their gravitational attraction pulls every satellite into an inclined orbit. The influence of the moon is about twice that of the sun. The ΔV for each satellite is about the same regardless of longitude, and the total perturbation, and therefore the required correction, are much greater than the east-west perturbation.

The average north-south change in inclination is 0.85° per year. North-south correction maneuvers require firing the RCS thrusters for typically 120–600 seconds every couple of months. The ΔV required for these corrections is about 45 m/s per year of satellite lifetime to keep a satellite within about ±0.1°of the plane of the equator.

If left uncorrected, a satellite's orbit would become inclined (Fig. 14.4). Because of this, as seen from an earthstation during one day, the satellite would appear to move up and down, above and below the Clarke orbit, and slightly east and west, describing a small "figure-eight" in the sky. The technical name for this "figure-eight" is an *analemma* (Fig. 14.5). The size of the northern and southern loops of the "figure-eight" depends on the inclination of the orbit: as many degrees as the inclination of the satellite's orbit. The east-west size is typically a few tenths of a degree.

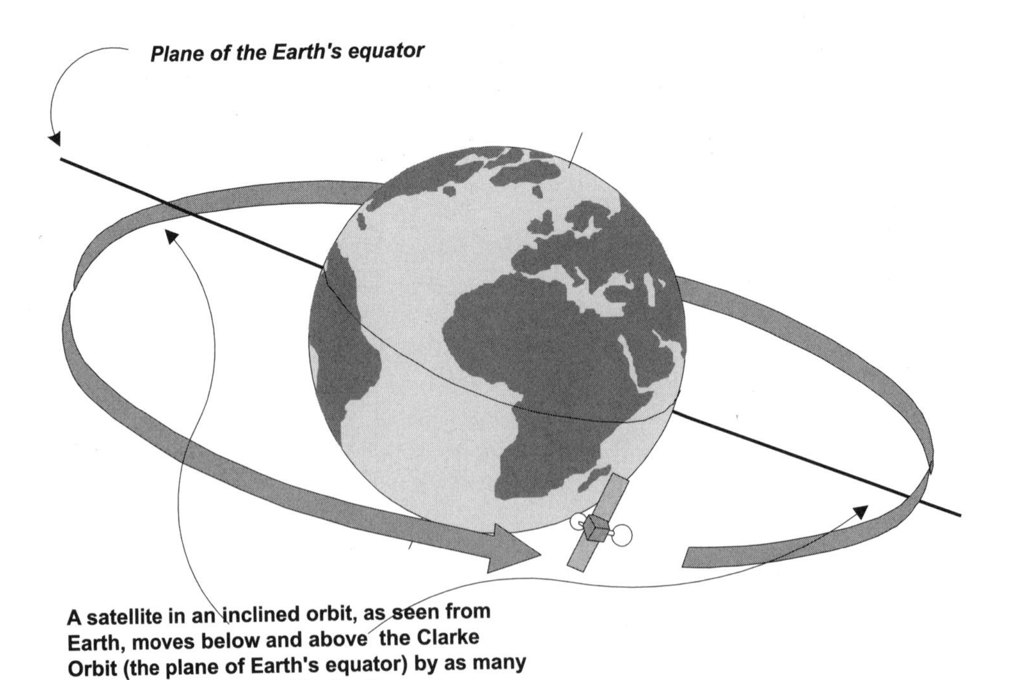

Figure 14.4 A satellite allowed to move into an inclined orbit moves slightly north and south of the celestial equator, by as many degrees as the inclination of the orbit. The satellite is still geosynchronous, but only approximately geostationary.

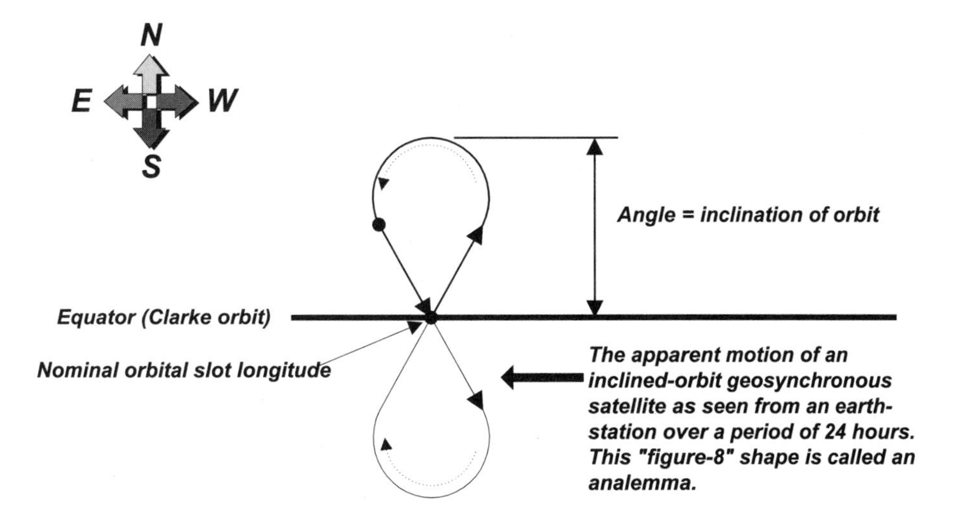

Figure 14.5 As seen from an earthstation, a satellite in an inclined orbit not only moves north and south of the equator, but also moves slightly east and west of its assigned longitude slot, typically by only a few tenths of a degree, over a period of one day. The "figure-8" apparent path of the satellite is called an analemma.

Satellites like this are said to be *inclined-orbit satellites*. Since they are no longer truly geostationary (although they are still geosynchronous), they may have to be tracked by the earthstations linking to them. Such a satellite must still be controlled to keep it within the east-west limits of its orbital slot assignment and within the maximum inclination decided on by its operators.

Whether it must be tracked by a particular earthstation will depend on the beamwidth of the earthstation (which we will see later depends on the size of the dish and the frequency in use). If the beam is wider than the excursions of the satellite, no tracking is necessary. Obviously, earthstations with tracking capability will cost more—often on the order of hundreds of thousands of dollars more—than truly fixed ones. Some satellite operators let their satellites go into inclined orbits to lengthen their lifetime, hence their revenue production. However, space segment charges for inclined-orbit satellites is typically only a third or less of the charges for a truly GSO satellite. Inclined-orbit satellites are often those already nearing the projected end of operational life, and may be used as a spare or backup.

14.1.5 Orientation and pointing

Another job for the reaction control system is to keep the satellite pointed at the earth. This must be done to an accuracy of better than 0.1°, especially if the satellite has narrow spot beams serving small areas of Earth. A 1° inaccuracy in pointing by the satellite would shift a spot beam by more than 700 km from where it was supposed to be pointed on the surface.

When a inclined-orbit satellite moves above and below the equatorial plane, naturally any antenna beam from the satellite moves north and south as well. If narrow spot beams are in use, this can move the spot beams completely away from their intended service areas. To compensate for this, Comsat Laboratories invented a patented *Comsat maneuver.* This technique simply tilts the antenna pattern down when the satellite moves up above the equatorial plane, and tilts it up when the satellite is below this plane. This tilting can be accomplished either by nodding the whole satellite, or by moving the antennas on the satellite if it has been designed to do so.

14.1.6 Relocation

Yet another way the satellite's precious fuel is used is in repositioning the satellite into another orbital slot. It takes fuel to get the satellite moving, and after it drifts slowly along the Clarke orbit, a second burst of fuel stops the satellite at its new slot. The faster the satellite is relocated, the more fuel it takes. Propelling a typical satellite to a speed of 5° a day and placing it in position can consume as much fuel as a year's worth of stationkeeping. In other words, relocating a satellite reduces its operational lifetime.

14.2 Satellite lifetime

It is important to emphasize that the operational lifetime of a GSO satellite is determined by its ability to remain pointed correctly at its orbital slot. This requires fuel, so the fuel capacity and efficiency of its thrusters determines the operational lifetime. More massive satellites will obviously require more fuel than smaller ones. Today, many GSO satellites are designed with a nominal lifetime of 12 to 15 years. (It is unlikely that we will ever deploy satellites with lifetimes designed to be much longer than 12 to 15 years, since their technology would be obsolete after that much time.)

Correct operation of the satellite's RCS also affects lifetime. Fuel loss, unplanned orbital maneuvers, or partial failure to correct a satellite's position and orientation can have serious business consequences. In one case, a satellite made use of both hydrazine and ion thrusters, and the amount of fuel on board was calculated to allow for a revenue lifetime of 12 years. After some time on-orbit, the ion thrusters failed. Without this system, the hydrazine RCS had only enough fuel for about 6 years, thus halving the lifetime of the satellite, causing its owner to both renegotiate the long-term transponder leases of its customers and restate its projected future earnings. (Operations insurance partially covered this loss.)

At the end of its life, a GSO communications satellite is typically empty of fuel, but still has most of its communications capacities operational. A couple of months before the fuel tanks are dry, the satellite's communications systems are turned off and the last of the fuel is used to boost the satellite a few hundred kilometers outward from the Clarke orbit, freeing up an orbital slot for future satellites. This move into a "graveyard orbit" is now required by the ITU of all GSO satellites. This, of course, costs in terms of the lifetime of the satellite. Roughly speaking, for every 100 km you wish to boost the satellite out of the way, you need to use enough fuel for a month of stationkeeping. (Several older satellites that reached their end-of-life before this requirement was established are still slowly wandering around the Clarke orbit, uncontrollable.)

Lower-orbit satellites, if they do not succumb to air drag and fall back to Earth on their own, are usually deliberately de-orbited to reduce the amount of orbital debris. This, of course takes fuel to accomplish, and is more difficult for MEO satellites than for ones orbiting closer to Earth.

Figure 14.6 In addition to controlling the satellites themselves, the overall operations of a satellite network need to be managed as well. This can include such tasks as assigning transponders to users, setting up cross-strapping if available, performing turnaround functions of specific signals, and any other items that control the flow of information through the system. This photograph shows the Network Operations Centre of the SES Astra fleet. (Photograph courtesy of SES Astra.)

14.3 Communication operations

The services a satellite is intended to provide dictate the design of the communications subsystem. The design and operation of the satellite dictate how much remote control of its operations is necessary. These tasks are usually the purview of a *Network Operations Center, NOC*. Figure 14.6 shows the NOC for the SES Astra satellite fleet.

In a case of a satellite that simply relays the feeds of many programming providers through the satellite to cable head-ends, not much change occurs from day to day. Typically, each transponder carries the same traffic for years. While there may be minor traffic changes, such as differing users for subcarriers alongside the video traffic, these usually do not require any changes to the satellite (the partitioning of the transponders' capacity is done at the uplinking earthstations).

At the other end of complexity, some satellites have been designed to act as versatile "jack-of-all-trades" communications systems; these may change the users of some transponders every few minutes, and if the satellite is so designed, require occasional cross-strapping of transponders. In such a case, an operations center is much more busy responding to the continually changing demands of customers, such as assigning transponders and timeslots to various users.

Satellites with onboard processing or switching capabilities may need more control from an NOC on Earth. This may include such tasks as switching on and off a cross-strap from one transponder to another, adjusting a matrix switch controlling the routing of input transponder channels to output transponder channels, or configuring data flow in a satellite to adapt to changes in the network that it serves. Other, more sophisticated satellites with OBP may be almost autonomous, preforming these tasks on-board.

Now we must leave the "Buck Rogers" part of the business, with its satellites and rockets, and get down to Earth.

Part 4

The Ground Segment

Chapter 15

Earth Stations: Types, Hardware, and Pointing

Earthstations are the mundane side of satellite communication. In the old days, they were typically huge, expensive, remote dishes that served as gateway points between the satellite systems and the public-switched telephone network, the PSTN. In the dawning days of satellite communication, they were so expensive that it was thought that each country would probably have only one or two (remarkably paralleling the predictions a century and a quarter ago that each city might have only one telephone!).

The very first commercial communications satellite, Telstar, was not in geostationary orbit. Thus, the huge antennas had to track the satellite accurately from horizon to horizon in less than half an hour to maintain communications through it. Requiring a dish to track a satellite greatly raises the cost of the dish. The faster the satellite moves across the sky and the more accurately the dish must track it, the more expensive the dish.

Thus, the industry migrated to geostationary satellites a few years after Telstar to make the dishes cheaper and simpler. This brings up what might be called the fundamental economic law of satellite telecommunications: Every dollar a satellite costs gets divided by the (potential) number of users on the ground, while every dollar a terminal costs gets multiplied by the number of users. This leads to designing more expensive satellites and less expensive terminals, a trend that has continued for more than three decades.

As antenna and receiver technology has gotten more sensitive during this time, we are now in an era in which direct contact is possible between individuals with handheld terminals and satellites, without the intermediary of huge gateway stations. More and more earth terminals look less and less like the "classical" image of a dish-shaped antenna. As an aside, while everyone commonly refers to the large dish as the antenna, this part of the station is actually nothing but a big reflector, and the true antenna is a small wire located at the focus of the dish.

(A note on terminology: North Americans tend to call these devices "antennas" while the British usually refer to them as "aerials.")

15.1 Types of earthstations

One way to categorize earthstations is by the diameter of the dish. While these classifications are not officially established or rigid, in general we can list the sizes as

- microterminals or miniterminals, used for DBS and VSAT services, with dish sizes ranging from around 0.5 m to 3 or 4 m in diameter;
- small dishes, typically 3–7 m in diameter;
- medium-sized dishes, typically 7–15 m in diameter;
- large dishes, 15–30 m in diameter.

(We might also mention the nondish antennas, such as the whip antennas used for some low-datarate applications, and the helical and patch type antennas used to receive GPS signals.)

Another way to categorize stations is by whether they are fixed or transportable. Most dishes in the world are fixed. The transportable stations range from half-meter DBS dishes on mobile homes to 1-m whip antennas for use with Little LEO systems, to 1- to 2-meter satellite newsgathering dishes on trucks to laptop-computer-sized terminals for use with Inmarsat satellites, to 3- to 4-m dishes used to broadcast events from stadiums.

It is also convenient to categorize earthstations into three major functional types, depending on their usage. While the categories have somewhat fuzzy boundaries, they serve as a useful way of thinking about the stations.

15.1.1 Single-purpose stations

The majority of the tens of millions of earthstations around the world are only trying to do one thing at a time. The major characteristic of such a station is that it typically is used to perform a single type of link with a satellite. Some may simultaneously be linking a few signals to a single satellite, but most do not. Such stations may be only uplinking (transmitting), only downlinking (receiving), or both.

The majority of antennas in the world are receive-only, such as for backyard television receive-only (TVRO, Fig. 15.1), direct broadcast services (DBS, Fig. 15.2), digital audio radio from satellite (DARS, Fig. 15.3), or reception by cable-television head-end stations (Fig. 15.4). Other examples of such stations might be those receiving a corporate teleconference or teletraining service, or receiving stock market ticker data for display in a brokerage office. Such antennas typically range from a few-inch-long rod on the roof of a car, to dishes or flat-plate (phased array) panels half a meter to a meter in size, up to several-meter-wide dishes for TVRO and cable head-ends.

A very few earthstations are transmit-only, most commonly for remote telemetry services such as monitoring pipeline flow, utility stations, or even reading

Figure 15.1 A mesh dish of a type which might be used to receive TVRO for a home or SMATV for an apartment building.

Figure 15.2 Small offset home dish for DTH video reception. (Photograph courtesy of SES Astra.)

electrical meters from satellite. Such stations often don't look anything like the general notion of a satellite earthstation.

Another example is the very small aperture terminals (VSATs, Fig. 15.5) used for point-of-sale systems for retail firms, hotels, and gasoline stations. Two-way corporate television would also fall in this category.

With the advent of the Inmarsat system in the early 1980s came the first direct satellite-to-end-user terminals, at first on ships. With technology improvements to both satellites and terminals, we have come to briefcase-sized individual terminals that

Figure 15.3 A car-mounted antenna for reception of SDARS. (Photograph courtesy of XM Satellite Radio.)

Figure 15.4 Typical receive-only dishes that might be used at a television station to pick up contribution feeds.

are self-contained with antenna, transmitter, receiver, telephone and data equipment, all in one small package designed specifically to use Inmarsat satellites only.

Handheld satellite telephone services, from such firms as Iridium and Global Star (Fig. 15.6) are available. Again, these are single-purpose terminals designed to work only with a single satellite constellation. The smaller data-only or navigation systems, such as OmniTRACS, are also single-company low-speed data systems direct to the end user. Low-speed SCADA systems don't even look like satellite equipment at all.

Figure 15.5 Offset VSAT two-way dish used at a retail store for point-of-sale communications with a corporate hub.

15.1.2 Gateway stations

Gateway earthstations (Fig. 15.7) still serve as the interface points between terrestrial networks and satellites. Often such earthstations consist of many, perhaps dozens, of dishes, and are connected to terrestrial networks by a variety of transmission technologies, including wires, coaxial cables, waveguides, optical fibers, and microwave towers. They may also serve as transit and relay points between satellites.

In gateway stations, the major activity is signal processing. A single gateway earthstation is typically at the confluence of many, many terrestrial signals simultaneously arriving from perhaps a whole country. Some are analog, some digital; some are telephone calls, others television signals, still others are datastreams. They are intended for destinations all around the globe, so some need to be directed to one satellite, others to other satellites. They arrive in various formats, various levels of multiplexing, and using different telecommunications standards (such as NTSC or PAL or SECAM television) and need to be converted, demultiplexed, deformatted, and converted to other standards appropriate to the satellite system they will go on and their destination. Thus, it is signal manipulations, not just uplinking and downlinking as for a single-purpose station, that is the focus of activity at gateway stations.

Some such stations are part of, and sometimes owned by, the satellite system owner, or its cooperating firms. Some are independent. Antennas linking with a particular satellite system operator's fleet of satellites must be designed, manufactured,

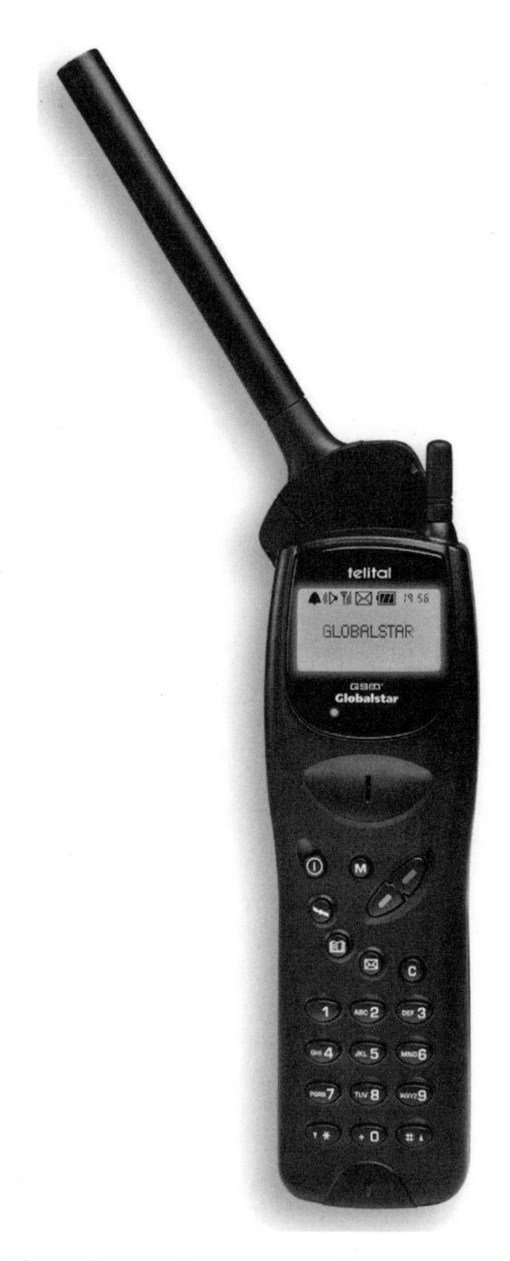

Figure 15.6 A GlobalStar satellite telephone. (Photograph courtesy GlobalStar.)

and operated in conformance to certain standards promulgated by the satellite fleet owner, and come in a range of sizes and capacities called *station types*. Many manufacturers can produce "type-approved" earthstations that are capable of operating properly with the specific satellite system. Thus, you will see mention in advertisements for such things as an "Intelsat Standard-B" earthstation, or an "Inmarsat Standard-C" terminal, or GlobalStar-approved satellite telephone.

Figure 15.7 The "antenna farm" of a gateway and network control dishes for the SES Astra satellite fleet. (Photograph courtesy of SES Astra.)

15.1.3 Teleports

A variant of the gateway station is the teleport (Fig. 15.8). These are typically operated by firms not specifically owned by or part of a specific satellite system. Teleports are a way of aggregating a market for satellite telecommunications, and of avoiding the difficulties of linking to satellites from congested urban business sites.

Teleports began mostly as a bypass technology, a way to go around the monopolistic terrestrial telephone networks. They were also useful for firms that could not justify the expense of putting in their own dishes, perhaps because their need for communication was not great enough, or perhaps because the line of sight from the office building to the satellite they wanted to use was blocked by a mountain or another office building. Teleports are often, but not always, located on the edges of cities, and are away from the interference of other earthstations, microwave towers, and similar interfering signals. Often, some point within the nearby urban business center will be set up as a communications hub for users within that area. If a company in the city needs links to satellites, it need only get

Figure 15.8 A major teleport with dozens of dishes, capable of linking to a wide range of satellites. Such teleports also offer value-added services such as format conversion, turn-around, and more. (Photograph courtesy of Verestar Washington Teleport.)

its signal to the hub, from where it is then relayed by fiber or microwave to the teleport outside the city.

In the early days, anyone who set up two or more dishes could call themselves a teleport, and often did. As earthstation equipment became smaller and easier to place right at a firm's office building, and the telecommunications industry became liberalized, some of the smaller teleports went out of business, while the larger ones grew. Today, teleports range in size and capacity from a few dishes to dozens of dishes that range in size from fractions of a meter to many meters in diameter to accommodate a wide range of demands from users. They often have dishes conforming to the standards of many satellite operators so as to be as versatile as possible.

Teleports also earn a lot of money from "value-added services," such as format conversion, encryption, production and post-production, turn-around services (receiving at one frequency from a satellite, and linking back up to another satellite in a different band), and other tasks. Some teleports offer such additional services

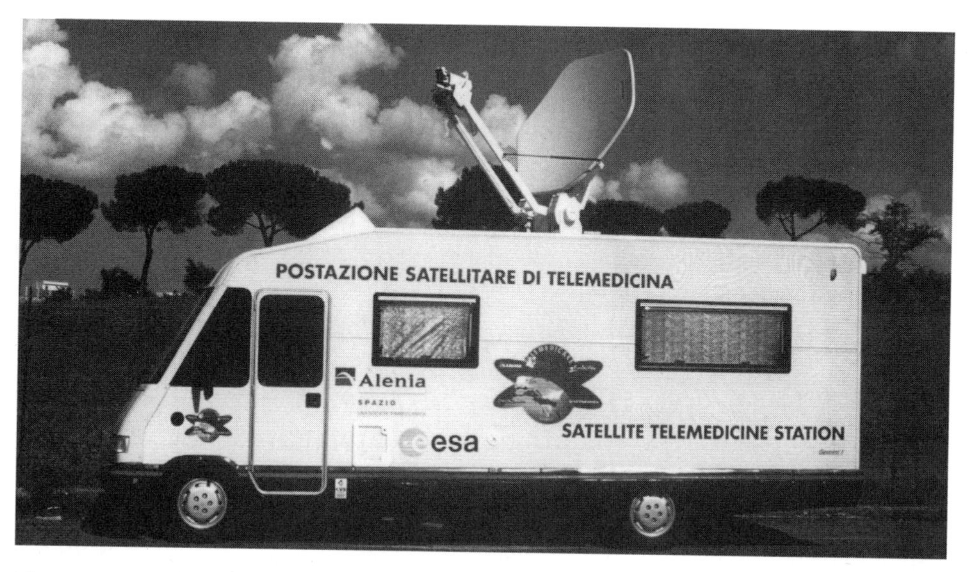

Figure 15.9 A transportable dish mounted on a van to provide telemedicine services in rural areas. (Photograph courtesy of Alenia Spazio.)

as leasing SNG trucks or services, transportable uplinks (Fig. 15.9) for temporary events, or even full production facilities.

15.2 Environmental effects on antennas

Most earthstation antennas are located in the open air. They are subject to various environmental conditions depending on their location. Such effects include temperature, solar radiation, wind loading, and possible seismic events. In extreme climates, particularly in polar areas, they may be enclosed in radomes made of radio-transparent material to protect them from the elements.

The major constant environmental factor for unprotected dishes is the wind. The dish surface is basically a huge sail, and can be subjected to tons of force in strong wind conditions. A 5-m (diameter) dish facing into a 40-mph wind feels a force of almost 1400 pounds. In a high enough wind, the dish can be destroyed. At lower windspeeds, this dish may point slightly away from the satellite.

The stability of a dish is very important, and the larger the dish and the higher the frequency band, the smaller the beamwidth of the dish, so the more accurate the pointing must be. In light and moderate winds, the main effect is to vibrate the dish and cause it to point slightly away from the satellite in use. If a dish wobbles in the wind, this reduces the signal to and from the satellite, and potentially increases interference transmitted and received. (You can actually see this in action sometimes when a news reporter covers an approaching storm and happens to have the uplinking dish in the field of view. You will see the dish vibrate, and the quality of the picture on your television change accordingly.)

Typically dishes are rated for pointing accuracy during operations in windy conditions. Every dish is also rated for some maximum wind velocity, and this number figures in the insurance policy issued for the dish. There are limits on other environmental conditions as well. For example, a typical x-y mount 9.3-m diameter dish for C-band use might specify the environmental conditions such as in Table 15.1.

Table 15.1 Typical environmental requirements for a 9.3-m dish antenna.

Operating Temperature:	–40° to 125° F
Operational Wind Load:	45 mph (72 km/h) steady; 65 mph (105 km/h) gusts.
Survival Wind Load:	125 mph (200 km/h) in any orientation
Seismic forces:	1g vertical and horizontal (≈ Richter 8.3)
Rain:	4 inches/h (102 mm/h)
Solar Radiation:	360 BTU/sq.ft./h (1135 W/sq. M.)
Atmosphere:	moderately corrosive coastal and industrial environment.

In the tropics, where rain is frequent and dishes are pointed almost vertically, some provision must be made to allow rain to flow out of the dish. This may be as simple as using a perforated dish surface, or may be made more complex with actual drains.

In latitudes with real winters, snow can accumulate on the surface of the dish. One effect of this is to alter the reflectivity properties of the dish, and thus the signal quality. A more catastrophic effect is to build up a mass of snow that actually causes mechanical failure of the dish or its mounting.

To combat this, dishes in cold climates are equipped with deicing methods. You can accomplish this by such means as blowing hot air over the dish, shining infrared lights on it, or embedding heating pads in the dish itself. Sometimes a plastic tarp is stretched across the aperture of the dish to keep out both rain and snow.

15.3 Antenna pointing

Except for omnidirectional antennas, an earthstation must be pointed at the satellite to which it is linking. Terminals communicating with NGSO satellites typically use omnidirectional antennas, so pointing is not an issue. But for GSO satellites, the dish (or flat plate) must be oriented so that it faces the satellite and is held there accurately.

To determine how to point a dish to link to a geostationary satellite, you need two numbers to either plot on a graph or plug into formulas. To see the geometry of the problem, it helps to be able to envisage what the Clarke orbit—the geostationary arc—would look like if you could see it in the sky from your earthstation.

Some years ago, a "space sculptor" (whatever that is) proposed placing 360 brightly lit satellites spaced every degree of arc around the Clarke orbit. This bad idea was quickly shot down by astronomers, the satellite communication industry, and the military services. But suppose for a moment that he had accomplished this grandiose celestial spectacle. If you went outside at night, what would you see?

From a site in the northern hemisphere, you would see the first bright point of light on the horizon just south of due east. The line of lights would rise upward and toward the right—toward the south—reaching a maximum angle above the horizon due south of you, and then declining to the right—westward—reaching the horizon again a few degrees south of due west. This is the GSO arc (Fig. 15.10). (In the southern hemisphere, the line would start a few degrees north of east, rise to the left culminating at due north, and decline to hit the horizon a few degrees north of due west.)

If you were at the equator on Earth, that line of lights would pass directly overhead. If you were at one of the poles, it would pass below your horizon and be invisible. Thus, the farther you are from the earth's equator, the lower the arc will cross your sky. A rough approximation of the maximum angle you can ever see a

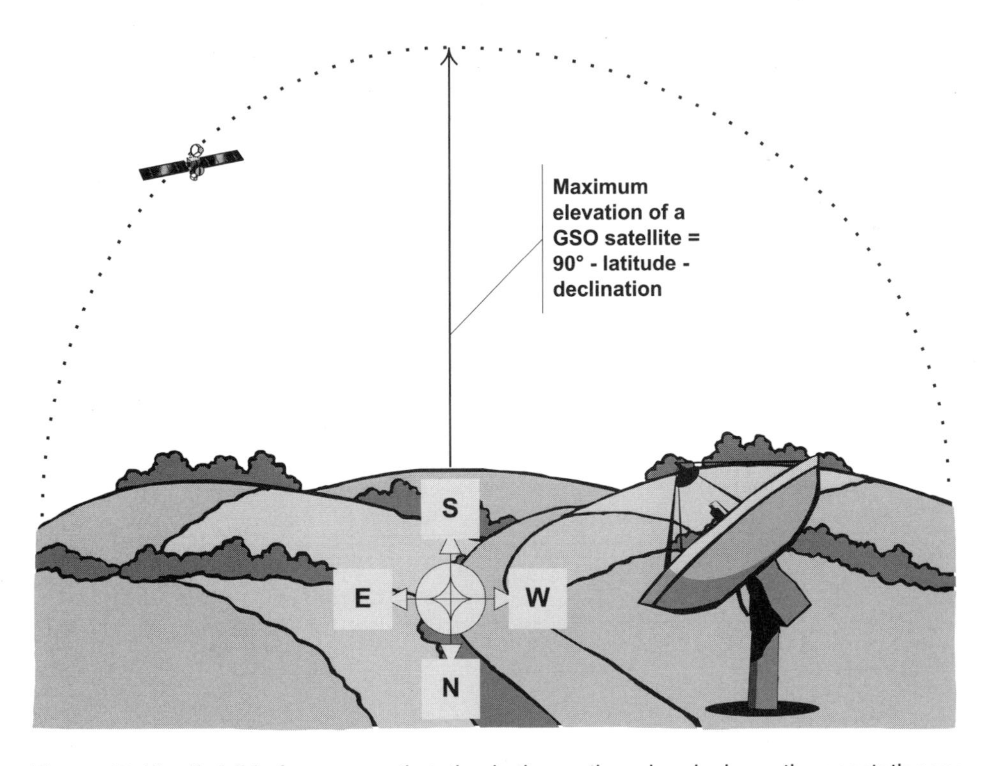

Figure 15.10 If visible from an earthstation in the northern hemisphere, the geostationary arc would touch the eastern horizon a few degrees south of due east, rise upward to the right, reaching a maximum elevation above the horizon due south, and slope downward to the west, meeting the horizon a few degrees south of due west. The higher the latitude of the earthstation, the lower the maximum elevation. The directions given here would be opposite in the southern hemisphere.

GSO satellite from your location is calculated simply by subtracting your latitude from 90°. Thus, say for an earthstation at latitude 40° (either north or south), the highest you could ever see a GSO satellite is 50° above the horizon.

Actually, because GSO satellites are "only" 36,000 km from Earth, and not at infinity, they really appear a few degrees lower in the sky than this. This small permanent equatorward tilt (southward in the northern hemisphere; northward in the southern hemisphere) is called *declination*, and is a function of your latitude (Fig. 15.11). At the equator, the declination correction is 0°; at latitude 40° it is 6.3°, so at 40° latitude, the highest you really ever see a GSO satellite is 43.7°; at latitude 60°, the declination reaches 8.2°. Thus, GSO satellites can never actually be seen from either pole on Earth; in fact, they are below the horizon above latitudes about ±82°.

When earthstation engineers install a dish, this declination adjustment is set into the mounting permanently, and you don't have to be concerned about it again (unless you move the dish to another latitude).

Since all GSO satellites are over the earth's equator, their latitude is zero. You can look up a satellite's orbital slot longitude. You then need to know your earthstation's latitude and longitude to calculate the difference between the satellite's longitude and your earthstation's longitude. (Check that both longitudes are expressed the same way, such as east longitude or west longitude.)

The two numbers you need to point a dish are the compass direction toward the satellite, called the *azimuth*, and the satellite's angle above the horizon, called the *elevation*. Elevation is a number of degrees between 0°, at the horizon, and 90°, which is directly overhead at the *zenith*. (Of course, you will never see a satellite at the zenith unless you are at the subsatellite point along the equator.) As we will see later in Chapter 18, the elevation is important because it determines how much air your signal passes through on its way to and from the satellite. Depending on frequency, the atmosphere can have severe negative effects on the quality of your signal.

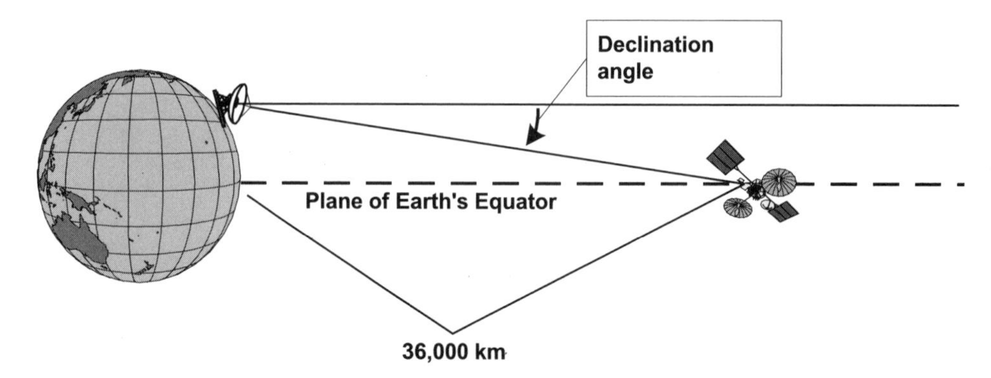

Figure 15.11 Declination is the slight equator-ward shift necessary in aiming an earthstation due to the fact that Clarke orbit satellites are only 36,000 km away, not at infinity. The declination angle is a function of earthstation latitude, ranging from 0° at the equator to about 8° at the highest latitudes that can see the Clarke orbit.

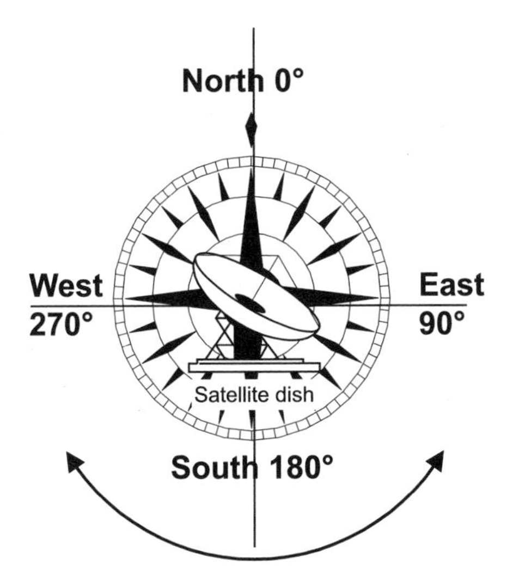

Figure 15.12 True azimuth is measured from true north, through east at 90°, then south at 180°, through west at 270°, and back to north at 360°. If pointing angles are specified using true azimuth, for earthstations in the northern hemisphere the angle will always be between 90° and 270°. For earthstations in the southern hemisphere, the angle will be between 270° and 360° or 0° and 90°.

Azimuth can be measured in two ways. One is called *true azimuth*, and is the method a geographer or navigator would use (Fig. 15.12). True azimuth has 0° at north and goes clockwise (as seen from above) to 90° at east, 180° at south, around to 270° at west, and up to 360°, which is the same as 0°—back at north again. So, if your earthstation is in the northern hemisphere, a GSO satellite's true azimuth will always be a number between 90° and 270°. (If you are in the southern hemisphere, it will always be either between 270° and 360° or between 0° and 90°.)

An alternative way of measuring direction is called *relative azimuth* (Fig. 15.13). In the northern hemisphere, the zeropoint for measurement is due south; in the southern hemisphere, it is due north. Using relative azimuth, the value will always be less than 90°.

Uplinks to satellites closer than 5° to the horizon are strongly discouraged because of the greatly increased probability of the earthstation's transmission interfering with terrestrial receivers.

In addition to just pointing at truly geostationary satellites, some earthstations must be designed to work with satellites that are in slightly inclined orbits. During one day, such a satellite swings slightly above and below the line of the equator in the sky. Further, unless you are looking due south (in the northern hemisphere; the opposite in the southern hemisphere), the line of the equator is tilted as seen from an earthstation. Therefore, both the elevation and azimuth of the satellite slowly change by several degrees during each day.

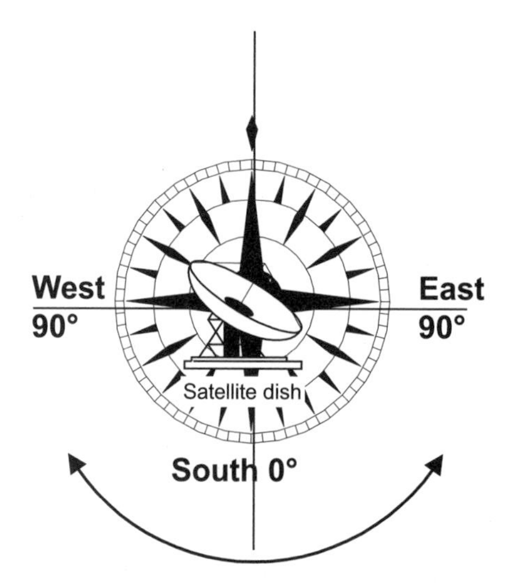

Figure 15.13 Relative azimuth for pointing a dish is measured relative to south for earthstations in the northern hemisphere, and relative to north for earthstations in the southern hemisphere. The angle is always within 90° of south or north.

The antenna mounting for such earthstations must have the capability to track such satellites accurately. As seen from an earthstation, the "figure-eight" diurnal path of the satellite is not oriented vertically, but is tilted. Thus, any tracking antenna must move slightly both side-to-side and up-and-down. Tracking is usually done automatically by having circuits monitor signal strength and adjusting the pointing of the dish to maximize it. This of course increases the complexity and cost of the earthstation. For a medium-sized earthstation, it can add tens of thousands of dollars to the cost of the station. The tracking must be accurate to less than a tenth of the beam size of the dish. As we will see in the following chapter, this beam size depends both on the physical size of the dish and on the frequency in use.

15.4 Antenna mountings

The support structure for an antenna can be either fixed—when only one satellite is of interest—or moveable—when the earthstation needs to look at more than one satellite. Even moveable mounts usually do not allow a full "horizon-to-horizon" pointing for the dish because this is not usually necessary.

A fixed mounting, such as shown in Fig. 15.14, can be anything that holds the dish pointed firmly at the satellite. Such systems range widely in complexity, but all must be able to hold the dish accurately even under the effects of wind. Fixed mounts include such things as small dishes strapped to a chimney, attached to

Figure 15.14 An earthstation dish mounted on a fixed mounting. The dish is aimed at the satellite it is to link with, and fixed in position.

metal poles in the ground, bolted to a windowsill, or just propped up against a building. Larger dishes have better mounts that are better at resisting the much greater effects of wind and are easier to adjust.

Antenna mounts intended for installation on a roof can be classified as "penetrating" or "nonpenetrating," depending on whether they are intended to be bolted into the roof structure. Nonpenetrating mounts may be held down simply by the weights of concrete blocks or sandbags. Over the past few decades, owners of office buildings have found a new source of revenue renting out roof space by the square foot for building tenants who want an antenna.

Moveable mounts have the ability to aim the dish anywhere along all or part of the Clarke arc. There are two basic ways of accomplishing this.

15.4.1 Elevation-azimuth mounts

This type of mounting works much like the familiar camera tripod: it has two axes of rotation that can move the dish up and down in elevation and side-to-side in azimuth. They are sometimes called *x-y mounts* or *elevation-azimuth mounts*.

There are some variations in design that have the horizontal axis contained within the vertical axis or vice versa, but they are operationally similar, and work equally well. For antennas that are infrequently repointed, the shift may be accomplished by manual movement. For more frequent movements of the dish, each axis is

Figure 15.15 A large earthstation dish with an elevation-azimuth, or xy mount. To look at different satellites, the dish can move in elevation around a horizontal axis and move in azimuth around the vertical axis. (Photograph courtesy of SES Astra.)

connected to a motor and geartrain that is controlled from the station's control room. Changes around both the azimuth and elevation axes are needed when shifting the view from one satellite to another, since the GSO arc crosses the sky at an angle.

Figure 15.15 shows an example of such a mounting.

15.4.2 Equatorial mounts

These mountings take advantage of the fact that all GSO satellites are over the earth's equator. Such a mount has one major axis of rotation, called a polar axis, which is exactly aligned parallel to the earth's axis. If the dish is then attached perpendicular to this axis, and then adjusted for declination, it will automatically point along the Clarke arc in the sky. Thus, when moving from one satellite to another,

Figure 15.16 A limited-motion equatorial mounting for a dish. The metal tube just beneath the dish that is angled from lower left to upper right is the polar axis, and is aligned carefully to be parallel to the axis of Earth's rotation. A dish attached to this axis at right angles, with the proper declination adjustment for the latitude of the earthstation, will thus point toward the Clarke orbit. The motor and screw mechanism below and to the right of the polar axis allows the dish to scan along a part of the arc.

turning is necessary only about this single axis, making the pointing a bit simpler. Figure 15.16 shows a typical equatorially mounted satellite dish.

If the dish is to be used with an inclined-orbit satellite, it must also have the ability to "nod" up and down several degrees perpendicular to the GSO arc to follow such a satellite over a day's movement. This adds a slight complication and increased expense to the dish.

In the next chapter, we will explore the properties of the dish itself that contribute to its ability to link to a satellite.

Chapter 16

Earthstations: Antenna Properties

The way satellite "dishes" handle incoming and outgoing radio waves has a major effect on the signal strength and interference levels we deal with when sending and receiving signals. The properties of dishes apply both to earthstations and antennas on the satellites. This chapter will consider how the mechanical properties of the dish control the ability of the antenna to transmit and receive radio waves. The following chapter will then deal with what happens electronically to the waves before and after they have bounced around the antenna structure itself.

16.1 Dish antennas

The most common type of equipment used at earthstations and on the satellites to receive and to transmit radio waves is the "dish." Three other kinds of hardware, the flat-plate (phased array) antenna, the horn antenna, and the omnidirectional antenna, will be discussed later.

Mathematically, the shape of the dish is a paraboloid of revolution; that is, a cross-section of the dish through its center would have the shape of a parabola (Fig. 16.1). This is the only geometric shape that has the property that parallel waves coming into it are reflected and brought to a *focal point*, also called the *focus*, or *prime focus*, in front of the dish. Since the geometrical properties are symmetrical, any waves sent out from the focus toward the dish are reflected and beamed outward into a cone-shaped region of space. We can use the analogy of an ordinary flashlight, which takes the small amount of light coming from a bulb and concentrates it into a small beam. An analogy for a receiving system is the reflector telescope commonly used by both amateur and professional astronomers.

The dish is not the antenna itself, although very frequently it is referred to as such. The dish part is just a reflector and must be manufactured to high accuracy. The surface accuracy of a dish should deviate from a perfect parabolic shape by no more than 1/50th of the wavelength in use. That corresponds to an accuracy of better than half a millimeter for dishes working in the Ku-band. The dish must maintain this accuracy no matter which way it is pointed in winds up to 50 mph.

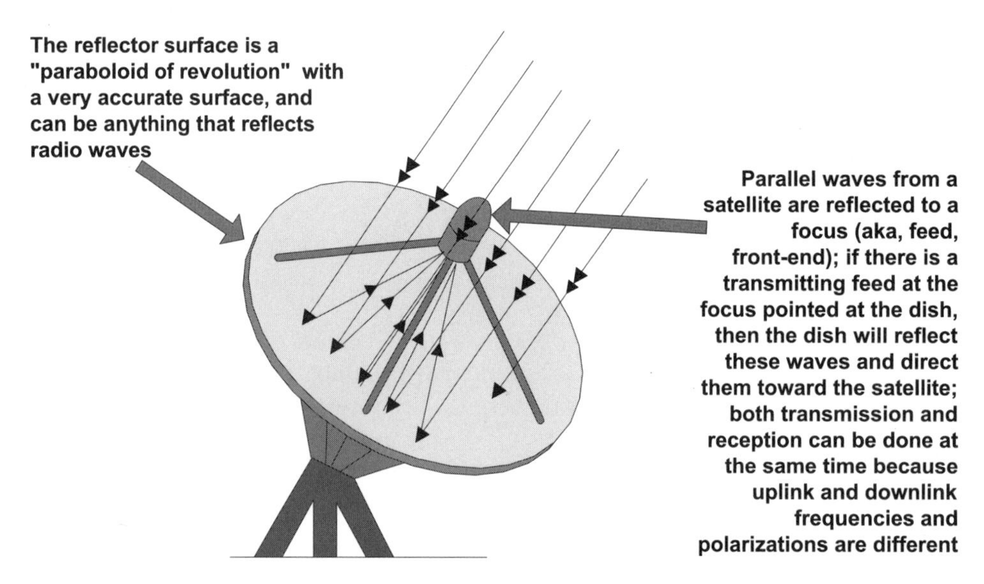

Figure 16.1 The shape of a dish antenna is a paraboloid. It must be accurately shaped and made of a material that will reflect radio waves. The waves are reflected to a focus in front of the dish, where the actual antenna is located. Because the received signal strength even after focusing is only a few trillionths of a watt, a low-noise amplifier is also located there to boost the power enough to send it inside to an office or home. The larger the dish, the more waves it collects, producing what is called dish gain.

The real antenna, the electronic component that actually receives the waves coming in or transmits those going out, is typically a small piece of wire located at the focus. It is usually located in a shielded box to keep it from receiving unwanted signals from other directions, and is sized in proportion to the wavelength being used. (For small dishes, another box for a high-gain amplifier is located just next to this antenna wire, and the whole package of equipment at the focus is called the *front end* or the *feed*; but more on this in the following chapter.)

The material of the surface of the dish can be anything that reflects radio waves. Some dishes are solid metal, often aluminum for lightness. Often the metal is perforated to reduce the mass and wind resistance of the dish and to let rain drain off. The size of the holes has to be kept much smaller than the wavelength in use or the dish will become less efficient. Some dishes are made of a screen-like metal mesh; others are simply metallic paint sprayed on a fiberglass or another substrate.

The accuracy with which the dish is made contributes to its *efficiency* as a reflector. Dishes made for use with high frequencies (i.e., short wavelengths) must be made to a higher geometrical accuracy than those for lower frequencies. No dish is perfectly efficient, and a usual figure for a dish efficiency is 55% (also written as 0.55). Most dishes have an efficiency that ranges from 30% to 80%.

16.1.1 Feed configuration

The way the waves coming in and going out "bounce around" is called the *feed configuration.* Many dishes use only a single bounce, with incoming waves reflecting off of the dish surface to the focus in front of the dish, where the antenna is located. Such a configuration is called a *prime-focus configuration.* When the dish is used to transmit, the transmitting antenna at the focus beams waves toward the dish, bouncing them off to space. This is the simplest arrangement.

Many dishes have the waves make more than one bounce. This is generically called a *folded*, or *folded-optics*, system. (It is called "optics" because the systems originated with optical telescopes for astronomers.) The advantage of this is that the whole dish+feed system is more compact. There are several different folded configurations, but all have at least one *secondary reflector*, also called a *subreflector*, located out in front of the dish to redirect the waves.

A common dual-reflector antenna, called *Cassegrain* (after the French astronomer who invented such a system for telescopes) has a convex subreflector positioned in front of the main dish, closer to the dish than the focus (Fig. 16.2). (If you want the geometrical details, this small reflector is a hyperboloid.) This subreflector bounces the waves back toward a feed located on the main dish's center, sometimes behind a hole at the center of the main dish. Sometimes there are even more subreflectors behind the dish to direct the waves to the feed for convenience or compactness.

Another common feed configuration is the *Gregorian* (named after another astronomer). This system has a concave (ellipsoidal) secondary reflector located just beyond the prime focus (Fig. 16.3). This also bounces the waves back toward the dish.

Figure 16.2 In a Cassegrain dish, instead of having the front-end electronics at the main focus, a convex secondary reflector, or subreflector, is placed in front of where the focus would be. This reflects the waves back toward the dish, so that the heavy electronics can be mounted on or even behind the dish. (Photograph courtesy of SES Astra.)

Figure 16.3 In a Gregorian dish, a concave subreflector is placed near the focus of the dish, reflecting the waves back toward the dish. The front-end is seen between the subreflector and the main dish. (Photograph courtesy of Alenia Spazio.)

In the configurations mentioned above, the waves come in (or go out) roughly parallel to the axis of revolution of the main dish. Technically, this is called an on-axis design, but the term is rarely used. What you will hear about are designs called *off-axis*, or *offset*, in which the waves are made to bounce at an angle to the axis by tilting the secondary reflector (Fig. 16.4). This often results in a more compact design, and is common for such small dishes as those used for VSATs, satellite newsgathering, and DBS.

Figure 16.4 An offset dish. The front-end is not aligned with the axis of the main dish. Such a configuration results in a more compact structure for small dishes.

16.2 Horn antennas

For some simple, low-gain applications, antennas may be what are called horns. There are two types, those that are pyramidal in shape, and those that are conical. Often, the feed for larger dishes is a horn antenna pointed at the dish's reflecting surface. They have also been used for global beams (those from a satellite covering the entire visible Earth) and other special purposes. There are also horn-reflector antennas for special uses.

16.3 Phased-array antennas

Some antennas use what is called a *phased-array.* They are also sometimes used on board satellites to provide highly defined beams to Earth. Phased arrays work by electronically combining the outputs of many small antennas in such a way that the signals coming into each are combined to reinforce each other.

One example of such an antenna is a flat plate antenna. It has gain and directionality characteristics comparable to a dish antenna. (In fact, a 1-m square flat plate antenna has about 2 dB more gain than a 1-m-diameter dish because it has a larger area.) These antennas are often used to receive DTH broadcasts.

Phased-array antennas do not have to be flat. They can be made "conformal" to the curved surface, such as the fuselage of an aircraft, or as a "patch" built into a small receiver.

Most phased-array antennas must be pointed toward their targets, just like common dishes. There do exist electronically steerable phased-array antennas that can sweep a beam over a wide range of directions by altering the phases of the signals in the tiny component antenna cells. Because of the cost of such systems, these technologies are mostly confined to military communications and on some satellites with directional steerable beams, but we might expect some such systems to enter the civilian satcomms hardware as the cost decreases.

(A small linguistic note: When small flat square aerials were being developed for British DBS services, they coined the term "squarial." Thankfully, this usage seems to be waning!)

16.4 Yagi antennas

Another type of antenna sometimes used for communication with a satellite is the Yagi (also called the Yagi-Uda). It consists of a number of rods, sized proportionally to the wavelength in use, set perpendicular to the direction of reception. They are most commonly used for frequencies below C-band. (In everyday use, they are the most common type of outdoor television antennas.)

Figure 16.5 A large dish capable of linking to many satellites simultaneously over a wide arc, up to 60°, of the Clarke orbit. The horizontal cross-section of the dish is a sphere, while the vertical cross-section is the usual parabola. This focuses the beams from different satellites along the arc to different locations in the curved structure mounted in front of the dish. In this structure there would be electronics to transmit or receive from each desired satellite. This particular dish is called a "Simulsat" by its manufacturer. (Photograph courtesy of ATCi.)

16.5 Multibeam antennas

There are times when a company needs to link to several satellites simultaneously, but the organization may not have the ability to have multiple dishes dedicated to specific satellites. The answer to this problem is a *multibeam dish* such as the one shown in Fig. 16.5).

A multibeam dish is typically wider than it is high. The geometrical cross-section taken perpendicular to the GSO arc is the usual parabola; the other cross-section is a circle. This allows satellites in different directions to bounce waves off of the dish to converge to several foci at different points in front of the dish for each satellite, with the trade-off of having a slightly less efficient dish. By placing a feed at each of these various foci, the single dish can work with many satellites. Of course, separate electronics are needed for each satellite's signal. Several firms make such multibeam dishes that simultaneously work with satellites along as much as 60° of the GSO. The most common users of such dishes, especially in receive-only use, are television stations and cable headends. Another common use for multibeam antennas is for TT&C earthstations, monitoring several satellites at once.

16.6 Nondirectional antennas

Although directional antennas have the ability to concentrate their powers in a narrow range of directions (a property called gain, to be discussed below), there are circumstances in which this gain must be sacrificed for the ability to pick up and send signals to a wide range of directions.

One such case is onboard satellites. Most of the time their high-gain antennas are pointed at Earth. But during orbital insertion, some orbital maneuvers, and possibly during emergencies involving the orientation control, the satellite may not be pointed at Earth, yet it is vitally important to get signals to and from it for telemetry and control. Thus, all satellites have an (approximately) omnidirectional antenna, commonly of what is called a bi-cone design. These TT&C links usually operate in the VHF region of the radio spectrum.

Another case in which reception from a wide range of directions is important is in the case of mobile satellite services into handheld terminals. It would be impractical to require that a user continually reorient the terminal's antenna toward a moving satellite, especially considering that more than one satellite may be used during a single call. Like the cellular telephone handsets before them, these satellite handsets have omnidirectional antennas.

16.7 Dish properties: directionality

An antenna does not send its radiation outward in a perfectly cylindrical beam like a laser. Instead, the beam covers a range of angles to either side of the exact direction that the antenna is pointed. The direction of the center of the beam is often called the *beam axis* or *boresight*. This range of directions on either side of the beam axis is called the *main beam* or *main lobe* of the antenna. You can see this effect by shining a flashlight on a wall: the spot will be brightest in the center, fading off to the sides.

Furthermore, because of a physical phenomenon called diffraction, every antenna also sends some small amounts of radiation into directions outside the main lobe; these smaller beams are called *sidelobes*. In fact, there is some radiation actually sent in directions almost opposite to where the antenna is pointed; these are called *backlobes*. One of the tasks of the antenna designer and manufacturer is to minimize these sidelobes and backlobes, although it is not possible to eliminate them totally. Figure 16.6 shows the typical directionality characteristics of a dish.

The number measuring the directionality of a dish, i.e., the width of the main beam, is the *half-power beamwidth,* or *HPBW.* It is the total angle, centered on the boresight direction, within which the sensitivity of the dish is at least half of that in the boresight direction. In other words, within the range of directions measured by the HPBW, the dish is up to half as good at picking up signals from the sides, and thus susceptible to interference. For a transmitting dish, within this HPBW angle, the dish is sending out a signal more than half as powerful as the one aimed at the intended satellite. Again, shine a flashlight on a wall, and measure the angle within which the light is at least half as bright as in the center of the beam, and you are measuring the HPBW. See Fig. 16.7.

A simple formula to calculate this angle is given in Appendix E.9. Without going into the mathematical details, what you should remember is that the gain of a dish is higher 1) if the dish is bigger, and 2) if the frequency is higher.

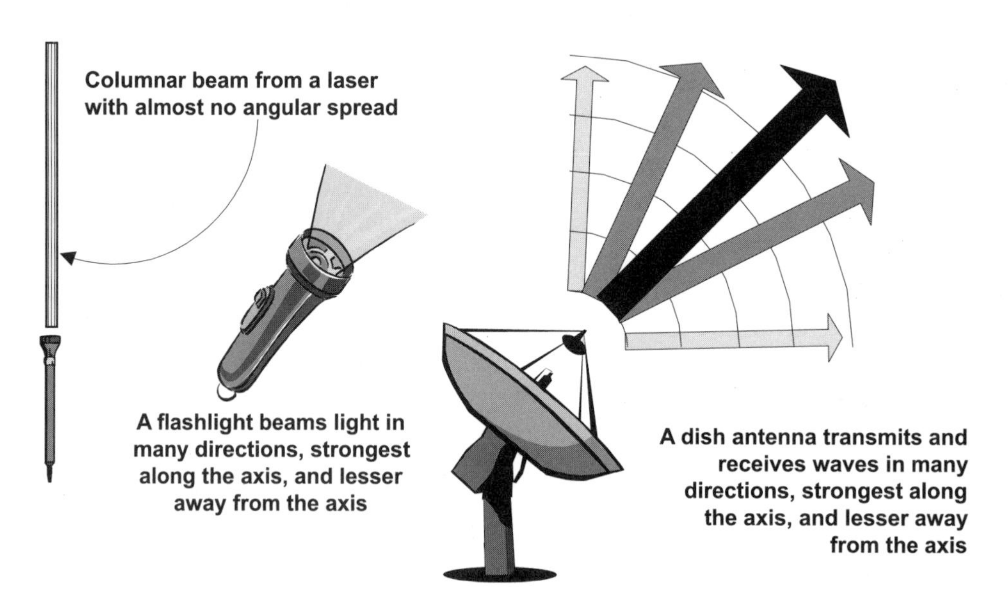

Figure 16.6 Like a flashlight, a transmitting dish sends most of its energy toward where it is pointed. But some goes off to the sides in what are called sidelobes. Not only do sidelobes waste energy, the also can cause interference with nearby receivers. When receiving, the sidelobes make the dish slightly sensitive to signals coming in from sources other than the satellite it is pointed at, causing this earthstation to pick up interference. Thus, a major dish design goal to is minimize sidelobes.

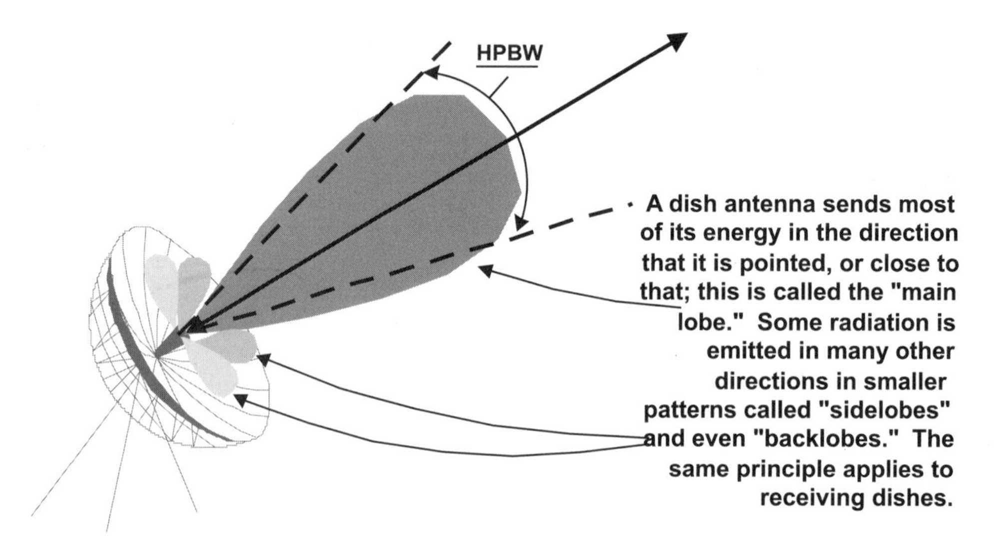

Figure 16.7 The measure of the spread of the main beam of a dish, either receiving or transmitting, is the angle over which the dish is at least half as powerful or sensitive as it is on-axis. This angle is called the half-power beamwidth, or HPBW. The HPBW decreases as dish size and frequency increase.

Thus, the problem caused by the breadth of the main lobe, and the existence of the unwanted sidelobes and backlobes, is interference. When the dish is receiving, it is not only picking up the wanted signal from the satellite that the dish is aimed at, it is also picking up small amounts of unwanted signals—interference—from other directions. In these other directions may be nearby satellites, other nearby earthstations, nearby microwave towers, or other electronic equipment. There will also be trees, buildings, clouds, people, and even the rooftop or ground that the dish is mounted on. All of these create interference, raising the noise level, *N*, to reduce your signal quality measurement, *C/N*. For an earthstation, this is called *downlink interference*.

Furthermore, when the dish is transmitting, it is sending unwanted radiation in these other directions, potentially causing others interference. These others may be adjacent satellites in orbit or nearby terrestrial equipment. This is called *uplink interference*. This latter problem becomes greater the lower the elevation to which the dish is pointed.

16.8 Dish properties: gain

One of the most important operational properties of a dish is its ability to intercept a lot of the weak radio signals coming from a satellite and concentrate them at the focus of the dish. This ability is called the *gain* of the dish. This is necessary because the signals are so very weak, typically trillionths of a watt. (After this, the electronics of the front end take over and amplify the signal still more, as we will detail in the next chapter.)

A dish acts just like a pan in the rain: the bigger the pan, the more rain it collects, i.e., the higher the gain. Another analogy is the use of a magnifying glass to burn a hole in a piece of paper. When simply held up to the sunlight, the paper may get warm but will not burn. However, if you capture a lot of sunlight by using the larger surface of a magnifying glass, and concentrate all of this energy in a small focus, you intensify the light and heat and can make the paper burn. The flashlight analogy is also apt.

Do the following thought experiment, shown in Fig. 16.8: Pretend you are at the center of a perfectly spherical room holding a light source that shines in all directions. Such a source is said to be *omnidirectional*, or *isotropic*. If you could measure the brightness produced by the lamp on the walls of your spherical room, it would be the same everywhere. Now, take a big, flat mirror and hold it just to one side of the lamp. What happens to the light in the room? It is no longer isotropic. On the side behind the mirror the light goes away; and where does it go? It is reflected to the other half of the room, doubling the brightness over there. Talking in decibels, we have achieved a gain in brightness of 3 dB! Also note that we have reduced the area that the light covers.

To produce the same doubled brightness without a mirror, you would have to use an isotropic source twice as bright. That is why the gain of a dish is measured in units sometimes called *dBI*, meaning dB compared to an isotropic source.

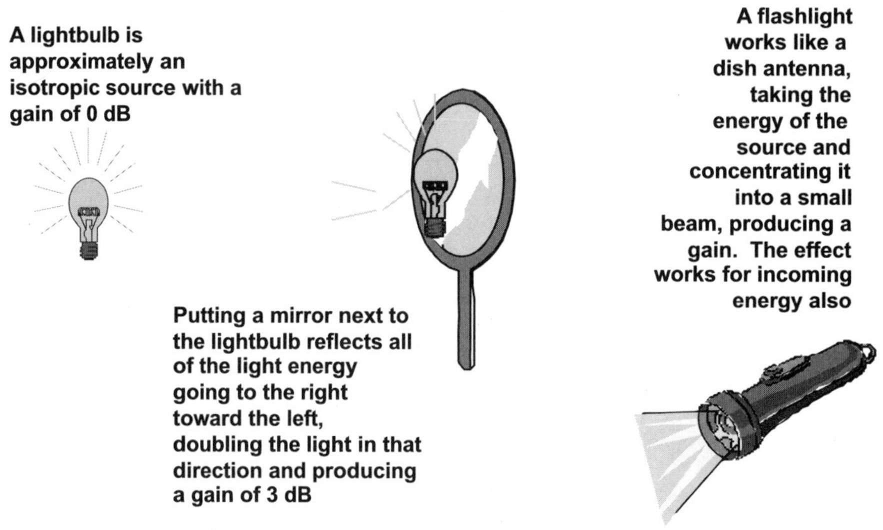

A lightbulb is approximately an isotropic source with a gain of 0 dB

Putting a mirror next to the lightbulb reflects all of the light energy going to the right toward the left, doubling the light in that direction and producing a gain of 3 dB

A flashlight works like a dish antenna, taking the energy of the source and concentrating it into a small beam, producing a gain. The effect works for incoming energy also

Figure 16.8 The concept of dish gain. A light bulb shines in all directions, isotropically. Simply putting a flat mirror next to a bulb directs all of the light that would have gone to one side to the other side, doubling the brightness on that side, for a gain of two-times, or 3 dB. Further curving the mirror focuses the same amount of light into a smaller cone, thus making the beam brighter, or higher in gain.

Now continue our thought experiment: What would happen if you started curving the mirror, making it concave toward the light source? First, the area illuminated would get smaller. Also, as would be expected by concentrating a fixed amount of light into a smaller and smaller area, the illuminated area—the spot beam—would get brighter.

Here we see another important trade-off for antennas: The smaller the beam, the higher the gain; conversely, wide-beam antennas have inherently low gain.

It is important to emphasize here that there is nothing electronic going on—yet. Gain is simply the result of the geometry of the dish—it catches a lot of waves.

The formula for calculating the gain of a parabolic dish and some typical values are given in Appendix E.9. For real dishes working in the radio region of the spectrum, gain is defined not only by dish size, but by frequency. The thing to remember is that the gain of a dish will go up as the size of the dish increases and as the frequency increases. Thus, a 1-m dish working at Ku-band (around 12 GHz), will have a higher gain than the same dish working at C-band (around 4 GHz).

The amount of gain provided by a dish can be quite large, and is an important component of the link budget we will later calculate in Chapter 19. Figure 16.9 is a graph of gain for various frequencies and dish sizes.

16.9 Limits on sidelobe gain

As we have seen, not all of the energy sent out by a dish goes in the direction that the dish is pointed: some of it goes into sidelobes. Conversely, when a dish is used to

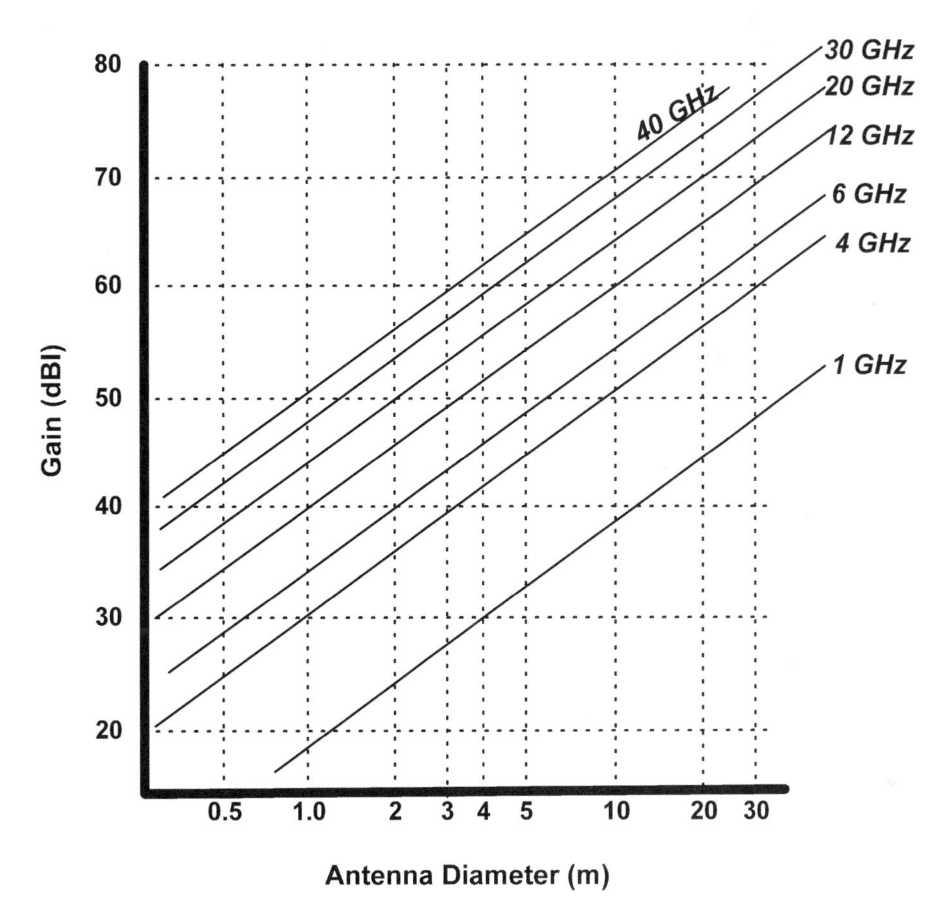

Figure 16.9 Dish gain is a function of both dish size and frequency. This graph shows the gain of typical 55% efficient dishes of various sizes and frequency bands. The higher gain at higher frequencies for small dishes is one of the reasons why engineers like to use the higher frequency bands.

receive, it is sensitive not only to the sources that it is pointed toward, but is sensitive to the sides as well. Thus, these sidelobes can both cause and pick up interfering signals.

For this reason, the regulatory authorities have set limits on the maximum off-axis gain of any dish that will be used to transmit. There is no such legal limit for receive-only dishes, but a user would certainly want to minimize the interference coming into a dish.

This legal limitation is called the *antenna gain rule* or *sidelobe gain rule*, and is defined by a mathematical formula (actually a set of formulas) that specifies how strong the signal from a transmitting dish can be in directions off of the boresight. The most commonly-used formula is given in Appendix E.9. It is illegal to use an antenna that does not conform to this rule for transmission. Frequently moved antennas that operate in changing surroundings, such as portable uplink dishes and satellite newsgathering vehicles, must regularly test their emission for conformity to this rule.

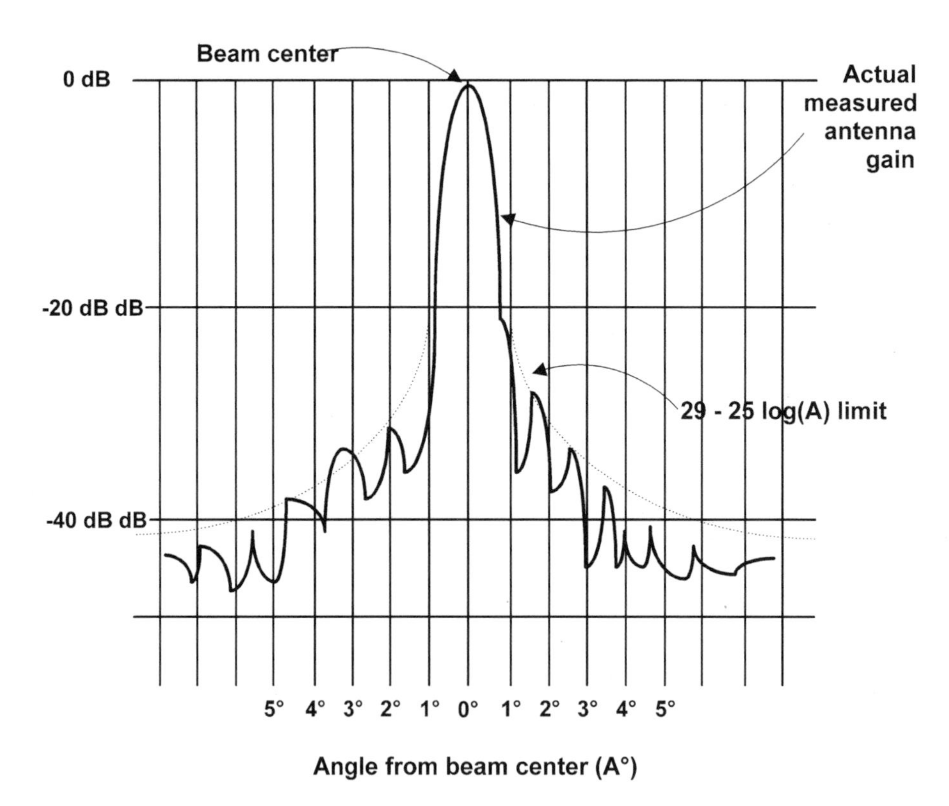

Figure 16.10 Because sidelobes can cause interference to others, regulators limit how much power can be projected into the sidelobes. This is called the antenna gain rule or sidelobe gain rule. For angles more than 1° from the beam center, the maximum output is limited by a formula, plotted in the figure as a dotted line. The solid line is the actual gain of a hypothetical dish.

Figure 16.10 shows a possible antenna pattern with the formula plotted as the smooth line superimposed on it. National telecommunication authorities may impose tighter requirements in some cases.

The topics we have covered in this chapter have thus far dealt with only the hardware of the antenna, but not with any associated electronics. Of course, the purpose of the dish is to increase the efficiency of the transmitting and receiving electronics and improve the signals to and from the dish. Thus, in the next chapter, we will consider the electronic components used to transmit and receive, and determine how these work with the dish to provide a useful radio link with the satellite.

Chapter 17

Earthstations: Signal Flow, Electronics, EIRP, and G/T

The functions of an earthstation and the electronics that accomplish those functions may be conveniently grouped under three major tasks:

- Contacting the satellite using the assigned frequencies in the radio-frequency band;
- Signal processing, which may operate at one or more intermediate frequencies;
- Interfacing with the sources and destinations of the signals at whatever baseband or broadband frequencies are inherent to the nature of the communication.

A small, single-purpose terminal will have a lot less complexity than a hub or gateway station. (*Hubs* are earthstations that serve to control and route traffic in networks.) Small, usually single-purpose earthstations and large gateway stations operate quite differently because of the different quantity of information they handle in the signal. In a small station, the electronics are simpler and usually less demanding in terms of noise and signal processing. Gateway stations, because they handle perhaps hundreds of thousands of signals of various types simultaneously, are more complex and need higher-quality equipment.

In this chapter, we will examine how signals flow into, through, and out of some typical earthstations. We will see how some of the electronics are used to accomplish these tasks, and we will elucidate the important ways that the electronics work together with the dish gain to allow links with the satellite.

17.1 The length of the link

The elevation angle of the satellite above an earthstation's horizon determines the exact distance between the earthstation and the satellite, called the path length. For an earthstation at the subsatellite point—directly under the satellite and thus located on the earth's equator and at the same longitude as the satellite—the satellite has an elevation of 90° and is 35,788 km (22,229 statute miles) from the earthstation. The lower in the sky the satellite appears—i.e., the farther from the subsatellite point the earthstation is—the farther away the satellite is. If the earthstation is as far away as it is physically possible to see the satellite, the two will be 41,679 km (25,888 sm) apart.

In the satellite business, the distance between a satellite and an earthstation may sometimes be measured in units of kilometers, while others use nautical miles, statute miles, or even the radius of the earth (which is 6378 km).

The distance is important for several reasons. For one, it controls how long the radio wave will take to travel between them when traveling at the speed of light. This travel time is important in connection with two items involving satellite links.

The first is delay, the fact that your signal takes a measurable time from satellite to Earth. The delay may produce an echo on two-way voice channels, which we will discuss later. The second item is that if several earthstations are using high-speed time-division multiple-access systems (to be discussed more fully in Chapter 20) to send packets of data through a satellite, each station must know precisely how far it is from the satellite to time its transmission such that its packet does not collide with those of other stations.

The second important reason for knowing the distance between earthstation and satellite is that, as we saw in Chapter 5, radio waves spread out and lose strength with distance. This decrease in signal strength is called the *space loss*.

Appendix E.10 gives the formula for calculating the space loss. The salient concepts to remember are that space loss increases as distance increases and as frequency increases. Figure 17.1 is a graph of space loss for various frequencies.

To show you the size of the space loss, here are the approximate values for space loss (uplink and downlink have been averaged) for some satellite frequency bands now in common use: L-band, 188 dB; C-band, 198 dB; Ku-band, 206 dB; and Ka-band, 212 dB. Recall that a space loss of 200 dB means that of only 10^{-20}th of the transmitted signal, or one one-hundredth of a billionth of a billionth, gets to the receiver!

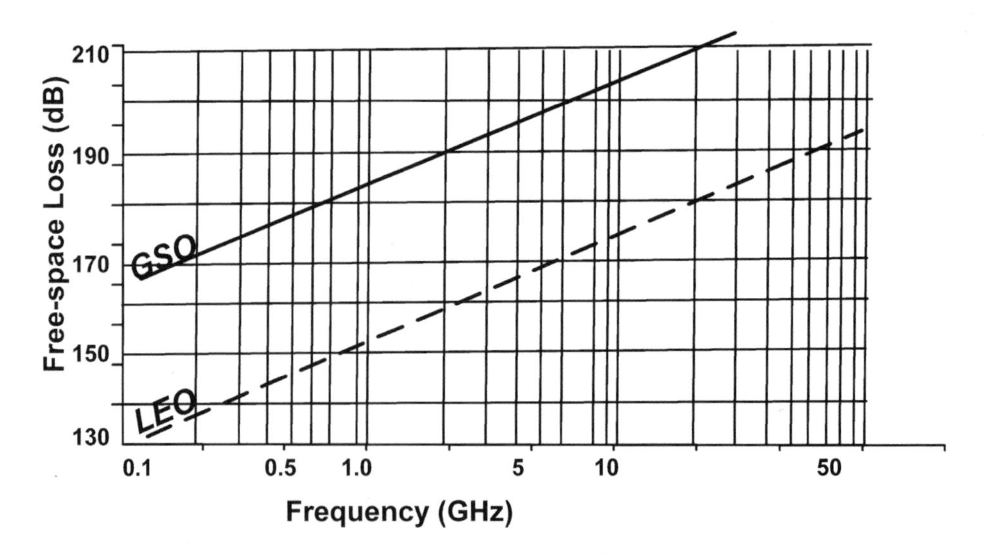

Figure 17.1 Space loss, the amount of power lost in the link between a satellite and an earthstation, depends on both the distance and the frequency. This graph shows the amount of space loss for GSO satellites and a typical low-Earth orbit satellite.

The difference in space loss from a satellite directly overhead and one on the horizon is about 1.3 dB.

17.2 Electronics

Transmitters are devices that send signals to other devices. They consist of electronic equipment that accepts low-power signals from a source, modulates a carrier frequency generated in the transmitter and boosts the power. In the process, transmitters may incorporate such other functions as frequency converters and modulators. The output of a transmitter is connected to an antenna, which may be a simple piece of wire sized according to the frequency in use or a more complicated array of electronic devices.

An amplifier has several major operating properties. First is its sensitivity to the signals that are input into it; this is the input *threshold*, the minimum level an incoming signal must reach to be recognized by the equipment. The maximum input signal the amplifier can handle is called the *saturation level*. Second, amplifiers are primarily characterized by their gain (the amount by which they increase the signal power), usually measured in decibels. Finally, amplifiers work within a defined bandwidth. An important operating characteristic is how uniformly different frequencies within the operating bandwidth are amplified. This ability to handle signals of various powers and frequencies is called *linearity*.

17.2.1 HPAs

Amplifiers may work with signals of all levels, depending on where they are in the signal chain. One type of amplifier takes low-power signals and increases them greatly in order to send the signal over a wire or from a dish. This is often called a *high-power amplifier, HPA*. These range in output power from a few watts to tens of kilowatts. One is shown in Fig. 17.2.

Typical hardware used for earthstation HPAs is similar to that used aboard the satellite for the transponders, but they typically operate at much higher powers. They also don't need to be as sturdy and reliable (and expensive) as space-qualified HPAs.

Traveling-wave tubes (*TWTs*), and their cousins *klystrons*, are used in all sizes and powers of earthstations to give the final boost to the signal heading for the satellite. TWTs are wideband devices, capable of amplifying a bandwidth of 500 MHz or more, and could amplify the whole bandwidth of a satellite; klystrons are narrower-band, typically 40 MHz or so, and thus are used to amplify the signals to individual transponders.

Smaller, low-power earthstations and the very small mobile terminals typically use solid state power amplifiers. They are typically capable of only a few watts to a few tens of watts of output power.

Figure 17.2 The feed of a medium-sized dish used at a television station. Side by side at the focus are an LNB for receiving the downlink, and a small solid-state high-power amplifier for uplinking to the satellite. Since uplink and downlink frequencies are different, there is no interference.

Some amplifiers are designed to accept very, very weak signals and boost their power to some moderate value to send the signal on to further electronic devices. Because signals from satellites are very weak, and it does not take much interference to degrade the wanted signal, amplifiers used at antennas must be designed not only for high gain, but also so that they themselves do not produce much competing electronic noise. For this reason they are called low-noise amplifiers.

17.2.2 LNA-B-Cs

The term *LNA*, or *low-noise amplifier*, is the most generic term for a high-gain, low-noise amplifier designed to interfere as little as possible with the desired signal. Figure 17.3 shows what one looks like. LNAs also have some operational bandwidth over which they work. The quality of an LNA is measured by its amplification or gain, in decibels, and by its noise, which can either be specified as a noise figure, NF, in decibels, or a noise temperature, T_N, in degrees Kelvin. The lower the noise figure, or the cooler the noise temperature, the better the LNA—and the more expensive. Inexpensive LNAs, perhaps used for DBS dishes, have noise temperatures in the hundreds of degrees; better LNAs, for systems such as VSATs or corporate earthstations, may have noise temperatures in the 30–120 K range; large gateway stations, where every tenth of a decibel of noise is important, may use cooled LNAs with noise temperatures in the single digits.

There are also two variations of the LNA, called the LNB and LNC. As before, the "LN" part stands for "low noise." The "C" stands for "*converter*," a device which shifts, or converts, the input frequency to a different output frequency. There are both *up-converters* and *down-converters* used in earthstations and satellites. The converters integrated into the LNAs are all down-converters, shifting the original frequency received from the satellite to some lower frequency. This simply makes it easier to pipe the signal into your office or home via coaxial cable, since lower frequencies travel through cables with less resistance.

Figure 17.3 This photograph shows some options for the feed of a DTH consumer dish, shown on the left with a single LNB. On the right are two optional LNBs. The black one is a dual-LNB setup, capable of receiving simultaneously from two satellites closely spaced in orbit. In the background is an LNB with the feedhorn removed. While we often loosely call the whole dish apparatus the "antenna," the actual antennas are the two small wires seen in the central hole. Note that they are oriented horizontally and vertically to receive the corresponding polarization. (Photograph courtesy of SES Astra.)

(Larger earthstations may use devices called *waveguides*, which are basically hollow pipes designed to carry radio waves within them, to carry the signals between the equipment room and the electronics at the dish itself.)

An LNC can typically be tuned to amplify and then down-convert the bandwidth of a single transponder, whatever that bandwidth might be. As mentioned earlier, 36 MHz is a common transponder bandwidth that typically carries a single analog television signal. If you need only to use the signal from a single transponder, then an LNC will do the job.

The only difference between an LNC and an LNB is the bandwidth it converts. The "B" stands for *block converter*, meaning that it handles the (typically 500 MHz) block of frequencies arriving from all of the transponders on the satellite. This is useful if you need signals from more than one transponder simultaneously. See Fig. 17.4.

An LNB maintains the 500 MHz bandwidth, amplifies the signal, and shifts the whole block of frequencies down to some more useful range. There is no industry-wide standard as to what range of frequencies it is shifted to: as long as the output frequency of the LNB at the dish is matched to the input frequency of the receiver in your office, it will work fine. For example, if your dish is receiving a downlink signal from a C-band FSS satellite (no matter which one—they all use the same band), then the LNB will be receiving a range of frequencies between 3.7 and 4.2 GHz, which is a bandwidth of 0.5 MHz = 500 MHz. Depending on the brand of the LNB and the receiver that you are using, the LNB may shift the band

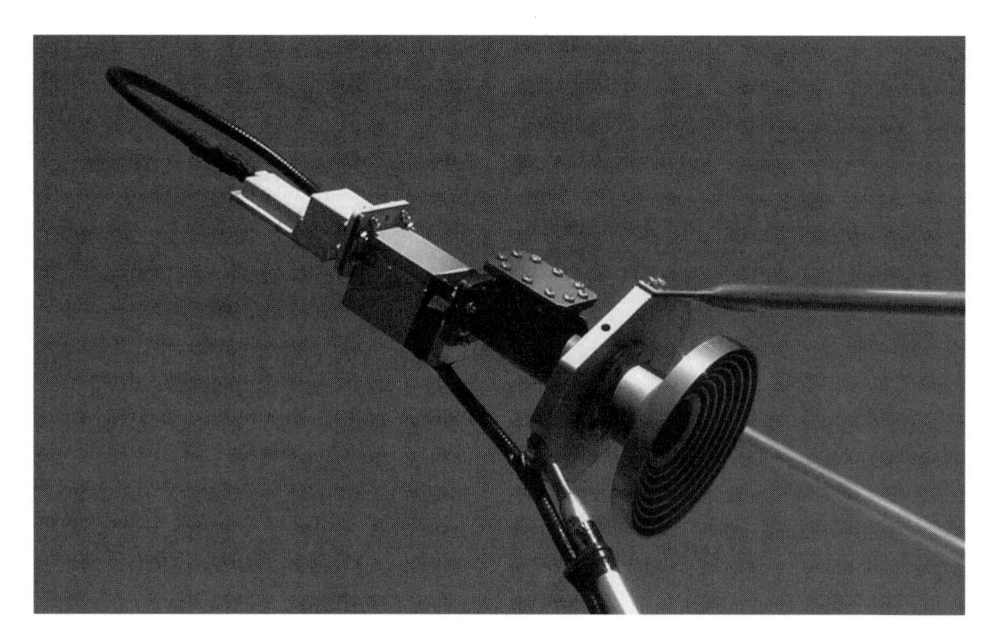

Figure 17.4 Close-up of a receive-only feed assembly. The dish, out of the picture down to the right, focuses the waves toward the feed. The structure with concentric circular rings is a field restrictor, to keep the actual antenna inside the black box from picking up stray waves from the surface the dish is mounted on, such as a roof or ground. The boxes behind this contain the electronics which first amplify the very weak signal to a useable level, then convert it to a lower frequency for ease of sending it into a home or office through the coaxial cable seen emerging from the back of the LNB.

down to 800–1300 MHz, or to 950–1450 MHz, or any other lower-frequency 500-MHz bandwidth range of frequencies.

LNAs are located near the focus of the dish. Immediately in front of them are polarizing components that separate the uplinked and downlinked beams by their polarization.

17.2.3 Receivers

A *receiver* is a device that accepts an incoming signal and processes it to produce a useful output, such as a datastream, a telephone call, or a television picture. Receivers also have some threshold and a bandwidth of frequencies over which they work. They will also often incorporate frequency converters, amplifiers, demultiplexers, error correctors, decoders, decrypters, and demodulators. A demodulator separates the signal itself from the radio wave carrier.

The details of a receiver depend on the nature of the signal. Data terminals, fax machines, telephones, and television sets are all very different examples of receivers. Figure 17.5 shows a typical DBS receiver.

Figure 17.5 A typical DTH receiver for home use. The electronics receive the signal from the LNB at the dish, demodulate it, check to see if the receiver is authorized to decode the signal, and then convert the signal to an analog form to send to the television for viewing. (Photograph courtesy of SES Astra.)

17.3 Earthstation functions and signal flow

Let us take a look at the signal flow and electronics of several typical kinds of earthstations. They vary greatly in complexity and cost because of the wide range of requirements. DBS "earthstations" are off-the-shelf components that can be bought at a local store for a few hundred dollars and installed yourself. Major gateway stations cost millions of dollars to buy and build, and take teams of engineers and more millions of dollars to operate.

17.3.1 Small receive-only earthstations

Most earthstations around the world are receive-only, such as the ones for DBS and TVRO television services. Figure 17.6 shows a basic functional diagram of a typical home DTH setup.

The very faint signal received from a satellite is brought to a focus by the geometry of the dish, as explained in the previous chapter. The actual antenna, a small piece of wire, is located at the focus. To keep the antenna from "seeing" beyond the edge of the dish where it might pick up interference, there may be a device called a *field limiter* which acts as blinders for the antenna. The detection of the proper polarization for the signal being received can be accomplished either by orienting the antenna physically, or by phasing the antenna electronically.

Even after being boosted by the dish gain, the signal strength reaching this antenna is still extremely small, typically a few trillionths of a watt. Since small dishes are not big enough to support the weight of lots of electronic devices at the focus, there is typically a small, lightweight, high-gain amplifier directly attached to the antenna at the focus of the dish, connected by as short a connection as possible.

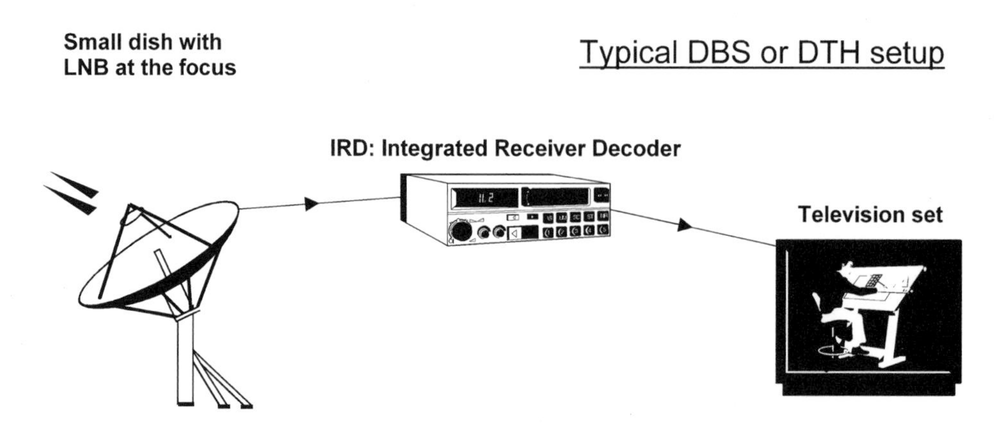

Figure 17.6 A simplified diagram of the signal flow of a typical DTH setup.

This is the LNA, LNB or LNC. It amplifies the weak signal and sends it into the receiver in the office at some intermediate frequency.

17.3.2 Two-way earthstations

VSAT earthstations and those providing such services as Internet-via-satellite for a small business or residence may be two-way links to satellites. Thus, in addition to the receiving components and functions just noted, they must also have a (relatively) high-power amplifier, HPA, to transmit a signal to the satellite. This is usually located at the focus of the dish, next to the LNA. Recall that the uplink and downlink frequencies are different to minimize interference, and the uplink signal and downlink signal are usually cross-polarized. Part of the antenna feed is a device called an *orthomode transducer*, which combines the uplink and downlink signals electronically split between the transmitting and receiving functions of the dish.

This HPA, which is typically around a few watts or tens of watts for small earthstations, is fed by a low-power signal at the intermediate frequency from equipment inside the office and carried to the feed by coaxial cables and/or waveguides. Figure 17.7 shows a diagram of a typical two-way VSAT station.

17.3.3 Large earthstations

Figure 17.8 is a block diagram of the basic components of a gateway station earthstation, and shows the three frequency regimes that are used. First is the equipment that interconnects the earthstation to the terrestrial sources and destinations of the signals. This can generically be called the interface function, and may be simple or complex. This equipment operates at the frequencies of the incoming signals.

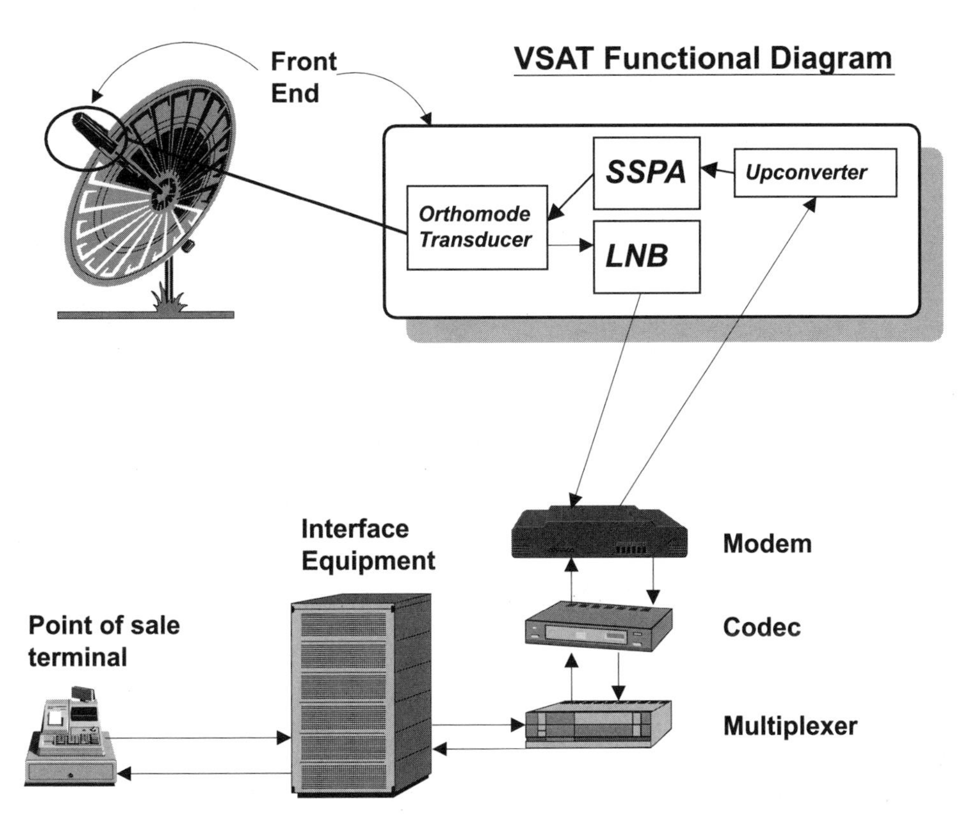

Figure 17.7 Simplified functional diagram for a typical two-way point-of-sale VSAT. The orthomode transducer separates the transmitted and received signals.

Since there may be many signals coming from different sources, of different kinds, and going to different transponders, demultiplexing ("demuxing") and deformatting equipment takes apart any grouped signals and recombines them in the ways the stations or satellites need them to be. Further signal processing may occur, such as a format conversion from one telephone trunking standard to another, or one television standard to another, multiplexing, digital compression, addressing, etc. Other miscellaneous processing that may occur here includes encryption and error control.

The incoming processed signals are then shifted ("converted") up to some convenient *intermediate frequency*, called the *IF*, for ease of further electronic manipulation. One standard IF for earthstations is 70 MHz. This is done simply for practicality. (Your home radios and televisions do something similar.) This IF is modulated by the signals, and then this modulated signal is further up-converted to the radio frequency, or RF, which will be used for the uplink to the satellite.

This signal may (or may not) be combined with others, and then goes through the high-power amplifier, the HPA, which boosts its power to that necessary to perform the uplink. At this stage, the earthstation may apply some more processing to allow multiple access to the satellite along with other earthstations. The signal is

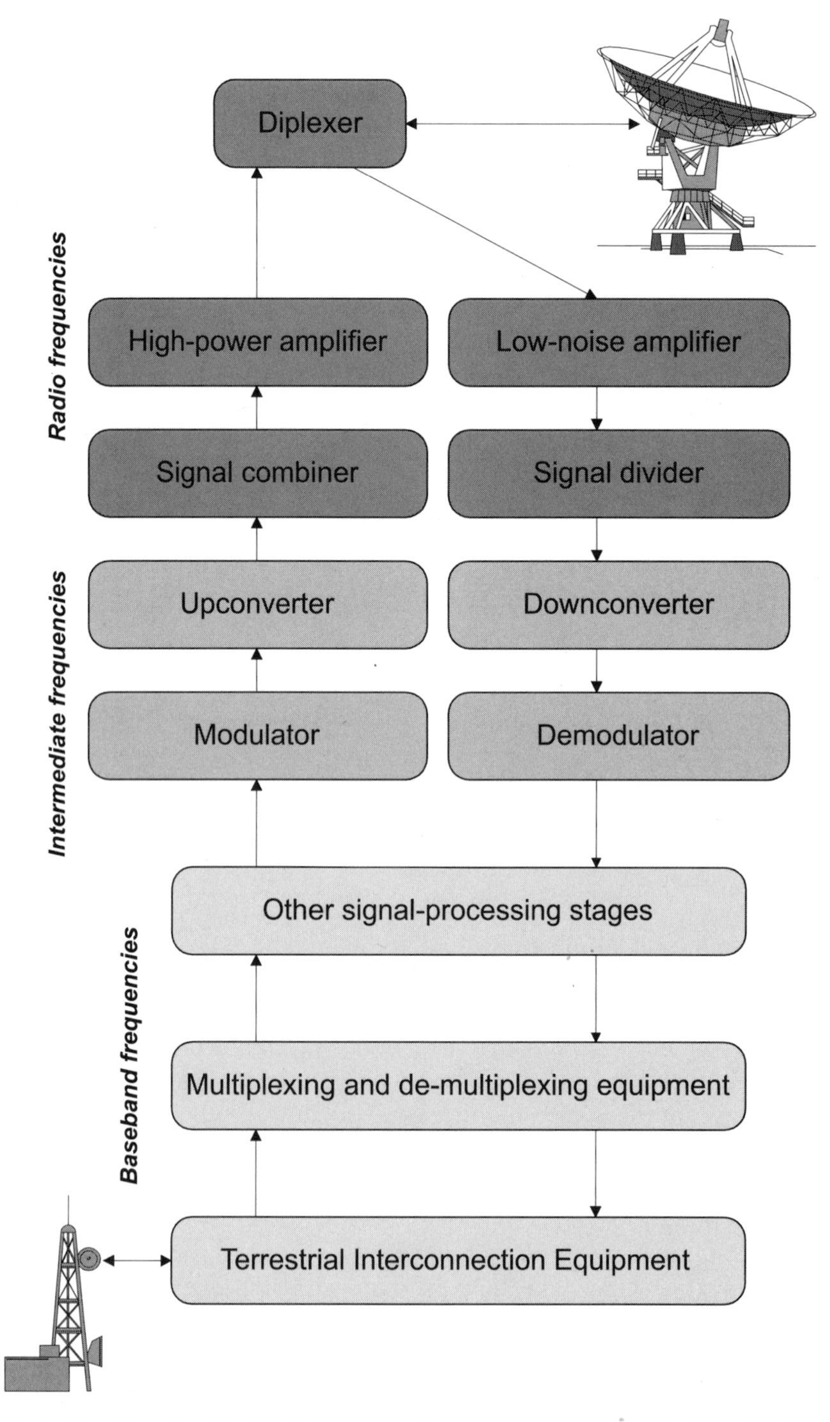

Figure 17.8 Simplified functional diagram of a large gateway earthstation. Here the focus of activity is interconnecting the (possibly tens of thousands) of incoming and outgoing signals with the satellite links. The terrestrial connections may include fiber optic cables, coaxial cables, microwave relays, and links from other dishes at the same earthstation.

then sent out to the antenna, where a *diplexer* may be used to combine sent and received signals. This high-power signal is then beamed at the dish, and when enhanced by the gain of the dish, provides the uplink signal sent to the satellite.

Similarly, a signal downlinked from the satellite is gathered by the dish and focused onto the receiving feed. A diplexer may again separate sent and received signals. The received signal is still very weak, and needs to be amplified before being sent on to later stages of processing. This boosted RF signal may also then go through a divider to split off individual signals contained within it.

The RF signal is then down-converted to an IF, demodulated, and converted to the baseband signal. After this, more signal processing may occur, such as more format conversions, decryption, etc. If the signal is multiplexed, it may then be "demuxed" and reformatted before being connected to the terrestrial network through appropriate interface equipment.

17.4 Other signal processing

In addition to amplifying the faint received signal, converting it to a lower frequency, and then putting it through a receiver that turns the modulated carrier back into the original form of the signal (a telephone call, audio feed, television picture, datastream, etc.), some other signal processing steps may be necessary, depending on the detailed nature of the signal. Some of these may be value-added services provided by the earthstation for the customer.

For television feeds, the most common processing is format conversion from a SECAM picture into an NTSC picture, for example. If the signal is scrambled to prevent unauthorized reception, it must be descrambled. If it was digital, it must be converted back to analog. Digital compression and statistical multiplexing with other video feeds is another common processing step for DBS service distributors. Scrambling (encryption) is another.

For telephone calls, another type of processing deals with the differences between how trunked telephone calls are multiplexed in Europe, North America, and Japan. This is another example of a format conversion.

One service common in the satellite industry is what is called a *turnaround.* A turnaround occurs when the earthstation downlinks a signal from a satellite in one frequency band and uplinks it to another satellite in a different frequency band. There is usually no change in the nature of the signal, just a relay function from one satellite to another.

There are a couple of common situations where this might be needed. One case involves a television station that needs to receive a feed from a transportable terminal or SNG truck out in the field at the site of some event, and the station does not have a terminal that can receive at the frequency band of the remote terminal. For instance, if the SNG truck has Ku-band equipment, and the station only has a C-band system, then the SNG truck could link to a satellite in Ku-band, to be received by a teleport at the Ku-band downlink frequency and uplinked to a C-band satellite that the station is capable of receiving.

Another case requiring a turnaround might be one where the origin and destination earthstations are too far apart on Earth to see a common satellite. An uplink from the origin station to a satellite could be received by some intermediate teleport, which relays the signal to a satellite visible to the destination earthstation.

17.4.1 The echo problem

Another common bit of signal processing applicable to telephony is *echo correction* to reduce or eliminate the irksome return of your voice to you after you have spoken. This is caused by the delay time in the round-trip link from earthstation to satellite to earthstation. It occurs in every telephone call—satellite or terrestrial—because of what are called "impedance mismatches" between pieces of equipment at each end of the phone call. Every phone call you make, even the ones carried along earthbound transmission media, has some echo, and the telephone companies automatically put echo correction devices into your voice circuit without you knowing it. The problem is worse with satellite calls because the round-trip time for the echo to get back to you is that half-second, due to the distance to GSO satellites and the speed of light. For NGSO satellites, the problem is much less, and usually negligible.

(As an aside, there are some "telephone calls" in which you do not want echo corrections devices: those that are calls not between humans but between computer modems. In such cases, the shrill whines and burbles you may hear as your modem connects tell the terrestrial telephone equipment not to insert echo correction into the call.)

There are two kinds of echo-correction devices. The older and simpler are called *echo-suppressors*. They work like a voice-actuated microphone, sensing when you are speaking and cutting off all signals coming in to you. The problem here is the same as with a speakerphone: if both parties talk at the same time, no one hears anything! In techspeak, the suppressor has turned a full-duplex circuit into a half-duplex circuit.

The newer, and much preferred technology, is called an *echo canceler*. This is a digital device which, as you speak, makes a digital "mirror image" of your voice. When the echo of what you said gets back to you, the device combines this mirror-image with the returning echo to cancel it out, bit by bit. This maintains a full-duplex circuit. (Remember, however, that only the echo is canceled, not any delay due to the round-trip transit time.)

Digital signals may be encrypted, or use spread spectrum or some other specialized technique that must be undone at the receiver end to obtain the original signal again.

In a gateway station, the earthstation itself is usually not the intended terminus of the transmission. Thus, there may be quite a lot of equipment devoted to connecting the incoming and outgoing signals to terrestrial networks, or even to other satellite networks. Combined with the other requirements, this is why the main focus of activity at such a station is signal processing, not just linking to and from satellites.

17.5 Dish + electronics

When an earthstation transmits to and receives the radio links from a satellite, what is important is not just the gain of the dish, as defined and explained in Chapter 16, but how this gain works together with the transmitting and receiving electronics to create a usable signal. At the transmitter, the appropriate parameter is called the EIRP. For reception, the number we are interested in is called the figure of merit.

17.5.1 EIRP

Recall that the gain of a dish depends on its efficiency, its size (basically its surface area), and the frequency being used. Remember, also, that gain is strictly a property of the dish and has no electronics involved.

When a dish is being used to send a signal (either from an earthstation or from a satellite), it works together with the high-power amplifier of the transmitter to produce a radio wave directed toward the destination. What is important for the link is how the HPA and dish work together to produce a signal.

Transmitter powers can be measured in watts, and typically range from a few watts in handheld satphones, to thousands of watts for large gateway stations. We also know that we can equivalently measure wattage in decibels compared to one watt, or dBW. Thus a 100-W HPA could also be called a 20 dBW amplifier; a 2000-W HPA would be rated at 33 dBW.

As a measurement of the actual effective radio power coming out of the top of the dish, we define an important figure called the *equivalent isotropically radiated power*, or *EIRP* (Fig. 17.9). It is defined simply as the sum, in decibels, of the transmitter power and dish gain:

$$EIRP = HPA\ power + dish\ gain.$$

This is one of the most important numbers that will go into our link budget in Chapter 19.

For example, if we have a 100-W HPA feeding a dish with a gain of 30 dBI, then the EIRP will simply be 20 + 30 = 50 dBW. This is the equivalent amount of radio wave power leaving the dish. It is the signal strength you would measure if you sat atop a transmitting dish with a field-strength meter. (But don't, because it is dangerous, as EIRPs can be in the billions of equivalent watts for large earthstations.)

A good analogy is a simple flashlight. A very small lightbulb can be made to produce a much brighter beam (over a limited area, its beamwidth) by focusing it with a concave mirror, which is the analog of the earthstation's dish. The wattage of the lightbulb is the equivalent of the power of the HPA, and the focusing power of the flashlight's mirror is its gain.

Values of EIRP for each transponder are typically part of operational specifications for each satellite. They appear on a *satellite footprint map*. It is important

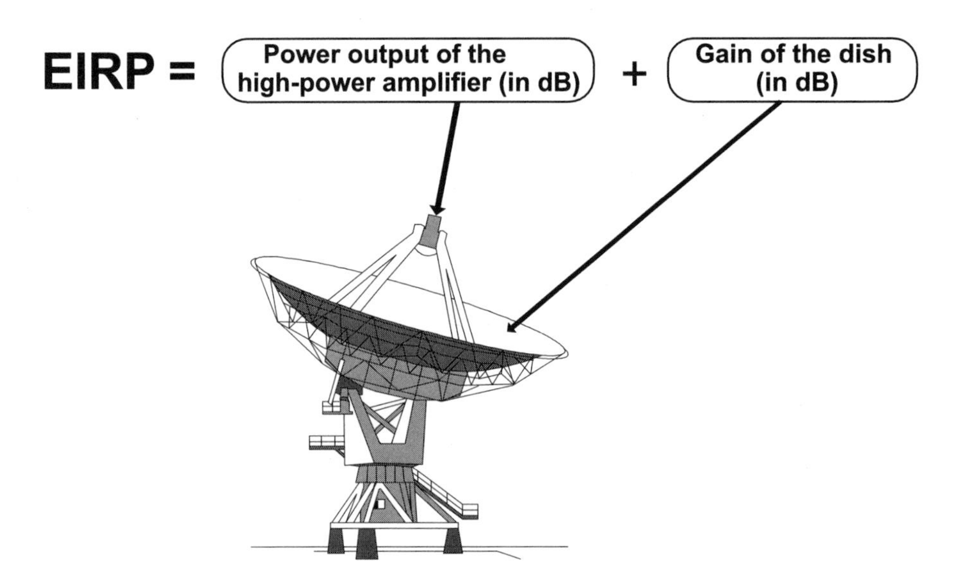

Figure 17.9 The definition of effective isotropically radiated power, EIRP. Expressed in decibels, it is simply the sum of the output power of the transmitter high-power amplifier and the gain of the dish. It is the amount of power you would measure just emerging from the dish.

to remember that EIRP is always measured at the dish, so if you see a footprint map with EIRP numbers for a given transponder on a satellite, that is the amount of power the satellite is sending down to Earth, measured as it is just leaving the satellite's downlink dish.

By the time the signal arrives at your earthstation, it has been reduced by the amount of space loss and other losses. We talked about space loss more fully earlier in this chapter, and Appendix E.10 gives a formula for determining the value. For now, just take an example: if the EIRP of a transponder is 45 dBW measured at the satellite, and the space loss at the frequency in use is 200 dB, then the actual amount of power that arrives into your earthstation's dish is 45 – 200 = –155 dBW.

Some satellite operators prefer to take this space loss into account when they publish satellite footprint maps for their users (Fig. 17.10). Thus, rather than showing the EIRP of a transponder, they will give the signal strength as it is received on the ground, corrected for the space loss. This number is called the *illumination level*, often symbolized by *W*, and is measured in dBW per square meter. It is simply defined as

$$W = EIRP - Space\ loss.$$

Some operators prefer to specify received power per unit bandwidth. Typically the bandwidth (*BW*) unit used is 4 kHz (because that is the bandwidth of a typical analog telephony channel). If this is done, the new value is called the *power flux density*, or *PFD*, defined as

$$PFD = EIRP - Space\ loss - BW.$$

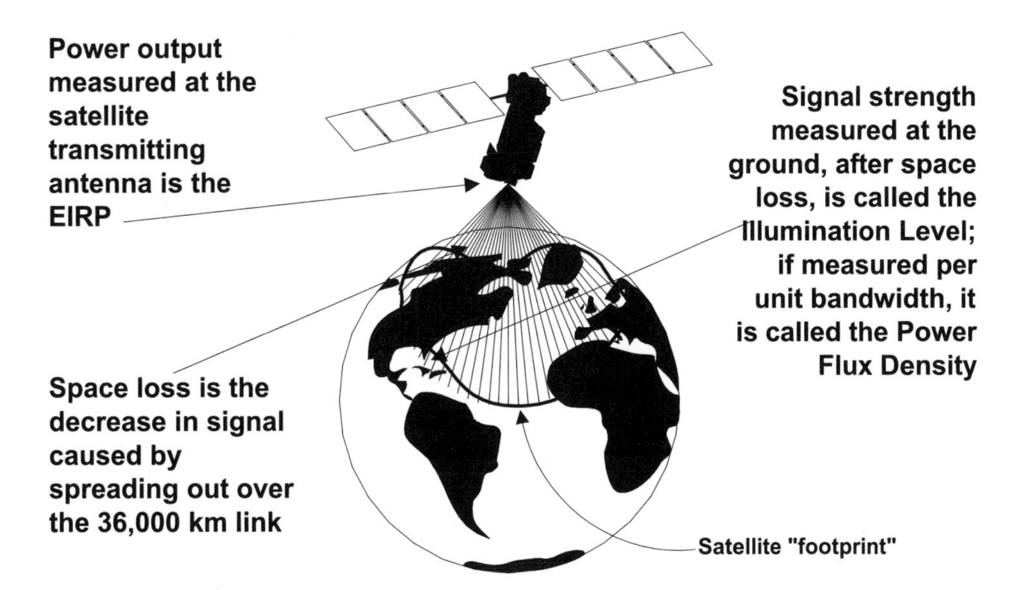

Figure 17.10 The EIRP transmitted by a dish is reduced in power by the space loss across the distance between sender and receiver. The received power is called the power flux density, or, if measured per unit bandwidth, the illumination level; both are measured in dBW per square meter.

17.5.2 Figure of merit

When a dish receives a signal, what counts is not just the gain of the dish, but how the dish works together with the receiving electronics to produce a useful signal. Thus, a receiving dish is like an optical telescope that gathers enough light for the eye to see it. What we are actually interested in is the sensitivity to weak signals of the dish+LNA combination.

The amount of radio power arriving at the dish is very tiny—a billionth of billionths of a watt. The dish amplifies this, but there is a problem with noise getting into the system. The noise comes from several sources, as we saw in Chapter 8. One of the major sources is often the high-gain electronics themselves, which amplify this very weak signal so it can be sent along to your receiver. These are the LNAs (or LNBs or LNCs) we discussed previously. The lower the noise produced by the electronics, the better your signal-to-noise ratio.

But there are other sources of noise that come from outside: other satellites, the sky, the objects surrounding the dish, and others. All of these add to the noise of the LNA to produce a figure called the *system noise temperature*. Again, the lower the better. This total noise interferes with your weak signal, so a good measurement of the efficiency of the combination of the dish and electronics is what is called the *figure of merit*, or *G/T*. This important factor is defined simply as

$$G/T = \frac{\text{Gain of receiving dish}}{\text{System noise temperature}}.$$

Both can be expressed in decibels. Thus, for example, if your receiving dish has a gain of 40 dBI, and the system noise temperature is the equivalent of 15 dB, then the G/T of this system is 40 – 15 = 25. (The units are technically dB per degree Kelvin, sometimes written dBK^{-1}.)

G/T is effectively a measurement of the sensitivity of the receiving dish to weak signals. The larger it is, the better. You can increase the G/T of a receiving system by increasing the gain of the dish (which depends on size and frequency) or lowering the LNA's noise temperature, or both.

Going back to our cocktail party analogy, when a transmitter sends to a receiver, it is similar to trying to hold a conversation. If the person you are speaking to has sensitive ears, you do not have to talk very loud; but if your audience is hard of hearing, you may have to raise your voice. Thus, EIRP and G/T cooperate to produce a useful communication, as we shall see when we put it all together to come up with a link budget in Chapter 19.

17.6 Using satellite footprint maps

You frequently see footprint maps of the EIRPs and G/Ts of satellite transponders. These are published by satellite system operators to help you figure out what you need to do to accomplish a successful link through the satellite. Simply put, the EIRP map tells you how strong a radio signal the transponder sends to various locations on Earth, while the G/T map tells you how sensitive the transponder is to uplinks coming from various locations on Earth.

From the earthstation's point of view, the EIRP map tells you how large your earthstation G/T must be to receive a useful downlinked signal, while the G/T map tells you how strong your earthstations's EIRP must be to successfully uplink to the satellite.

Consider the footprint maps for the Optus satellite. The EIRP map, Fig. 17.11, shows the strength of signals sent by the satellite to the ground, and gives you the data needed to determine what G/T your receiving earthstations must have. Conversely, the G/T map, such as Fig. 17.12, shows how sensitive the receiving antenna on the satellite is to signals uplinked to it from various locations on Earth, and gives you the data you need to calculate what your uplinking earthstation's EIRP must be to be received successfully by the satellite. Both EIRP and G/T are measured at the satellite.

A radio beam from a satellite is a bit like the beam of light from a flashlight: it is typically brightest (has the strongest signal) at the center of the beam, and decreases in brightness as you move in angle off from the beam center. If there are smaller regions where the beam is strong, these may be called *hot spots*. If the shape of the mirror of a flashlight, or the reflector structure of a dish, is altered a

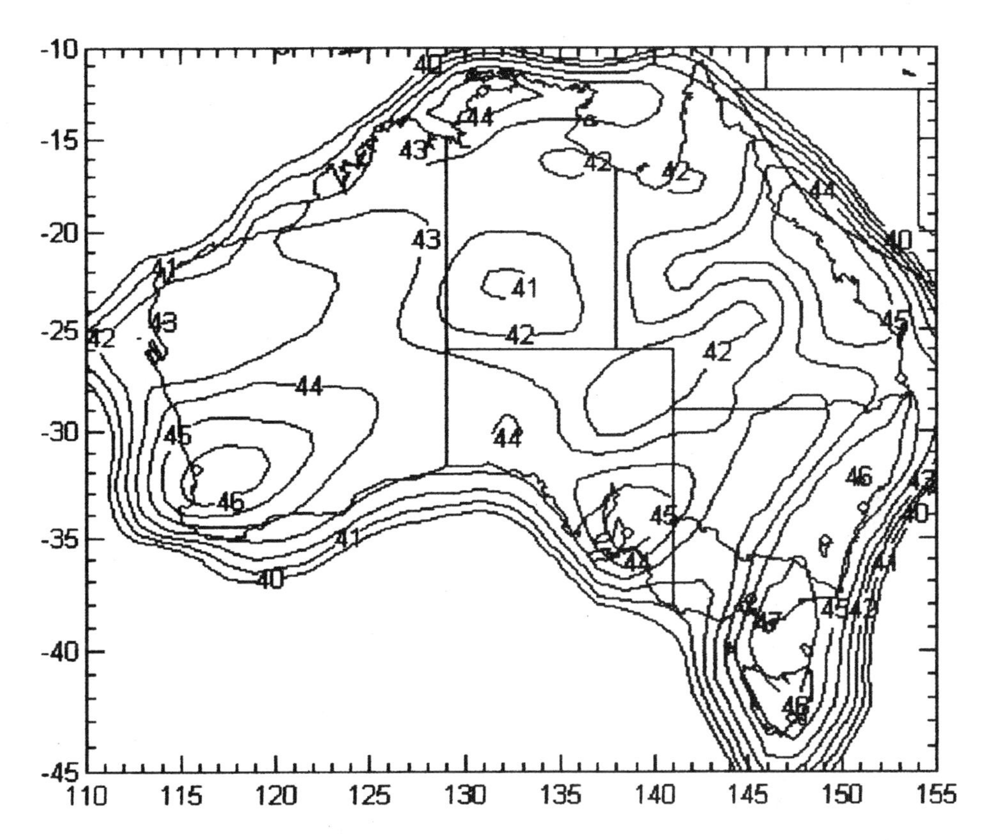

Figure 17.11 A typical computer-drawn EIRP service footprint for a satellite: this one is the Optus satellite serving Australia. Notice the contours of signal strength. Note, for example, that the power sent from the satellite toward the city of Perth (not marked, but in the lower left of the continent) is 46 dB, whereas the power sent toward Alice Springs, in the center of the continent, is only about 41 dB. Thus, a receiving earthstation in Alice Springs would, all else being equal, have to have a dish with a G/T 5 dB better than one in Perth to compensate for the weaker signal. (Image courtesy of Optus.)

bit, you can arrange for the light or radio waves to cover irregularly shaped patterns, called beam patterns. Most flashlights don't do this, but many satellites do shape the beam to cover specific areas such as a country, a continent, or a time zone. This concentrates the power of the transponder into the desired service area, and reduces the power sent to regions you don't want to cover.

Similarly, the area of reception for a dish is typically most sensitive right at the center of the beam, and is less sensitive at angles off from this beamcenter; but engineers can design the satellite's receiving antennas to be more sensitive (have a higher G/T) to some directions than others.

In summary, the satellite's EIRP contours tell you what you need to do to receive from the satellite, while the *G/T* contours tell you what you need to do to send to the satellite. Once you know the EIRP and *G/T* that your earthstation must have, you have reduced the problem to a shopping business decision. Suppose, for instance, your calculation shows you need an earthstation with a G/T of 23. From

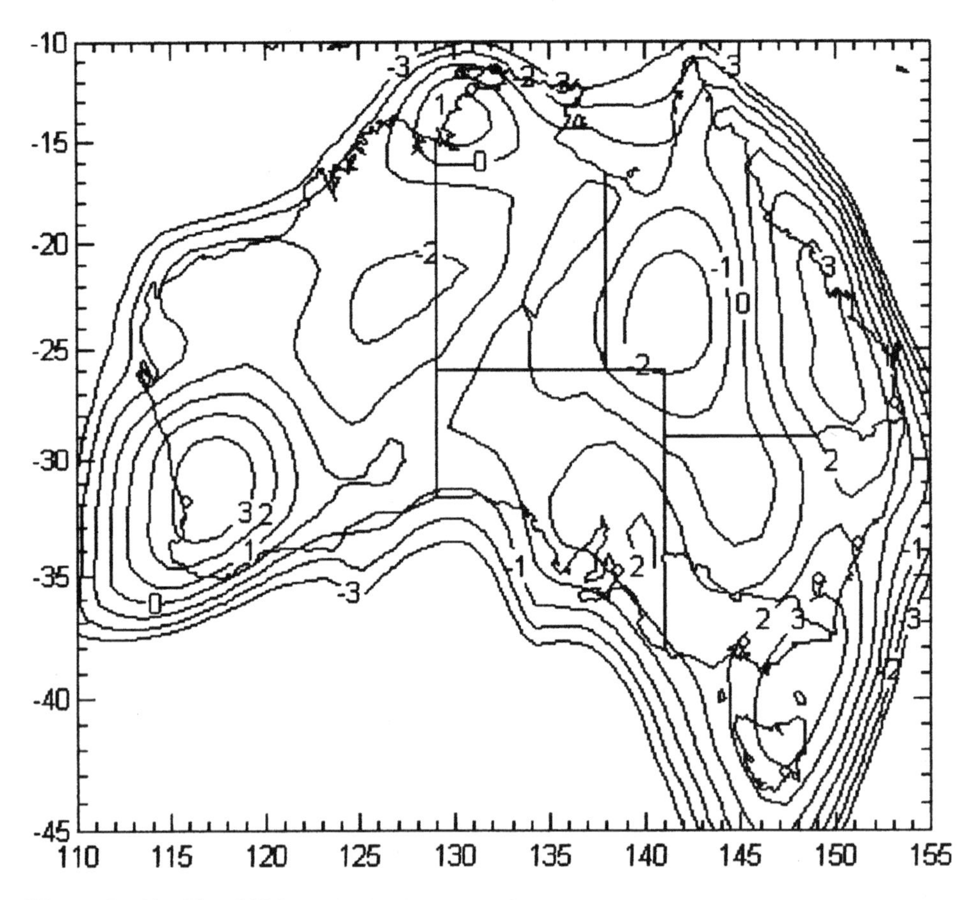

Figure 17.12 The G/T footprint for the same Optus satellite as in Fig. 17.11. Such a map shows the sensitivity of the satellite's antennas to signals being uplinked from various regions of Earth. The measurements are in dB/K. Note that the satellite is more sensitive to signals arriving from southeastern Australia than to those from the center of the continent, by about 5 dB. This means that, all else being equal, an uplinking earthstation in Alice Springs would have to have an EIRP about 5 dB stronger than one in Sydney to produce the same signal strengths in the satellite. (Image courtesy of Optus.)

an operations point of view, it does not matter how you achieve this value. You could opt for a small, inexpensive dish with low gain, but you would require an expensive LNA with low noise for the gain and the noise temperature to combine to a G/T of 23. Another possibility would be to buy a larger, expensive dish with higher G using a higher-T, less expensive LNA. Alternatively, you could adopt some intermediate values as long as the G/T is 23. Similar trade-offs between dish size and HPA wattage are made for the uplink to achieve whatever EIRP your calculations say you need.

Recall that once you have decided on the details of performing a link through the satellite, much is then fixed operationally. For instance, once you specify the earthstations, the bandwidth of the signal, and the frequency band you will use, the only variables that remain really variable—that is, that you can adjust—are the

EIRP and *G/T*. These are the two operational parameters most under your control when calculating a link budget.

So now it is time to begin to put all of these operational parameters together to produce signals linking earthstations and satellites. In addition to the signal loss caused simply because of the distance from earthstation to satellite, there is another major signal loss that occurs when the signal passes through the atmosphere. In the next section we explore that important subject and turn our attention to putting all of the elements together to produce a link budget and determine whether we can perform our communications task and with what level of quality.

Part 5

The Satellite ↔ Earth Link

Chapter 18

Atmospheric Effects on Signals

The atmosphere causes several serious negative effects upon radio signals, and these effects are very dependent on three things. The first factor is the frequency in use: some frequencies—the lower ones—are hardly affected; higher frequencies are greatly affected. The second factor is the elevation of the satellite as seen from the earthstation, which determines how much air the radio wave traverses. The third major factor is the climate of the location of the earthstation.

18.1 An optical analogy

A good way of gaining an understanding of the major effects that the atmosphere has on signals, and their frequency dependence, can literally be seen most days of the year. Just look at the sun—carefully!

On a clear day, you don't stare at the noonday sun: it is too bright and would hurt your eyes. It is high in the sky and its light is passing through only a few kilometers of air. However, just before the sun sets or rises, you can usually safely enjoy the sunset or sunrise, because the sun appears much less bright. This is simply because its light is now passing through more than a hundred kilometers of air, and the sunlight is attenuated in strength. Thus, we see the very important factor of elevation angle at work.

You can also note here the local effects of climate and environment. If you live in a desert region, the sun on the horizon will be much brighter than if you see it through a dense layer of urban smog.

Just as important as elevation angle is frequency dependence. The noonday sun is basically white, perhaps a bit yellowish, but the setting sun is typically orange or red. The light at the blue end of the visible spectrum—the high frequencies—is removed more than the reddish colors, leaving only the lower reddish colors to reach your eye. Here we see the frequency dependence of absorption. (As an aside, when the bluish colors are removed from the sunlight, the air molecules scatter them around the whole sky, which is why a clear sky is blue.)

The same kinds of effects happen to radio waves going between a satellite and your earthstation: they are attenuated by the atmosphere, and the amount of attenuation depends on the frequency of the link. Let's look at the details.

18.2 Elevation angle and path length

In Fig. 18.1, you can see that an earthstation linking to a satellite high in the sky—at a high elevation angle—sends a signal through a relatively short length of the atmosphere close to the surface. A satellite at a low elevation means that the signal passes through a much greater length of air. Since the effects on a signal depend on the path length through the atmosphere, they therefore depend on the elevation angle.

In the diagram is a line labeled "top of the atmosphere." Of course, there is no "top" because the air density slowly thins out as you go higher above the surface. There is no sign posted reading "Now leaving Earth—entering outer space." But there is an effective top to the atmosphere as far as signal effects are concerned, because most of the effects are due to water and water vapor in the atmosphere. This is several kilometers above the surface, depending on your latitude and the season. Thus, an earthstation looking at a satellite directly overhead would be looking through this much air, and thus, water vapor.

A major secondary cause of signal problems is oxygen. Water vapor and oxygen have different absorption spectra, and so affect different wavelengths differently. Both absorb only slightly at frequencies below 10 GHz. Water vapor is the major absorber between 10 and 40 GHz and above 80 GHz, while oxygen absorbs the most from 40 to 80 GHz.

As the elevation angle to the satellite approaches the horizon, the path length through the air increases, and increases especially fast once the elevation angle gets

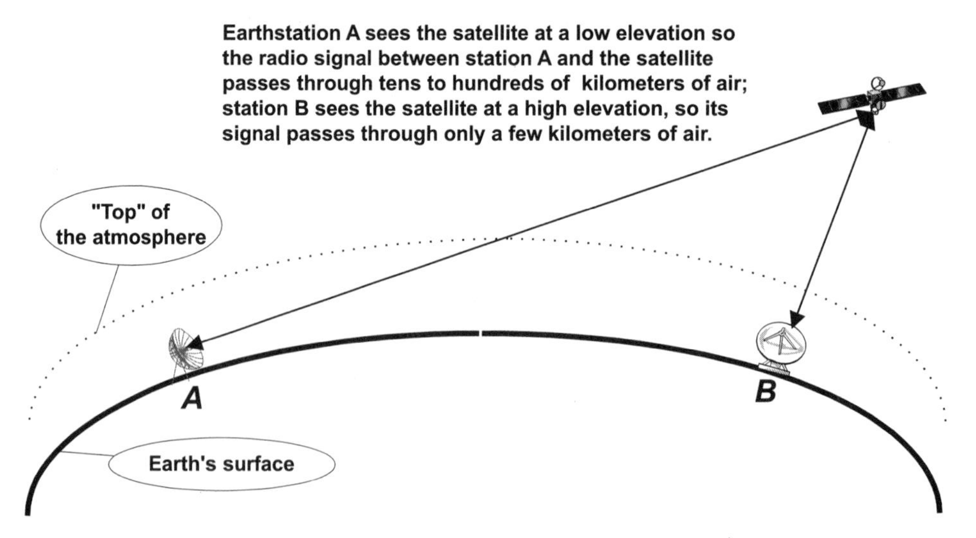

Figure 18.1 The importance of a high elevation angle is due to the fact that a link to or from a satellite near the horizon passes through more air than a link to a satellite high in the sky. Atmospheric degradation of the link is dependent on the path length through the air. The deleterious signal effects are moderate from the zenith down to about 10° elevation, and grow very severe below 5°, which is why guidelines suggest not receiving from satellites this low. If an earthstation is transmitting to a low-elevation satellite, it poses a much greater interference threat to nearby electronics and may not be able to be licensed.

below 10°. If you assume that the atmosphere is 10 km thick, then the path length for an elevation of 90° is 10 km. At an elevation of 10°, the path length is about 33 km, so the atmospheric effects will be a little over 3 times as severe. If you look just 5° above the horizon, the path length has increased to 51 km, so the effects are 5 times worse. And if you look at a satellite only 1° above the horizon, the path length is 126 km.

For this reason, there are guidelines for the minimum elevation at which one should receive a satellite signal. These are only guidelines, not laws, so no one will arrest you for looking at a very low satellite—you will instead pay a penalty of having to have much higher EIRPs and G/T and, consequently, more expensive equipment.

For FSS services from GSO satellites, the guideline recommends receiving from a satellite with a minimum elevation of 5°. For DBS services, which are at Ku-band and thus more affected by the atmosphere, the recommendation is to look no closer to the horizon than 20° in most areas, 30° above the horizon in mountainous areas, and 40° elevation in high rain regions.

For the case of a receive-only earthstation, the penalties of looking close to the horizon fall only on you and the quality of your signal. However, when you have a transmitting earthstation, you could interfere with other people. The sidelobes from your transmitting earthstation could lie close to the horizon and are potentially a source of interference for nearby receivers. Thus, uplinks to satellites less than 5° above the horizon are very strongly discouraged, because you are effectively turning your earthstation into a microwave tower that has a power thousands to millions of times stronger than typical terrestrial microwave links, possibly interfering with nearby terrestrial systems. In fact, you may not be able to obtain a station license.

18.3 Atmospheric effects

Recall that the measurement of signal quality is carrier-to-noise ratio, *C/N*. Anything that decreases the *C* or increases the *N* is bad for your signal. The water and oxygen in the atmosphere hit your desired signal with a "triple whammy." These effects are

- Water vapor and oxygen absorb some of your signal, thereby decreasing the carrier strength *C*;
- Various components in the air themselves emit noise, increasing the *N*; and
- Various components partially depolarize the carrier and its cross-polarized same-frequency signal. This both decreases *C* and increases *N* from interference from the cross-polarized carrier.

Thus, by the time the signal gets through the air, it has had two decreases to its *C*, two increases to its *N*, and so a great diminution of *C/N*, the signal quality. Note that all of this is highly dependent on frequency and climate.

18.3.1 Through a gas darkly

Water in the atmosphere comes in several forms, some more troublesome than others. The biggest problems are caused by liquid water and water vapor, because solid water has different electronic properties.

Even with a clear blue sky, there is some water vapor in the atmosphere. As always, the effects depend on frequency. In Chapter 19, we will calculate a link budget first for a clear-sky condition with an assumed average amount of attenuation based on the frequency in use. Only after a clear-sky link budget is calculated are the effects of inclement weather added in, and this correction depends on the climate of the earthstation.

Rain is the most serious atmospheric problem. The distribution of rain in both time and space varies greatly from one part of the world to another. Electronically speaking, the major effect of rain is that it simply absorbs some of the intended signal, reducing the carrier strength. Rain in the atmosphere also depolarizes both the desired signal and any cross-polarized component. Rain landing on dishes and radomes can absorb signals in the higher frequency bands. And, since rain is obviously not at a temperature of absolute zero, it also emits noise. The rain climate models we will examine try to characterize the absorbing effects of rain. In general, however, we should keep in mind as a rule-of-thumb that rain is of major importance to our link budget for frequencies above about 10 GHz, that is, for Ku-band and up.

Sleet is another form of water in the atmosphere, but it is solid water with much less severe electronic effects. However, sleet is formed when rain high in the atmosphere freezes as it falls through very cold air. Thus, while the signal may hit sleet at the earthstation, high in the atmosphere it may still be passing through rain.

Snow is yet another variation of water in the atmosphere and is complicated by the fact that it comes in many forms, from big, wet, sloppy, watery flakes to dry, hard pellets. Watery snow has more signal effects than dry snow.

Snow can also accumulate on a dish, with two effects. In a snowstorm, it builds up irregularly on the dish surface, and since it can reflect radio waves, it alters the curve of the dish, which reduces the efficiency factor of the dish, thereby reducing the dish gain and hence the signal quality. Further, snow can accumulate to the point that it causes the dish to collapse under its weight like a tree branch.

For this reason, in all but tropical climates, many earthstations have some form of deicing equipment. One method is to blow hot air over the dish surface. Another technique is to shine infrared lamps on the dish. Still another method imbeds heaters into the dish surface. In very inclement climates, particularly arctic regions, the dishes may be enclosed in heated radomes which allow radio waves in but keep the weather out.

Hail is frozen water. As such, it has few signal effects, but larger hail can damage the earthstation.

Water vapor in the atmosphere, other than the general background amounts that are always present, also comes in the form of clouds and fogs. The electronic effects vary widely and depend on both the average size of the water droplets and the number of those droplets per cubic centimeter. Clouds and fogs differ greatly, and so do their effects. Another rule of thumb, however, is that fog and clouds are of major effect only above about 20 GHz, that is, from Ka-band and upward.

18.3.2 Frequency dependence

Rain attenuation is highly dependent on the frequency being used. For example, at 4 GHz, the C-band downlink frequency, the clear-sky attenuation is only about 0.4 dB; when it is raining 10 mm/hr, the attenuation increases slightly to about 0.5 dB. Below C-band frequencies, attenuation is almost negligible. This is one reason that S-band, L-band, and VHF frequencies are used for mobile links and TT&C: they are almost immune to atmospheric problems.

Going above 10 GHz, however, the problems grow. At Ku-band, in the 10–18 GHz range, in clear weather the attenuation is less than a decibel. But when it rains, the attenuation jumps as high as 10 dB. If the rain rate is higher, the attenuation will be correspondingly worse. Thus, Ku-band is much more susceptible to rain fade than C-band.

The situation is worse still at Ka-band, in the 20–30 GHz region. Clear-sky attenuation is not too bad, typically a few decibels. But when it rains, the attenuation shoots up to 30–50 dB. (Recall that 50 dB attenuation reduces the signal strength to 0.00001 of its original strength.)

There is a huge peak of attenuation around 60 GHz caused mostly by oxygen. This tells us that these frequencies are useless for satellite links to earth, even in clear weather. While most atmospheric problems are caused by water, the peak at this frequency is mostly the result of the properties of molecular oxygen, which is why it shows up even when it is not raining. (As an aside, the impenetrability of the atmosphere at 60 GHz is exactly why some military satellites use these frequencies for satellite-to-satellite links: there is literally "no way on Earth" such signals can be intercepted from down here.)

This has implications for the future of satellite telecommunications. When we fill up the spectrum below about 50 GHz, the frequency range of about 50 to 70 GHz is not useful for space-to-Earth links, so the next useful frequency band will be above about 70 GHz. Radio equipment is much less advanced at those frequencies than down in C- and Ku-bands.

18.3.3 Rain fade and rain fade margins

When you are leasing satellite capacity, you are typically quoted what is called a *continuity of service* figure, which will be a percentage of time, usually per year, that your signal quality is guaranteed to be at or above some level of quality. The figure given is usually specified with the caveat that it is "less solar outages," because these are largely unavoidable. A continuity of service guarantee in a lease contract might quote you, "99.9% continuity of service, less solar outages." If this is the case, you can expect to have a signal of useful quality all but 0.1% of the time, averaged over a year. Continuity of service figures in the range of 99% to 99.95% are common in the satellite industry.

As there are 8760 hours per calendar year, 1% of a year is 87.6 hours, and 0.1% of a year is 8.76 hours. So the quality guarantee quoted above means that you are willing to accept an impaired signal for about 526 minutes a year. The impairment could be anything from a slight degradation in signal, to being completely "off the air."

Since you are guaranteed that you will have a good signal the rest of the time, the signal provider must be prepared to compensate for a rain attenuation of up to some amount to achieve the specified continuity of service. For instance, if the calculations determine that the attenuation may reach up to, but no higher than, 7.5 dB for 99.95% of the year, then you must be prepared to compensate for this amount of attenuation to achieve a 99.95% continuity. This attenuation figure is the amount by which your signal may be reduced by rain for that amount of time, compared to clear-sky conditions. This extra amount of signal that must be provided is called the *rain margin*, or *rain fade margin*, or just *fade margin*. The rain margin is the amount by which you must be able to turn up the power of your transmitter (*EIRP*), or increase the sensitivity of your receiver (*G/T*), to compensate for the effects of the rain. It is a kind of "padding the account," or insurance, to ensure the continuity of service level needed for the application.

Figure 18.2 is a flow chart showing the steps that go into an atmospheric attenuation calculation. First you need to know the kind of climate at your earth-station and what continuity of service level you or your customers need. These then give you a figure for how heavily it rains and how often. Rain rate and the frequency band in use go into standard formulas that calculate the expected amount of attenuation by the atmosphere, per kilometer of air your beam traverses. The path length through the air is determined by the elevation angle to see the satellite.

Since rain is the chief item that degrades the signal, you need to know how often and how heavily it rains to compute how much and how long your signal will be affected. Thus, you will use rain climate models.

18.4 Global rain climate models

Because the major culprit in damaging your signal is rain, radiowave propagation specialists have devised what are called *rain climate models* of the earth. These are maps, with accompanying tables of average rain rate, which characterize the occurrence of rain in different locations. The rain rate, in mm/hr, is given in each model's tables as a function of climate type and percentage of the year.

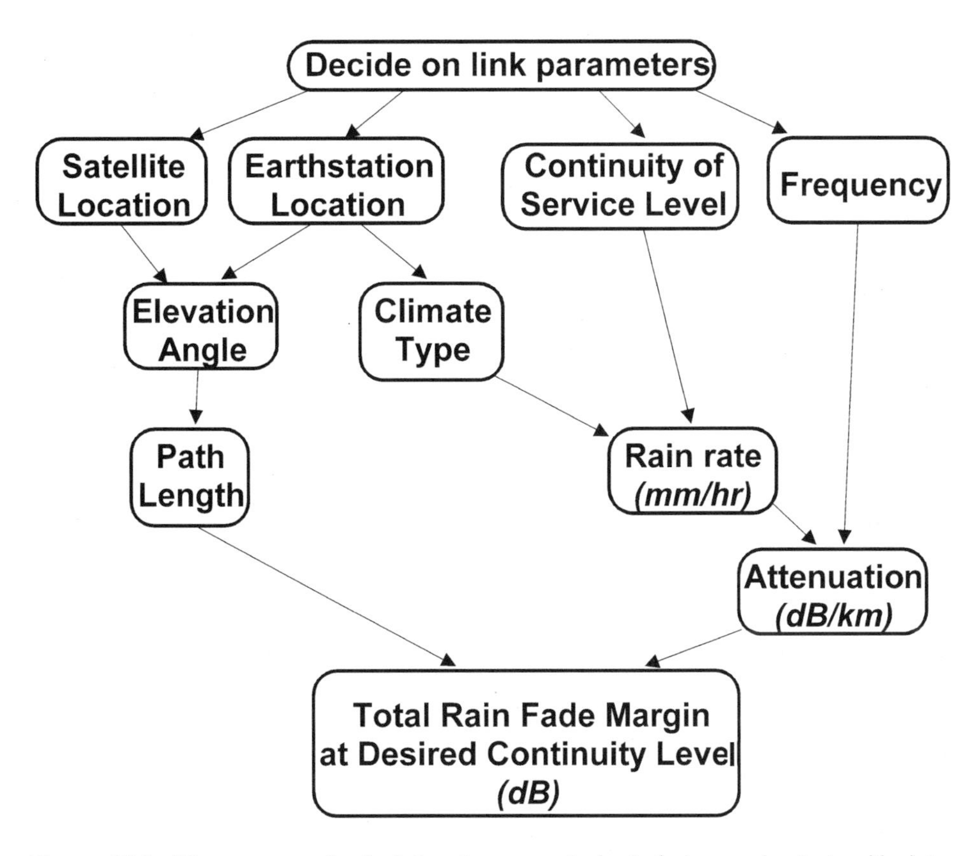

Figure 18.2 The process of calculating the amount of rain fade margin starts with determining the basic parameter of the link: satellite and earthstation locations, service level desired by the customer, and the frequency band to use. Following the flow in the diagram, these then determine the path length of the link through the air, and thus the attenuation amount to be expected.

While there are several such models, there are two that are most commonly encountered. These are the Crane model and the ITU model (sometimes still known by its earlier name, the CCIR model). It should also be noted that these are "broad-brush" models dealing with large areas of the earth, and do not take into account the variations in the microclimate of a particular earthstation's location. Thus, they are approximate. When accurate knowledge of the actual effects of rain and other atmospheric effects on a specific earthstation are needed, you must make actual measurements of the effects over a long period of time.

One word of caution: both rain climate models use letters of the alphabet to denote climate types. The Crane model uses types A through H; the ITU model uses types A through P (skipping letters I and O). There is no relationship between these models and their letter designations. Furthermore, the boundaries—imprecise as they are—are different between the two models. A "type F" climate in one model is not at all the same as a "type F" climate in the other model. So, be careful: when performing a link budget calculation and accounting for rain fade, you need to know which climate type each earthstation is in. If a client tells you her station is "in a type B climate," you need to ask which model she is using. Failure to use the right table will result in very erroneous results (and possibly lawsuits!).

18.4.1 The Crane model

One of the two predominant rain climate models is the Crane model, named for its originator, R. K. Crane.

Figure 18.3 shows a world map with the various climate types, denoted by letters. The climate types are as follows:

- Type A: Polar dry climate; tundra;
- Type B: Polar moderate climate; taiga;
- Type C: Temperate maritime climate;
- Type D: Temperate continental climate;
- Type E: Subtropical wet climate;
- Type F: Subtropical arid climate;
- Type G: Tropical moderate climate;
- Type H: Tropical wet climate.

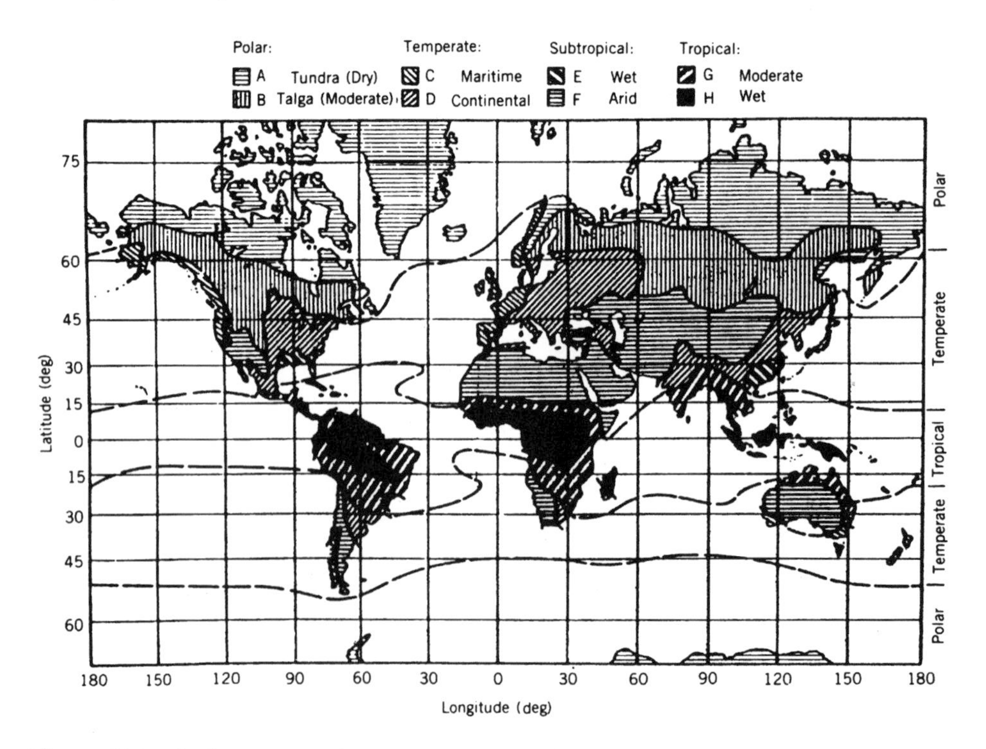

Figure 18.3 A climate map of the world using the Crane model. Eight types of climate are defined by the average frequency of rain rate, shown in Table 18.1. (Source: NASA.)

As you can see, most continental areas of the world are in B, C, and D climates. In fact, some of these climate types are subdivided into subtypes such as B1 and B2, and D1, D2, and D3. Figures 18.4 and 18.5 show more detail of climate types in North America and Europe. You can see the wet subtropical areas of the U.S. Gulf Coast, and the desert regions of Spain and North Africa.

Table 18.1 is the rain rate table to accompany the Crane Model maps. For example, if your earthstation is in a type D2 climate, you can expect that for 0.1% of the year it will be raining 15 mm/hr. In a type H tropical rainforest climate, during the same amount of time it will be raining a lot heavier, 51 mm/hr, thus requiring a much higher rain margin.

Table 18.1 Crane Climate model point rain rate in mm/hr, as a percentage of the year that the rain rate is exceeded. [From Ippolito, 1986]

% Year	A	B	C	D1	D2	D3	E	F	G	H
0.001	28	54	80	90	102	127	164	66	129	251
0.002	24	40	62	72	86	107	144	51	109	220
0.005	19	26	41	50	64	81	117	34	85	178
0.01	15	19	28	37	49	63	98	23	67	147
0.02	12	14	18	27	35	48	77	14	51	115
0.05	8	9.5	11	16	22	31	52	8	33	77
0.1	6.5	6.8	7.2	11	15	22	35	5.5	22	51
0.2	4	4.8	4.8	7.5	9.5	14	21	3.8	14	31
0.5	2.5	2.7	2.8	4	5.2	7	8.5	2.4	7	13
1.0	1.7	1.8	1.9	2.2	3	4	4	1.7	3.7	6.4
2.0	1.1	1.2	1.2	1.3	1.8	2.5	2	1.1	1.6	2.8

18.4.2 The ITU model

The rain attenuation model formulated by the ITU is similar to the Crane model, in that it partitions the world into a number of climate types with corresponding rain rates for various percentages of the year. The ITU model uses 14 climatic regions denoted by letters of the alphabet from A to P (leaving out I and O). To repeat the caution mentioned above: the letters for types in the ITU model are often very different from those of the Crane model. For example, The subtropical moist climate, such as in Florida and the gulf coast states, denoted as a type E in the Crane model, is called a type N in the ITU model.

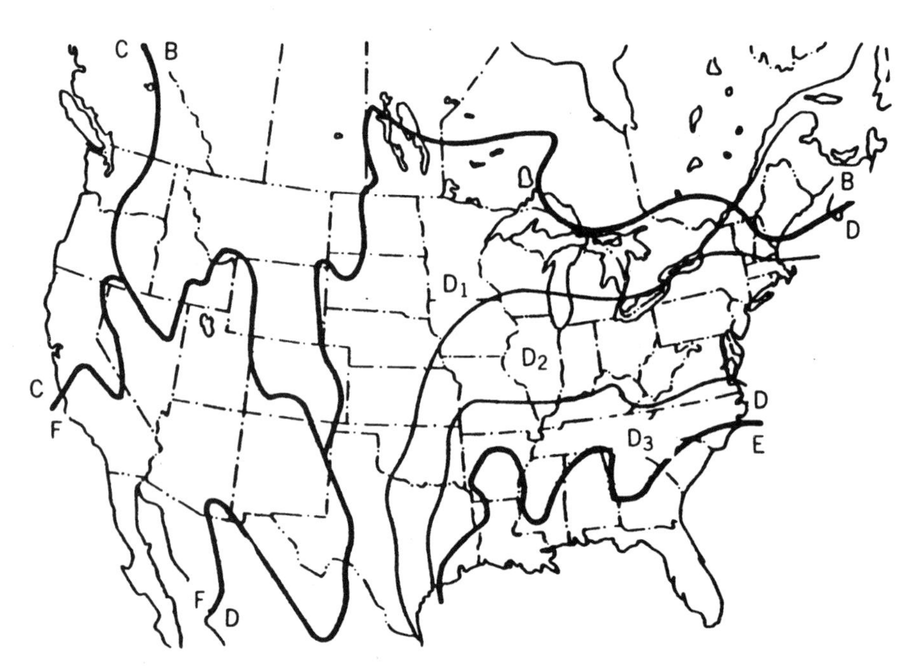

Figure 18.4 A more detailed climate map of North America using the Crane model. (Source: NASA.)

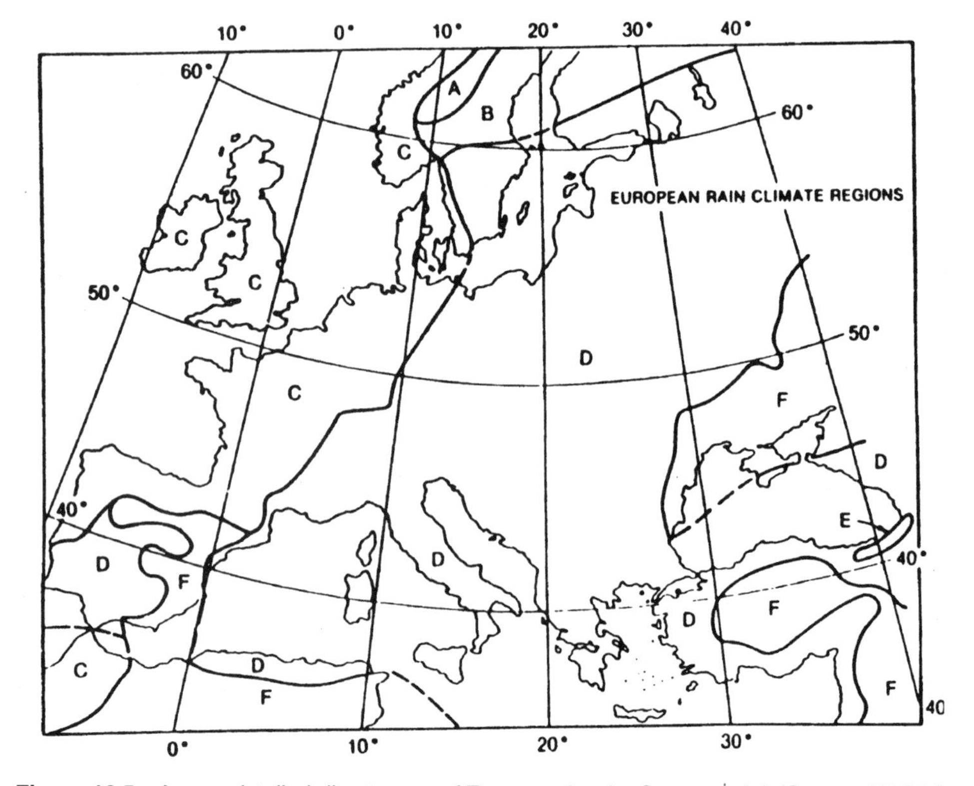

Figure 18.5 A more detailed climate map of Europe using the Crane model. (Source: NASA.)

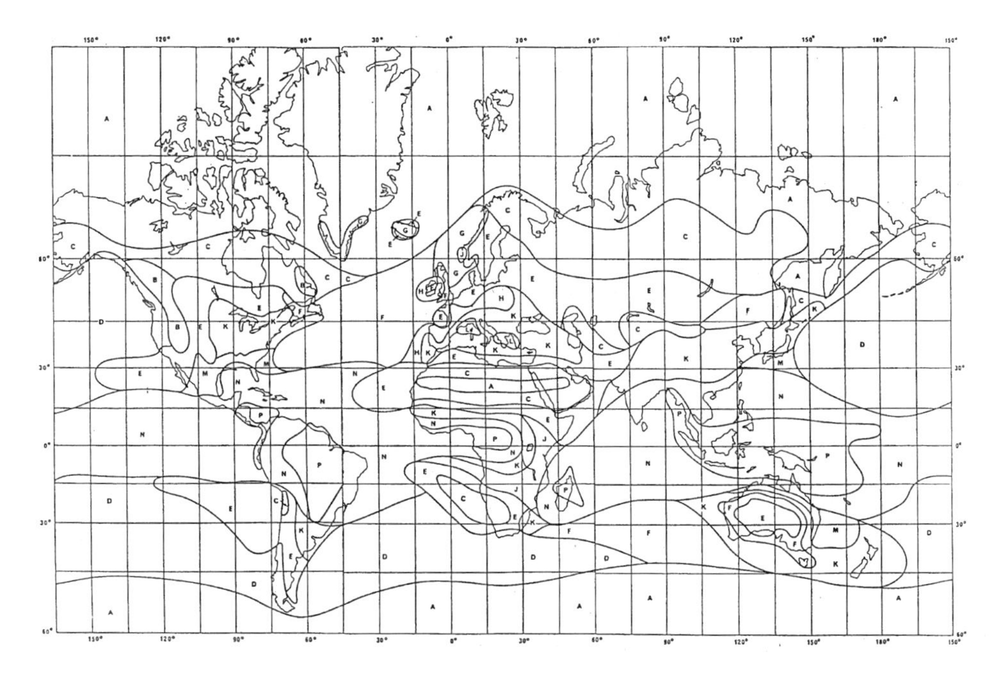

Figure 18.6 A global map of climate types using the ITU model. There are fourteen different climate types. Table 18.2 is used with this map to calculate rain fade. (Image courtesy of ITU.)

Figure 18.6 shows maps of the world with ITU climate types. Table 18.2 is the accompanying rain rate table. It is used in a similar way to the maps and table for the Crane model to estimate the percentage of time a certain number of decibels of attenuation will affect your signal.

Table 18.2 ITU climate model: percentage of the year that rain rate, in mm/hr, is exceeded. [From Ippolito, 1986]

Year %	A	B	C	D	E	F	G	H	J	K	L	M	N	P
1.0	0.5	1	2	3	1	2	3	2	8	2	2	4	5	12
0.3	1	2	3	5	3	4	7	4	13	6	7	11	15	34
0.1	2	3	5	8	6	8	12	10	20	12	15	22	35	65
0.03	5	6	9	13	12	15	20	18	28	23	33	40	65	105
0.01	8	12	15	19	22	28	30	32	35	42	60	63	95	145
0.003	14	21	26	29	41	54	45	55	45	70	105	95	140	200
0.001	22	32	42	42	70	78	65	83	55	100	150	120	180	258

18.5 Noise from the atmosphere

The second effect of the atmosphere on satellite links is the electromagnetic noise emitted by the molecules that make up the atmosphere and the material suspended in it. This is also very dependent on frequency.

The noise is measured in terms of noise temperature, which can be converted to a noise figure in decibels as we saw in Chapter 8. For C-band and below, the effect is slight, and there is little or no increase in noise when it is raining. For instance, at 4 GHz, the clear-sky noise temperature is about 25 K—corresponding to a noise figure of about 0.35 dB—and when it rains, this increases to about 35 K, an increase of only about 0.15 dB in noise level.

At Ku-band, the rain noise is much greater. In clear weather the noise temperature is only between 30 and 40 K; but when it rains, it soars to almost 200K, an increase of about 2 dB. The noise from the atmosphere rises to a bit less than 400 K—or about 3.7 dB—as we go to Ka-band and higher frequencies, and plateaus there.

18.6 Polarization effects

The third major effect that the atmosphere has upon signals is depolarization. Almost every satellite link is polarized—either linearly or circularly—to increase the number of channels available by frequency reuse and to reduce interference. Unfortunately, the atmosphere can partially depolarize the signals.

There are three major causes of depolarization. Once again, one culprit is water in the atmosphere. One cause is raindrops; a second cause is ice crystals. Crystals are not spherical and thus interact with waves vibrating in different directions differently. This effect is most important for frequencies above about 3 GHz, C-band or higher. The third cause is irregularities on the ground or equipment from which waves from your intended signal bounce off of, changing their direction of vibration so that they enter your antenna at various orientations in competition with the desired signal. This is called *multipath*.

18.6.1 Rain depolarization

Contrary to the typical depiction of raindrops seen in illustrations, they are not drops of water elongated and oriented roughly vertically. Left to themselves, raindrops would be perfectly spherical. But when they fall from the sky, the wind resistance they encounter on the way down flattens them out into a shape like a squashed beach ball, technically called an oblate spheroid, that is wider than it is high, with an axis of symmetry tilted somewhat to the vertical. Thus, radio waves passing nearby are influenced differently depending of how the plane of vibration is related to the tilt of the drop. This partially depolarizes the radio waves.

The details of the physics are much too much to get into. Radio propagation engineers measure the amount of depolarization by a number called the *cross-*

polarization discrimination. This value, abbreviated usually by *XPD*, is measured in decibels, and is the ratio of the signal that you want to the depolarized signal that you don't want. The bigger the number the better your signal. (Although a slight caution is warranted here: some people invert this ratio and call it the same thing; in that case, the smaller the value the better your signal.)

There are different XPDs for vertical waves, horizontal waves, and circularly polarized waves. Calculated values are always approximations since one can never know all of the details of the particular rainstorm you might be looking through.

There are, however, a couple of general rules to keep in mind:

- The amount of depolarization goes up as the frequency goes up.
- The amount of depolarization goes up as the rain rate increases.
- Most importantly, the amount of depolarization increases as the total amount of rain absorption increases.
- Circularly polarized waves are affected more than linearly polarized ones by as much as 15 dB.

Propagation experts usually claim that C-band links are rarely affected by more than a few decibels of depolarization. Ku-band and Ka-band signals, on the other hand, may degrade by 10–15 dB sometimes, as they are more susceptible to rain fade. At a given frequency, link engineers will usually calculate a value for depolarization as a function of rain attenuation.

18.6.2 Ice depolarization

Unlike rain depolarization, the depolarization caused by ice is not dependent on the total attenuation of the signal. It occurs when thin, roughly parallel ice crystals high in the atmosphere affect different waves. It can occur even in a clear sky because the ice crystals are not always visible. (It can even change rapidly, since such events as lightning strikes can change the orientation of the crystals.)

The details of ice depolarization are still not well understood. To take this effect into account, an ITU publication suggests that when calculating depolarization, one should lower the XPD by about 2 dB in North America, and about 4–5 dB in northwestern Europe. These values are, of course, an approximation. In general, though, ice depolarization is not the major contributor to the total depolarization effect on satellite links.

18.6.3 Multipath depolarization

This type of depolarization occurs because some signals do not enter your dish directly from the satellite, but may have bounced in after reflections from various objects on Earth. Thus, the signal may have taken multiple paths to your antenna. A similar effect happens when you listen to an FM radio station in your car: you encounter regions where both direct and multiply-reflected signals come to your antenna, causing the sound to flutter.

Most satellite links, except those for the mobile services into handheld telephones or other receivers, use highly directional antennas and link to satellites high in the sky. For such links, multipath degradation is not a big problem except in some urban areas where buildings make good reflectors. However, when you link to satellites at less than 10° elevation angle, the multipath problems get worse—another good reason to avoid low-elevation satellites.

Below about 3 GHz, multipath can be a severe problem. Since these are also the frequencies used for mobile services such as the Big LEOs and Little LEOs, multipath is of greater concern. The problem is complicated when such systems are used in the "concrete canyons" of urban areas or in mountainous regions.

Much more research is needed on these effects. The details of multipath effects are still little understood, and no good model for prediction exists. Fortunately, for C-band and above with links to satellites not too close to the horizon, the problem can be ignored.

There is one other problem caused by the atmosphere. It is called the *Faraday effect*, or *Faraday rotation*. This is the changing of the plane of polarization of a signal by the interaction of the ionosphere layer of the atmosphere with Earth's magnetic field. It is mostly a problem at VHF frequencies.

18.7 Scintillation

One final atmospheric effect on the signal is caused by differences in the density of the air and its components along the path from the satellite to your earthstation. This density change causes the amount of attenuation to vary rapidly with time. This is called scintillation, and is much the same effect as that which you see when a star twinkles. This effect tends not to be as severe as the water-related problems.

One source of scintillation is the rapid changes in the electrons in the layer of the atmosphere called the ionosphere, which is about 200–400 km above the surface. Sometimes the fluctuations can be 10 dB or more and occur for frequencies up to about 6 GHz. Fluctuations can be particularly severe at frequencies below 300 MHz, the frequencies most used for mobile terminals and TT&C. This effect is most common in equatorial and polar regions, and is less common in temperate latitudes. Because ionosphere scintillation is caused by charged particles in the atmosphere, it is also dependent on the electrical and magnetic storms that originate from the sun and hit our atmosphere. It can vary greatly at the same instant from one earthstation to another. In temperate latitudes, the effects occur most commonly around midnight and more often in the summer months.

Scintillation can also be caused by slight variations in the density of the lower few kilometers of the air, and these affect signals from satellites up to approximately 30 GHz. Low-elevation links are affected the most. Fluctuations can typically amount to several decibels.

18.8 Scattering

Scattering is a process that changes the direction of a wave and scatters it over a range of other directions. In other words, it takes part of the desired signal and spreads it around. (This was shown earlier in the discussion of blue-wavelength scattering from sunbeams to produce the blue of a clear sky.) In radio frequencies, a similar effect occurs. There is actually more than one kind of scattering, caused by different physical processes, but all result in reducing the carrier wave reaching the earthstation or the satellite. The two most common types of scattering affecting satellite links are Rayleigh scattering and Ricean scattering. If you want more details, consult a book on radio propagation.

18.9 Improving the quality of a degraded satellite link

For mobile satellite systems, the greatest problems users run into are typically caused by simple blockage: the loss of a line-of-sight between the mobile user and the satellite(s) by buildings, mountains, or foliage. Atmospheric problems are smaller, as the atmosphere has fewer effects at lower frequencies. For the higher-frequency bands, however, atmospheric problems dominate, and become increasingly important as we begin to use the Ka-band and eventually higher-frequency bands.

As we have seen, users of communications links want to be able to use the link at (at least) a specified level of quality (*C/N*, *BER*, *S/N*, etc.) during some period of time, often expressed as a percentage of a year. The determinant of signal quality is the link budget, which we will calculate in the next chapter. This calculation is performed first assuming clear-sky conditions, and then by taking into account the negative effects of all of the possible atmospheric problems to produce a fade margin, in decibels, that the link must be able to overcome to produce the continuity of service level desired. Many times the performance of the link is affected by degradations from a variety of causes. Sometimes each degradation may be minor, but in combination they can limit the performance of the link.

As the frequencies increase above Ku-band, atmospheric problems become more severe, so various ways of compensating for the problems are being invented. There are a number of ways to (partially) reduce the problems caused by the atmosphere; these methods are sometimes called *link restoration* techniques. Not all are ready for use in the field. Here is a sampling:

- *Uplink power control* monitors the fade in the downlink and calculates how much the uplink power should be increased.
- *Downlink power control* can be used with small beams from the satellite to minimize cloud attenuation.
- *Path diversity* can be used with NGSO satellites so that an earthstation receives from more than one satellite at a time (and perhaps from terrestrial repeaters, in the case of some SDARS systems).

- *Site diversity* is the act of using geographically separated earthstations so that both will hopefully not be affected by the same atmospheric conditions, particularly storms; one study suggested that 40 km would be a good separation for use with Ka-band systems.
- *Time diversity* works by sending the same information over the same channel at different times, hopefully avoiding losses; such a technique is useful for mobile users subject to scattering effects.
- *Frequency diversity* transmits the same signal over two different frequencies, either simultaneously all of the time, or occasionally by switching to another frequency when the main one is experiencing problems.
- *Adaptive processing* corrects problems "on the fly" by monitoring the quality of a channel and adaptively altering the datarate or the FEC rate to compensate.

None of these techniques is perfect, but they will become more important as demand for higher-speed services pushes satellite links into higher frequency bands.

Now that we have looked at the various things than control the strength of a signal linking a satellite to an earthstation, it is time to put all of the pieces together to produce a calculation that will tell us the quality of our signal: the link budget.

Chapter 19

Putting It All Together: Link Budgets

OK, this is where you become a satellite telecommunication engineer. (Well, not quite: we have left out a lot of nasty mathematics and physics and other details, but you have the general idea.) In previous chapters we have assembled all of the components that go into a link budget: we have compiled all of the things that either contribute to or detract from the quality of a signal sent between an earthstation and a satellite. While most of the other formulas mentioned in the book have been banished to Appendix E, the link budget is defined and discussed in detail in this chapter due to its importance.

When a customer has specified a desired quality of the final received signal, a satellite engineer can then calculate the necessary operational parameters to achieve that goal. This is the purpose of a link budget. It is the computation that tells us how well we can accomplish a link.

We are now going to calculate a link budget. While there are a lot of numbers in this chapter, don't panic. They are almost all simple additions and subtractions of items that have already been explained. The main purpose of this chapter is to show you what a link budget is rather than expecting you to do one yourself. Most importantly, you should take away from this chapter an understanding of what items in the link have the largest effects on the user's signal quality, and what you can and cannot adjust to make it better.

The analogy is the common household budget: add what comes in, subtract everything that goes out, and see what is left. That "bottom line" is a number, but you cannot say whether that number is good enough without knowing what you need. You'll need a better "bottom line" if you want to vacation on the Riviera than if you want to go camping at a nearby park, so knowing the user requirement is absolutely necessary to determining if the budget is sufficient for your purposes. If it is not, you have to make choices about increasing income or decreasing outflows.

The bottom line for a transmission is signal quality. As you will remember, the quality of a signal is usually measured by the value of the carrier-to-noise ratio, C/N, expressed in decibels. After we have calculated the C/N, we then examine the number and compare it to the requirements for the type of signal that we are trying to send, based on the (sometimes very messy) details of the coding, modulation, and multiplexing used with the signal and the desired subjective quality of the information that we are sending.

There are standard telecommunications references, some of them listed in Appendix C, which will tell you, for a given kind of signal and a calculated *C/N*, just what quality of signal you will receive. The details will, of course, depend on the nature of the information. References for television transmission will give you the correlation between *C/N* and on-screen video quality, references for telephone transmissions will correlate *C/N* with audio quality, and references for digital signals will correlate *C/N* with bitrate and bit error rate.

A link budget is calculated for each of the two links that make up a full communications path from sender to receiver, the uplink and the downlink. Later, we will see how to combine the two link budgets into an overall channel link budget.

19.1 The link budget

There are three major items that contribute positively to the *C/N*. (Recall that all values must be expressed in decibels.) These are

- EIRP: the effective strength of the transmitter, either at the earthstation or on the satellite;

  EIRP is the sum of the transmitting amplifier power and the gain of the transmitting dish.
- G/T: the sensitivity of the receiving equipment;

  G/T is the gain of the receiving dish minus the system noise temperature of the receiving electronics.
- a constant value of 228.6 dB that is added to the calculation.

  This is simply the decibel equivalent of a physical constant that comes in due to the conversion of noise temperature (in kelvins) to noise figure (in decibels). (For those with some physics background, this is Boltzmann's constant expressed in decibels; for those for whom this is gibberish, forget it!)

In addition, there are four factors which detract from signal quality, again all in decibels:

- Space loss: the amount of power lost because of the distance the signal must travel from sender to receiver.
- Atmospheric losses: the amount of power lost because of absorption and other effects when the signal passes through the air. The calculation is usually done first under clear-sky conditions, putting an average value in for this term that depends on the frequency in use. For C-band, 1 or 2 dB is common; for Ku-band, 5 dB may be typical; for Ka-band, 10 dB may be a good value for atmospheric loss.
- Miscellaneous other losses, such as from slight mis-aiming of the dish, or slight error in polarization orientation, losses in the connections between antenna and amplifier, etc. Usually from zero to a few dB are assumed here unless you know better.

- Bandwidth: the bandwidth of the signal in use expressed in decibels.
  This term appears as a reduction in signal quality because the wider the bandwidth your receiver accepts, the more interfering noise it receives from the environment.

Figure 19.1 shows these contributions to a link.

Here is the full calculation for a link budget (drumroll, please!):

$$
\begin{aligned}
C/N = &+ EIRP \\
&+ G/T \\
&+ 228.6 \\
&- \textit{Space Loss} \\
&- \textit{Atmospheric Losses} \\
&- \textit{Miscellaneous Losses} \\
&- \textit{Bandwidth Used}
\end{aligned}
$$

Let's try a couple of examples to see how this works.

Suppose that we are using a satellite to send an analog television signal to an earthstation in the C-band. Recall that the C-band downlink frequency is about

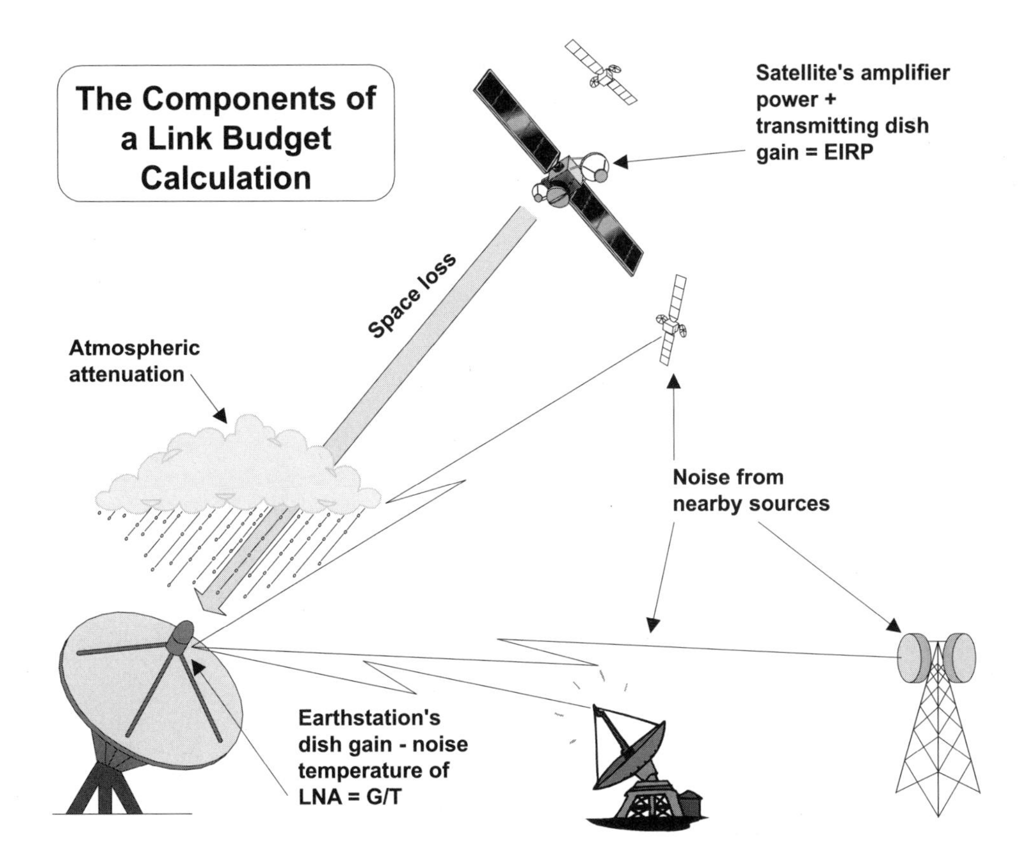

Figure 19.1 The components that go into calculating a link budget.

4 GHz. At the satellite, which is doing the transmitting, let's say that the satellite uses a 24-MHz bandwidth transponder, the SSPA of the transponder has a power of 25 W, and the downlink antenna on the satellite has a diameter of 2 m. From the formulas in previous chapters, we can calculate that the 25 W corresponds to 13.98 dBW. The gain of a 2-m dish at 4 GHz can be calculated to be 35.84 dBI. Adding those two numbers together gives us an EIRP from the satellite of 49.82 dBW.

Further, imagine that the receiving earthstation has a diameter of 3 m, and a system noise temperature of 720 K. The gain of this 3-m dish at 4 GHz is 39.37 dBI, and the noise is equivalent to 28.57 dB. Subtracting the noise from the gain, we arrive at a figure of merit, or G/T, of the receiving earthstation dish of 10.80 dB.

Under clear-sky conditions we'll assume an atmospheric loss of 2 dB, and miscellaneous losses of 3 dB.

The bandwidth contribution is the 24 MHz expressed in decibels, 73.80 dB-Hz. The space loss will be 196.22 dB at 4 GHz, and there is the constant to be added in.

Thus the calculation goes:

$$\begin{aligned} C/N &= 49.82 + 10.80 + 228.6 - 196.22 - 2.00 - 3.00 - 73.80 \\ &= 14.20 \text{ dB}. \end{aligned}$$

Is this good or bad? To find out, we have to go to some standard engineering reference books to find that this is a pretty strong, high-quality signal to deliver a television picture. For example, we know from experience that we should aim for a C/N of more than 6–10 dB in order to receive a good-quality signal if the signal uses frequency modulation.

Remember that this link budget was done under an assumed clear-sky attenuation. We could next subtract the likely rain attenuation based on the receiving earthstation's climate type and desired continuity of service. We see that even if the rain was fairly heavy, say with an attenuation of 5 dB, we would have enough power in the link to provide the service required.

Figure 19.2 is a graph of all of the ups and downs experienced by a signal sent from one earthstation to another.

19.1.1 Turning the formula around

We often want to know how powerful a satellite must be to perform some specified link. In that case, we know the *C/N* that we require to accomplish the link and the sensitivity of the receiving earthstation, so we turn the formula around and solve for the necessary EIRP of the satellite. When we have this number, we can then look in the satellite directories to see which satellites are visible from the earthstations that have the power (EIRP) required to give us the signal that we want. In our household budget analogy, this is rather like knowing how much our planned vacation will cost, and juggling the family finances to come up with the money needed.

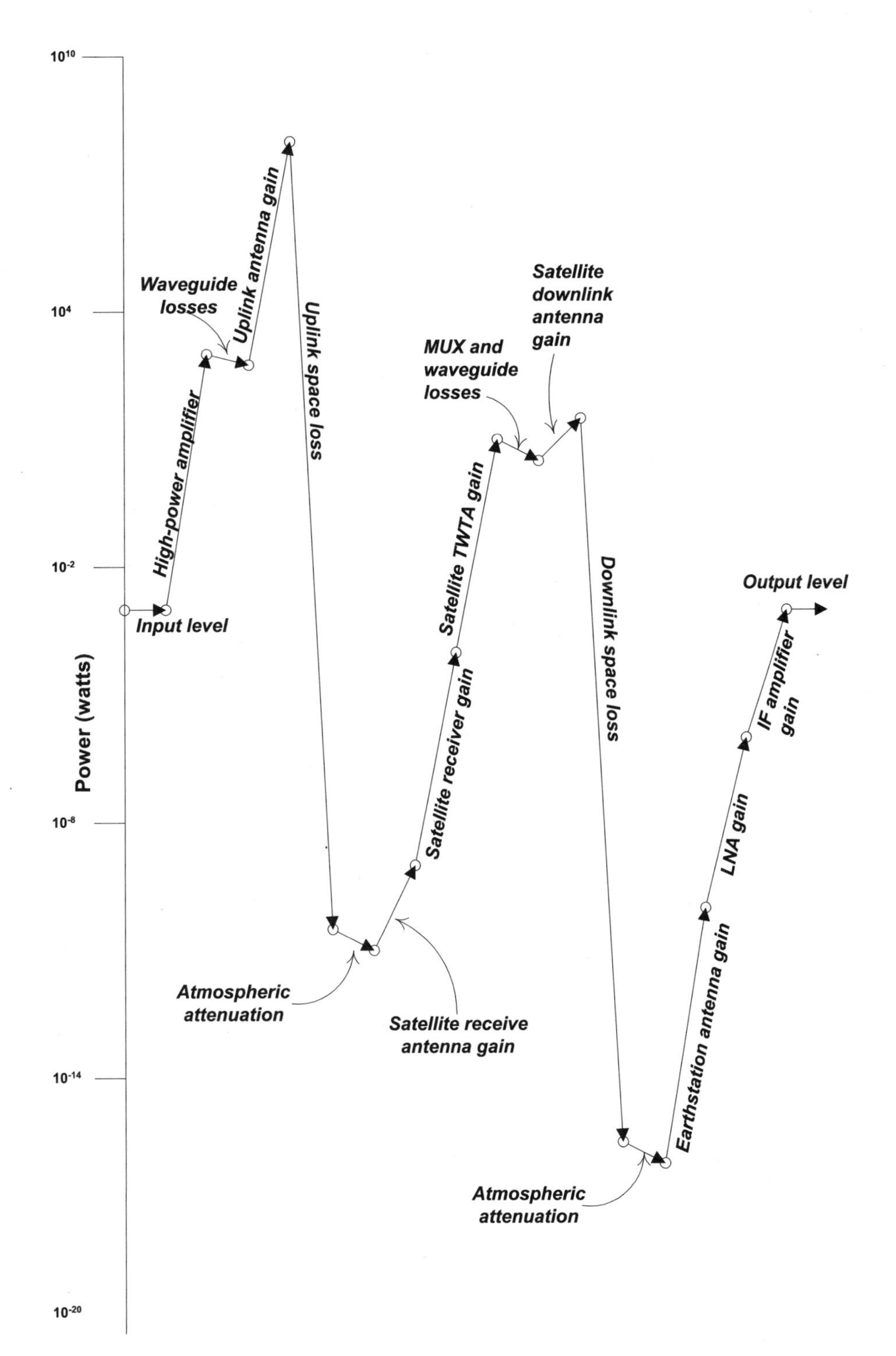

Figure 19.2 A flow chart of the changes in signal power in a link between a satellite and an earthstation. The signal power changes over a huge range of perhaps 10^{21}, which is a thousand billion billion times.

For example, suppose we know that we want to receive a 27-MHz-wide Ku-band downlink signal at an earthstation with a *G/T* of 23 dBK, and we want a minimum *C/N* for the link of 14 dB. Assume that the space loss is 206.4 dB. Since it is Ku-band, we will assume a clear-sky attenuation of 5 dB, and a miscellaneous loss of 1 dB. Then the calculation goes

$$\begin{aligned} 14 &= EIRP + 23 - 206.4 - 5.00 - 1.00 - 74.3 + 228.6 \\ &= EIRP - 35.10; \end{aligned}$$

or, solving for the unknown value, we see that the required downlink power of the satellite must be an *EIRP* = 49.1 dBW. We then consult a list of GSO orbital positions to see which satellites with at least that power are located in a part of the Clarke arc visible both to the receiving earthstation and also to the originating uplink earthstation.

Of course, a similar link budget must be calculated for the uplinking station, to make sure that the satellite receives the minimum signal strength it needs to receive the signal, called the *threshold.* If the uplinked signal is at least as strong as the threshold value, the satellite transponder can pick it up, amplify it, and downlink the signal. If the uplink power is too low, the transponder will not respond to the signal. Note that it is a case of raw power, not C/N, that is important in reaching the threshold.

A popular alternative is to calculate the signal quality per unit bandwidth. Often the unit of bandwidth used is 4 kHz, a relic of the old analog telephony channels. This is called C/N_0 and is called the *carrier-to-noise density ratio*. It is the same as C/N, but without the bandwidth term. Some engineers prefer to use C/N_0 in their calculations.

19.1.2 Digital signal quality

In the case of a digital signal, we might use the equivalent measure of quality called bit error rate, *BER,* as the desired end-user measurement, and the energy-per-bit-to-noise density, E_b/N_0 as our quality figure for the link. The BER will depend on the C/N, the error-correction technique in use, and the datarate. The formula for converting between them is given in Appendix E.11.

From the value of E_b/N_0, engineers can calculate the BER obtained using various types of digital modulation (BPSK, QPSK, etc.) and various error-correction techniques.

19.2 What's fixed and what's changeable?

Although it was mentioned in an earlier chapter, it is worth emphasizing here that some of the parameters going into the link budget are under your control, and some may not be. (The most obvious item is that the 228.6 value is fixed by the laws of physics!)

Take a look at the seven parameters that go into the *C/N* calculation. Once you have decided the kind of signal that you are sending, you have fixed the bandwidth to be used. Once you have chosen a frequency to use, you have fixed the amount of space loss and the amount of (clear-sky) atmospheric loss. The value of miscellaneous losses is an approximation based on your knowledge of the earthstation's operations.

Thus, the two main items that you can adjust are the *EIRP* and *G/T*. As we have mentioned several times before, if the transmitter is strong (large *EIRP*), the sensitivity (*G/T*) of the receiver can be low, and vice versa. Both items are partially determined by the frequency band you choose to use. Once that is decided, then *EIRP* is determined by the size of the dish, and *G/T* is determined by the noise environment and the quality of the LNA electronics. So these are the two operational parameters most often juggled to obtain a *C/N* high enough to give the quality of the signal you want.

Both the uplink and downlink will have a *C/N*. It is the overall C/N that controls the quality of the signal at the destination. If you are using a bent-pipe (nonprocessing) satellite, the C/N of each link must be combined to calculate the overall C/N. This is a bit messy mathematically, so the details are left to Appendix E.12. If you are using a satellite with onboard processing, that processing controls the downlink C/N and thus the quality of the signal at the destination. In this case, the averaging explained in Appendix E.12 does not apply.

We have now reached the climax of the technical part of the book. Once you understand what goes into a link budget, you have a grasp of the contributions of the transmitter, receiver, and intervening space. (If you don't want to do the calculations yourself, there are several commercially available link budget software packages available, as well as some interactive ones online.) So far, we have considered only the case of a simple link from one earthstation, through one satellite, to another earthstation. In the real business world of satellite telecommunication, however, we often have the need for many earthstations simultaneously linking to and from a single satellite. The techniques for doing this in an orderly manner are called multiple access, and are discussed in the next chapter.

Chapter 20

Multiple Access: Many Users on One Satellite

Satellites are capable of communicating with more than one earthstation at once. The suite of technologies, protocols, and procedures that make this possible—and that avoid (or minimize) the interference of possibly thousands of earthstations simultaneously linking to a single satellite—is called *multiple access*. Like every other technical decision, the choice of an appropriate multiple-access method depends on a detailed knowledge of the criteria of the communication being performed.

As we saw back in Chapter 3, there are many such criteria, such as power restrictions, interference restrictions, spectrum efficiency, reliability, timeliness, and cost, to name a few, that may be important to the people communicating. Again, as with everything, there are trade-offs: no technique can maximize all of the operational characteristics at the same time. In a way, it is rather like the (in)famous "Iron Triangle" of criteria for completion of any project: you can get the job done using three criteria—quality, speed, cost—but you can maximize only two of the three at any one time!

One of the most important issues that determines a good multiple-access design is the topology of the network: because different multiple-access methods have advantages and disadvantages relative to each other, different methods may be used on different links of the same communications applications. Furthermore, methods may be combined to increase the efficiency of communication.

20.1 Compact system descriptions

Sometimes satellite engineers describe their systems in a very obscure and compact way by a string of letters such as FM/SCPC/FM/FDMA. To the uninitiated, this arcane alphabet soup is acronymical overkill. But this is just a shorthand way of describing a signal flow from source to satellite. Each section is one of the blocks in the transmission of information, in the order the signal flows: 1) source modulation to produce a baseband signal; 2) baseband multiplex method to combine several users' signals into a broadband signal; 3) modulation method used to send the broadband signal to the satellite at radio frequencies; and 4) the multiple access method used by all of the earthstations simultaneously using the transponder.

Thus, the example given above tells us that we have some frequency-modulated original signals (probably telephone calls), each of which is assigned to a unique carrier, that the broadband combined signals are frequency-modulated onto the satellite link radio frequency, and that all of the stations linking to the satellite used frequency division to partition the capacity of the transponder in use.

It is worth emphasizing here that most communications satellites are simple bent-pipe relayers of radio signals, so the multiple-access techniques are performed by equipment at the earthstations linking to the satellite, not aboard the satellite itself. Later in this chapter we will discuss an example of multiple access using onboard processing.

There are four basic operational parameters that may be partitioned to allow multiple users to share a satellite or a transponder on a satellite. These are (1) space, (2) frequency, (3) time, and (4) code. In addition, to provide more control and flexibility, two (or more) of these can be combined.

We will continue to use the earlier analogy of a big cocktail party in the ballroom of the Embassy of the Duchy of Grand Fenwick. Space-division multiple access is analogous to each conversation being given a separate part of the room; frequency division is analogous to each speaker being asked to use a different tone of voice; time division is analogous to each user being told speak in turn; code division is analogous to having each speaker use a different language.

20.2 SDMA: space-division multiple access

SDMA is the physical separation of different signals. In our party analogy, it would involve separating conversations in the ballroom so that they are far enough that they are not heard by others. Another example would be the carrying of separate telephone calls in separate wires, all of which used the same cable duct to hold the wires.

Recalling that all FSS satellites in a given band (such as C-band) operate on the same frequencies, we see that the requirement for satellite spacing, explained in Chapter 11, is just an example of SDMA. Terrestrially, the same principle of physical separation and limitation of power allows radio and television stations in different parts of the continent, or cellphones in different areas of a city, to reuse the same frequencies.

In the satellite business, SDMA is more commonly referred to as *frequency reuse* because it allows almost identical satellites to operate independently by separating them along the Clarke orbit and limiting their power.

For a single satellite, SDMA, or frequency reuse, is accomplished by having distinct radio beams from the satellite pointed at different parts of the earth, as in Fig. 20.1. These may be categorized by the regions they serve, such as hemispherical beams, regional beams, national beams, or spot beams. If the beams do not overlap, then the same frequencies can be used in each beam, effectively multiplying the bandwidth available by the number of beams. For instance, if an FSS satellite over

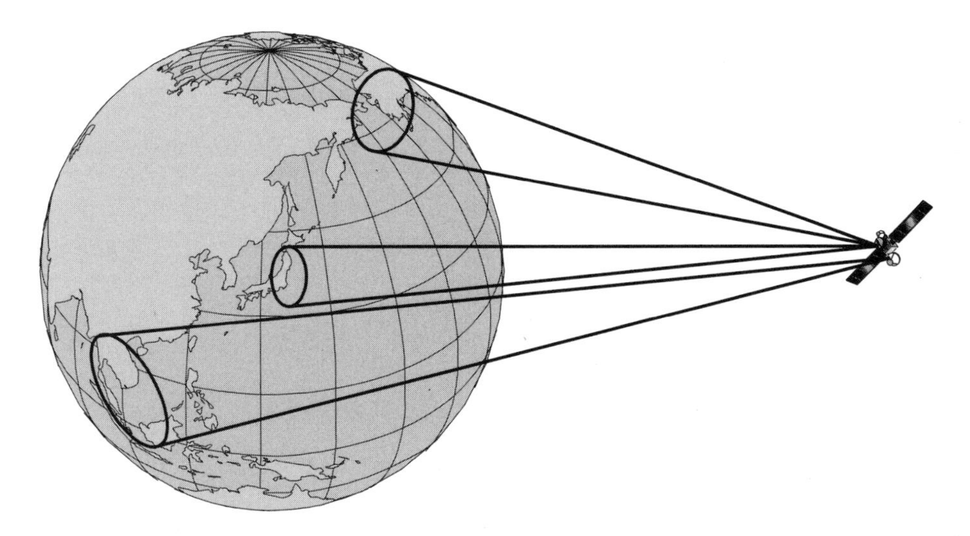

Figure 20.1 In space-division multiple access, SDMA, separate beams from the satellite partition the service area into separate regions. Since the beams do not overlap, the same frequencies can be used in every beam. If two beams should overlap, cross-polarization can be used to separate the signals. Thus SDMA is often called frequency reuse.

the United States had a separate beam for each of the four CONUS time zones, then the effective bandwidth of the satellite would be 2000 MHz, four times the official allocation.

We have seen that it is electronically impossible to make sharp boundaries for such beams, so how do we avoid the interference that would occur where the edges of such beams overlapped? The answer is polarization, which is actually a kind of SDMA. If the adjacent beams are cross-polarized, it will minimize interference.

Satellite examples are the NGSO telephony satellite constellations, such as GlobalStar. Here we have something like a cellular telephone network "turned on its head," with the cells defined by the beams from the satellite. But in this case, since the satellites are moving, so are the beams. Keep in mind that each of these cells is moving across the surface of the earth at several thousand mph. During any particular telephone call you may make using this system, you may be using several different satellites as the beams move to cover you, move away, and another satellite comes into view.

For GSO satellites, the ability to use SDMA depends on the use of the satellite. A satellite that is designed for mobile communication for users anywhere the satellite is visible must have a single global beam, 17.3° in diameter, that covers the entire region of the earth that the satellite can see. Since there is only one beam, there can be no frequency reuse except for cross-polarized beams.

In contrast, consider one of the New Skies Satellite's GSO. Figure 20.2 shows the beam patterns of this satellite. Note that three large beams cover three service regions in North America, northwestern South America, and a region in the middle of South America. Another regional beam handles traffic from Europe and a bit of North Africa. In this way, New Skies can try to accommodate the traffic demands

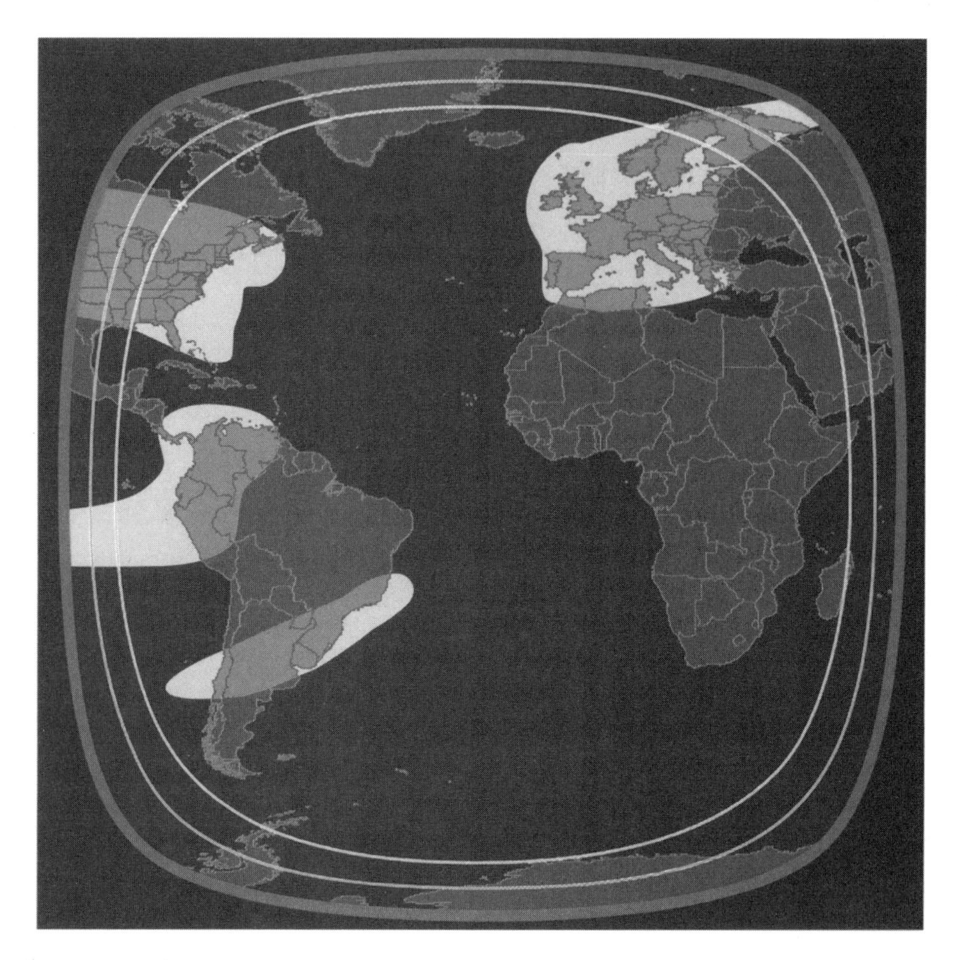

Figure 20.2 Satellite system operators often use regional beams to serve customers where the traffic is greatest. The light areas shown here are four regional beams. (Image courtesy of New Skies Satellite.)

that exist and are forecast for this part of the world during the lifetime of this particular satellite.

As another example, look at Fig. 20.3. Eutelsat's AtlanticBird 3 has several fixed beams, but this footprint shows several possible areas its steerable spot beam might point to, to serve customer demands. Because the beams are small, the EIRP at the center of the beam is quite strong.

Going back to our cocktail party analogy, you can see that as the room fills up, eventually people and conversations will be too close to one another to avoid mutual interference. Thus, you reach a point of diminishing returns if you try to pack too many separate beams into an area.

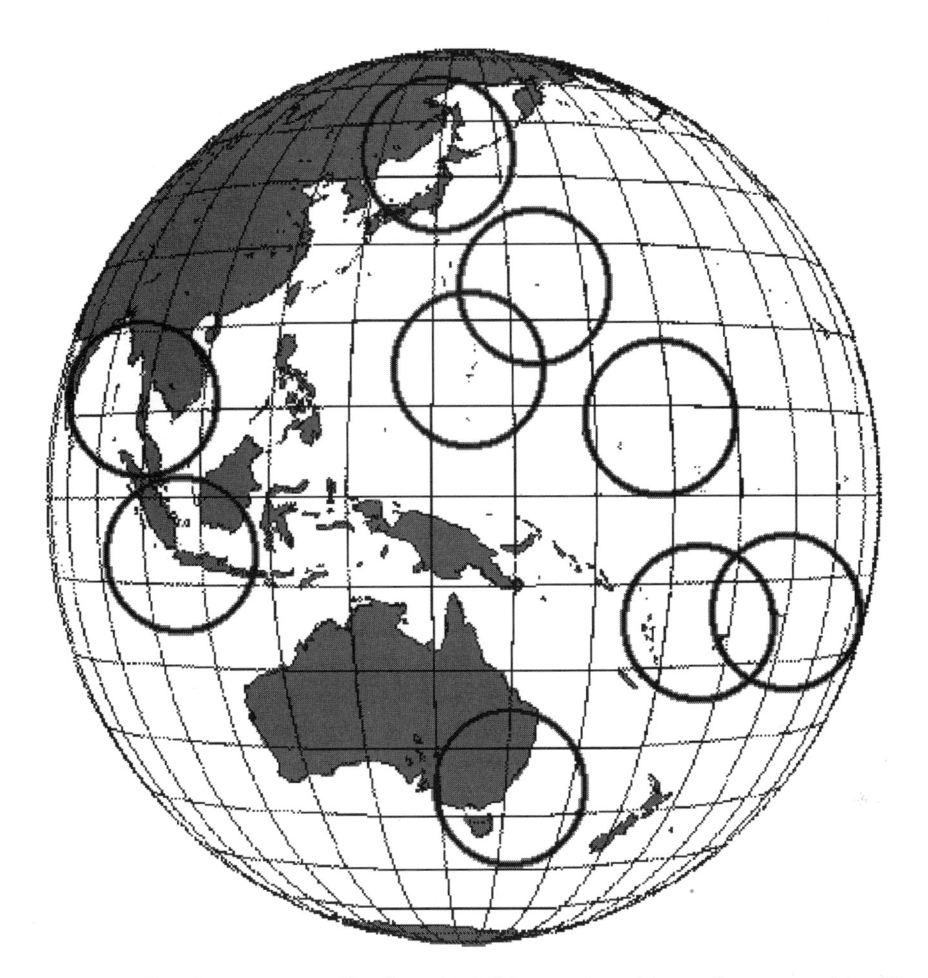

Figure 20.3 The Japanese satellite Superbird-C has a steerable spotbeam capable of being directed toward markets over a wide area of the western Pacific Rim. The circles shown are some typical possible areas of service. (Image courtesy of Superbird.)

20.3 FDMA: frequency-division multiple access

Another common way to subdivide the communication capacity of a satellite, and allow multiple users to use it simultaneously, is to have different frequencies for different users. Using our cocktail party analogy, suppose in an attempt to minimize interference, we insisted that each couple holding a conversation talk at a different pitch from others in the room. (It's not practical, but you get the idea.)

In fact, almost all satellites are inherently frequency-divided, for they take the allocated bandwidth and divide it into a number of narrower bandwidths called transponder channels. Early satellites typically had a dozen transponders; later satellites commonly had 24; today satellites have anywhere from a dozen to several dozen transponders. This partitioning of the satellite's total bandwidth is the concept of the frequency plan explained in Chapter 13.

The primary difference between the subdivision of the full allocated bandwidth of a satellite and the subdivision of frequencies within a transponder is that the former is wired permanently into the design and operation of the satellite, assigning each transponder a bandwidth and center frequency that usually cannot be altered. In contrast, all FDMA assignments subdividing the capacity of a transponder are done at the earthstations, and can be changed if needed.

Depending on the use, each transponder may or may not be subdivided. For instance, if the transponder's entire bandwidth is being used for a television signal, no subdivision is possible (except, perhaps, for inserting subcarriers in unused space beside the video carrier). However, if the traffic intended for the transponder consists of smaller bandwidth signals, the bandwidth of the transponder can be split up into narrower bandwidths. Figure 20.4 shows the concept of what is called generic FDMA.

To avoid interference among the individual signals, there are unused frequency spaces between signals called *guardbands*. Frequency partitioning a transponder is rather like taking a big sheet of wood and sawing it into slats: the more slats you make, the more wasted space from the sawing. To partition a 4-by-8-ft sheet of wood into two sheets requires only one cut about an eighth of an inch wide, and thus the waste of only a small amount of sawdust from the kerf. But to make the big piece into, say, 100 slats would require 99 cuts and waste a total of more than a foot of the width—a quarter of the capacity—of the sheet. Thus, you reach a point of diminishing returns when you try to subdivide a transponder bandwidth too narrowly.

One big disadvantage of FDMA is that since the transponder is a single electronic device, and you are putting multiple signals through it, they cause and are subject to the intermodulation distortion, IM, mentioned in Chapter 8. In fact, IM is usually the limiting factor on the number of simultaneous signals that can be put through a frequency-divided transponder.

Just as with frequency-division multiplexing, you can subdivide the allocated bandwidth into either a random assortment of sub-bands, or into a uniform group

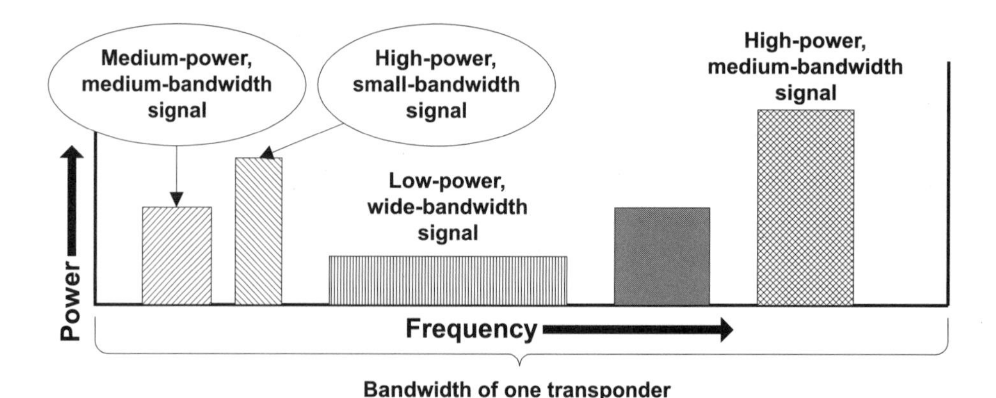

Figure 20.4 Frequency-division multiple access assigns different earthstations distinct separate frequency bands within a transponder. As with frequency multiplexing, the differing powers and bandwidths of the signals mean that they must be carefully controlled as to their placement within the transponder bandwidth.

of bands. If the transponder is subdivided into a number of identical frequency slots each carrying a single signal, it is called *single carrier per channel*, or *SCPC*. When more than one independent signal is sent on a single carrier wave, this is called *multiple channel per carrier*, or *MCPC*. Figure 20.5 shows one example of SCPC on a satellite transponder. Modern SCPC systems can typically carry 800 to 1600 voice or low-speed data channels on a 36-MHz bandwidth transponder.

Generic FDMA is used when different users have different, or perhaps changing, requirements for bandwidth. Some coordination scheme is needed to avoid interference. This may take the form of a pre-assignment of bandwidths if the signals are constant, or of a signaling channel if the bandwidth assignments change.

There are several variants of FDMA techniques. In one older system, analog telephone channels are combined by frequency-division multiplexing (FDM), and the resultant broadband signals are frequency modulated (FM) onto a carrier wave and sent to a satellite, with each separate signal using a different frequency slot within a particular transponder on a particular satellite. In condensed engineerspeak, this could be referred to as an FDM/FM/FDMA system.

Another variant might be dubbed a PCM/TDM/PSK/FDMA system for telephone channels digitized by pulse-code modulation, with the channels time-division multiplexed together and sent to the transponder as a phase-shift-keyed digital signal occupying part of a transponder.

SCPC is used when the traffic consists of approximately identical signals, such as datastreams or telephone calls. SCPC can be used either for analog or digital signals and is a good system for multiple users each of which has a low traffic load. A good example in the satellite area is an old technique called SPADE. It was invented in the early days of satellite communication by Comsat Laboratories for Intelsat to partition its transponders for telephony.

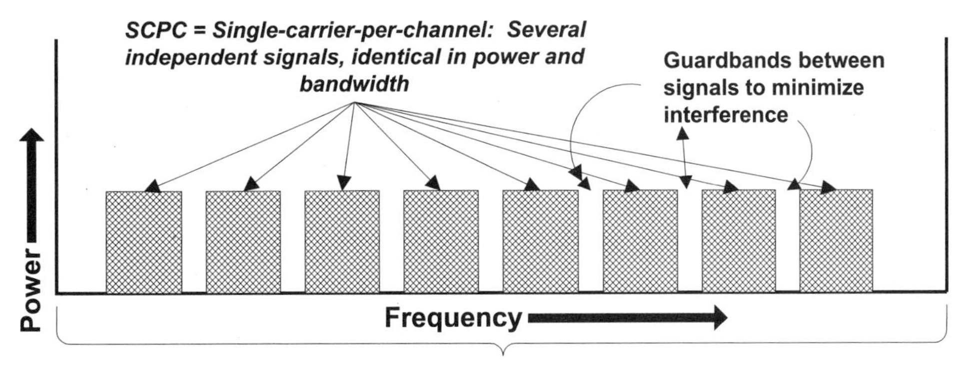

Figure 20.5 SCPC, single carrier per channel, is used when the traffic from different earth-stations consists of identical signals. Thus the bandwidth of a transponder of a satellite is thought of as consisting of a large number of identical power-bandwidth boxes, any one of which can carry many signals. The assignment of a specific signal to a specific box may done by consulting a control channel on the transponder or by assignment by a controlling earth-station.

20.3.1 SPADE

SPADE stands for "SCPC PCM multiple-access demand-assigned equipment." Looking at these terms individually, the SCPC part indicates that the system will assign individual signals to individual frequency bands within a transponder's bandwidth; PCM says that the telephone calls will be digital pulse-code-modulated signals (see Chapter 7); the multiple-access part says that many users can use the system; demand-assignment means that each user uses it as needed, on demand, but does not have a permanent assignment to some frequency slot. SPADE is obsolete now, but provides a good example of an FDMA system. (SPADE is an example of that horrid object, the second-order acronym: each letter itself stands for another acronym!)

Figure 20.6 shows how it works. A satellite has a number of 36-MHz bandwidth transponders. Considering only a single transponder, the 36 MHz is subdivided (by equipment in the earthstation) into 800 frequency slots, or channels, each 45 kHz wide, which was the bandwidth necessary in those days to carry a one-way voice signal. Channels are assigned in pairs to carry the two-way signals

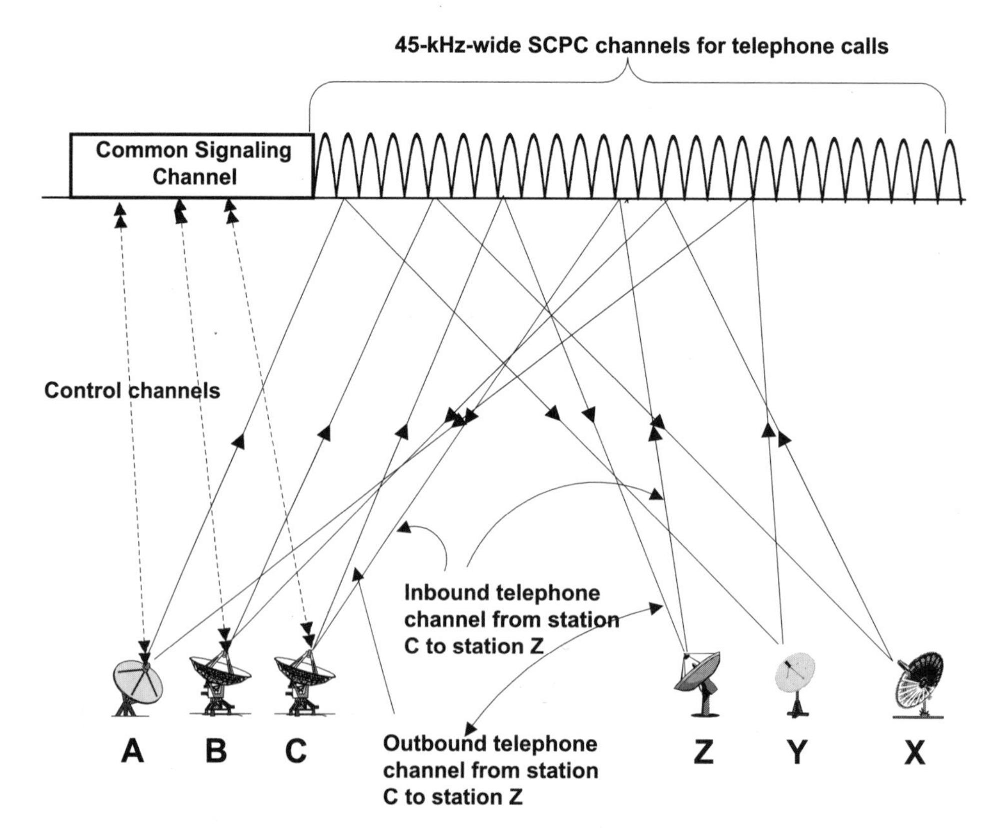

Figure 20.6 SPADE is an example of a frequency-division, demand-assigned multiple-access system. A single transponder is divided into 800 narrow SCPC channels. A common signaling channel keeps track of which earthstation is using which channels.

of a telephone call. A few channels at the edges and at the center of the transponder were left vacant as guardbands, so each transponder has the capacity to simultaneously handle 397 telephone calls.

To keep order, there is a separate "common signaling channel" band at one end of the transponder's bandwidth that keeps a record of the channels in use. When any earthstation needs to send a telephone call over the satellite, the earthstation equipment looks at this CSC for vacant channel pairs. If it finds one, it places the call over this channel pair and the CSC takes note of that. It does not matter where the earthstation originating the call was, nor the destination of the call, as long as both earthstations can see the satellite. Each telephone call was completely independent of all others. Thus, channel pair #345 might at one time be carrying a telephone call between New York City and London, while channel pair #346 was handling a call between Montreal and Paris. But once the NYC-London call hung up, that same channel pair became free, as noted on the CSC, and could be used then for a call between, say, Mexico City and Madrid. Thus, we see that the system was, indeed, demand-assignment and multiple access.

The biggest advantage of FDMA systems is their simplicity and the independence of each earthstation. Other than using separate channels, earthstations require no coordination among them. This keeps things simple from the earthstation's point of view. On the other hand, a separate modem is needed at the earthstation for each channel.

The biggest disadvantage of FDMA is the inherent intermodulation interference mentioned earlier, which requires backing off the power of the amplifiers in the system and the loss of useful bandwidth due to the increasing number of guardbands. This limits the spectrum efficiency of FDMA systems, and thus their capacity. Special techniques, such as using narrower bandwidths, can somewhat increase capacity, but FDMA systems, compared to most time-division systems, are lower capacity.

20.4 TDMA: time-division multiple access

Another way of keeping different signals separate is to assign each a different time slot in which to transmit. In our cocktail party analogy, we could ask each couple to speak for a few seconds, then stop so the next could speak, and so on. During each brief exchange, the speakers would have the undivided capacity of the room at their disposal, and the undivided attention of everyone in the room. (Of course, it would be polite to pay attention only to remarks addressed to you.)

Unlike FDMA, with *time-division multiple access*, *TDMA*, each earthstation briefly uses the full bandwidth of a transponder (or part of a transponder if it is partitioned into frequency bands). However, whereas in the case of FDMA each earthstation (at least for the duration of traffic) uses a part of the transponder continually, in TDMA, each earthstation only links to a transponder for brief periods of time, usually milliseconds long, and during those periods sends bursts of data often called packets. (While TDMA could be done using analog signals, it almost always actually uses digital signals.) Each packet contains header information as to which earthstation it came from, and to which earthstation(s) it is going. Every earthstation linking to the transponder is listening to all of the packets all of the time, but pays attention to the signal only when it sees its own address in the header information. It then processes the packet of bits and sends them to their destination. Figure 20.7 shows the basic concept.

In order to make this scheme work, each earthstation must transmit at such times as to interleave its packets smoothly with those arriving from all other stations. All of the packets come together at the transponder somewhat like a smooth shuffle of cards. To the satellite, it is just a continuous stream of signals. If two signals should arrive at the satellite at the same time, the satellite cannot distinguish between them, so both get relayed back to Earth as useless noise. This is known as a *data collision*. Each earthstation is monitoring the return of the packets it sent to the satellite, so it can detect if a collision occurs.

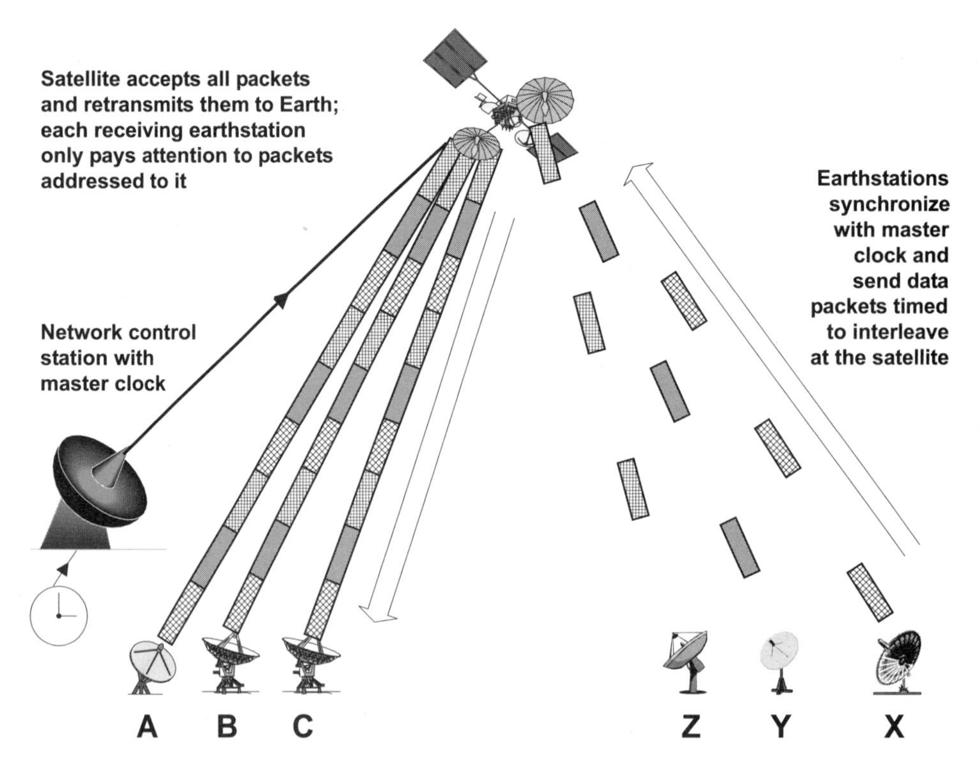

Figure 20.7 In time-division multiple access, each earthstation is assigned a timeslot in which to transmit its packet of data, timed precisely such that it interleaves with all of the other packets and does not overlap with any of them. The satellite receives all of them and relays them back to Earth. Since the packets are addressed to specific earthstations, a particular earthstation receives all the packets but only decodes those intended for itself. The major drawback of a high-speed TDMA system is that there must be a very precise master clock controlling all of the transmissions.

For high-speed TDMA, each station must "march to the same drummer" and take its timing cues from a master control station that sends out timing pulses to every earthstation. Furthermore, each earthstation must know precisely how far it is from the satellite, since this determines how long it takes for its uplinked signal to reach the satellite. The higher the speed of digital traffic through the transponder, the more precise the clock must be. Some of the highest speed systems use a master clock with accuracy better than one part in a billion.

The assignment of the time slots to each earthstation—the times when they may send packets of data through the satellite—may either be made as a fixed rotation among all of the earthstations, or in response to requests to transmit from an earthstation. This former method is called fixed assignment, while the latter is a bit more complicated but more versatile and sometimes referred to as a *demand assignment,* or *reservation system.*

Because each signal from each earthstation is a single frequency with a bandwidth filling the bandwidth of the transponder, we are not dealing with multiple frequencies and thus will have no intermodulation distortion problems. Therefore, the satellite amplifiers and earthstation amplifiers do not have to be backed off (i.e., turned down to avoid interference; see Chapter 8), and the full power of the transponder is available for use.

The major disadvantage of a high-speed TDMA system is the increased complexity and cost of equipment. There must be a master control station, timing equipment at each earthstation and possibly on the satellite, and typically more data manipulation and buffering hardware. The benefit of all of this is higher-speed traffic.

20.4.1 VSAT as an example of multiple access

One example of a low- to medium-speed TDMA system is common in the application often called *VSATs*, *very small aperture terminals*; another name for these is *microterminals*. These are small (typically 1–3 m in diameter) earthstations capable of sending and receiving digital data at bitrates ranging from the few tens of kilobits per second to a few megabits per second. They are often used to connect point-of-sale terminals in retail stores to central computers in company headquarters for inventory control and credit card authorization. In a typical application, when a cash register at a store rings up a purchase, the information describing the transaction is assembled, and the earthstation on the roof of the store sends a request for permission to transmit the data to the hub earthstation at headquarters. The controlling hub takes into account all such requests from all of the stores, and transmits a "reservation" back to the store telling it when to uplink the data on this particular transaction. The delay is typically only milliseconds to a few seconds. At the appointed time, the store's VSAT terminal sends the data. After processing, some seconds later the store receives confirmation of the purchase, perhaps approval of the credit card charge or other information, such as updates of inventory.

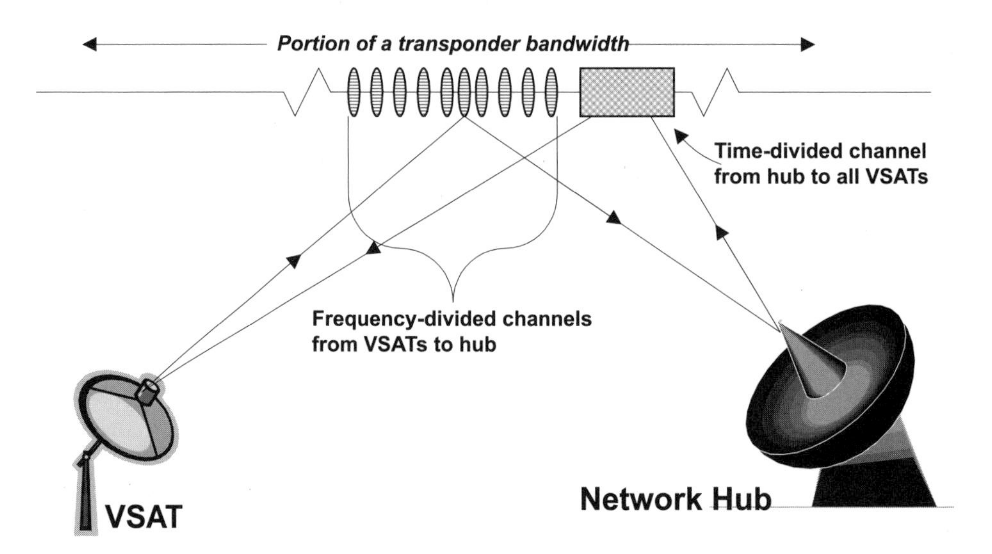

Figure 20.8 An example of one possible multiple-access scheme for a VSAT network. In this technique, the low-speed data from each store is sent to the satellite through SCPC channels on one transponder of the satellite, while the higher-speed data returned by the VSAT network hub uses TDMA in a separate part of the same transponder. By combining multiple-access methods, the efficiency of communication can be increased.

It is possible, even common, to combine TDMA and FDMA techniques on different parts of the link because the nature of the traffic on the inbound and outbound links are different. Figure 20.8 shows one example.

Much higher speed TDMA systems are operated by major carriers, and carriers' carriers, to trunk datastreams, telephone calls, and other traffic. Such systems can operate at hundreds of megabits per second datarates, or higher. Such systems require higher powers, *C/N*s, and larger, more complex earthstations.

20.4.2 Aloha

There is also a very low speed system in use that illustrates the TDMA concepts and makes the additional point that sometimes it is user friendliness, not cost or communications efficiency, that is the prime criterion of service.

Such a system is *Aloha* (which is thankfully not an acronym), invented for the University of Hawaii to connect personal computer users to a campus mainframe. The primary criterion was to give each user the feeling that his or her terminal was hardwired to the mainframe, even though it may be distant with no copper connection. To do this, the designers used knowledge of how PC users work: for instance, few users sit for hours and type 100 words per minute; instead, the demands of each user are brief, small, sporadic and unpredictable.

Each personal computer is connected to a small two-way earthstation. When the PC user types in something, such as a file request, the data goes from the PC to the earthstation, which assembles it into a data packet with the PC's user address in it, and immediately transmits it to the satellite and transponder being used. If all goes well, this signal is relayed back to the earth by the satellite and is received by the earthstation at the main campus. The data packet is decoded and sent into the mainframe computer, where the request is processed. The results are then assembled into a data packet with the PC user's address on it, and are immediately transmitted to the satellite. Every user's earthstation is continually listening to the satellite, but only pays attention when it sees its own address. The addressed earthstation receives the packet, decodes it, and sends the result to the PC user's terminal. Thus, the user gets the impression of a direct connection to the mainframe.

All of the transmissions are entirely random, since they essentially depend on when each PC user hits "Enter" on the keyboard. What happens if, by chance, two users send off a packet at once? If they reach the satellite transponder simultaneously, we get a data collision, which results in no useable data. However, since each earthstation is listening to the transponder, it expects to get back from the satellite the same datastream it sent, about a quarter-second later. If it does not, the earthstation realizes that a data collision has occurred. Built into each station is a circuit that, in that event, waits a random number of milliseconds and tries to transmit again. Usually this second try works, and if not, it tries until it gets through.

Such a system is very nice for PC users, but seemingly very bad from the telecommunications engineer's point of view. Because of the uncoordinated nature of the transmissions, the maximum efficiency of an Aloha system is only 18%! However, efficiency was not this system's design criterion.

The efficiency can be improved, actually doubled, by making a small change to institute some control on transmissions. If you have a master clock, and allow transmissions to begin only at a clock tick, the number of collisions is reduced. This system is known as *slotted Aloha*. Further controls on when earthstations may transmit can further improve the efficiency, but at a loss of the "interactive feeling" to the users. Such systems are sometimes called *reservation Aloha*, or reservation TDMA.

There is a trade-off between packet length and the delay each user will experience in sending a packet.

20.5 CDMA: code-division multiple access

CDMA is a digital-only technique which relies on some very complex mathematical properties and manipulations. It is particularly good for providing services to individual users (such as cellphone users and MSS satphone users); it is not used in trunking applications.

Basically, each user's datastream is a stream of numbers that represents whatever the signal may be. Before transmission, each such number stream is multiplied in a coding device by a carefully chosen encoding datastream based on a mathematical formula. This results in a coded datastream. Each user's datastream is encoded with a unique coding formula. The coded datastreams from all users are simply added together and transmitted. Because there may (or may not) be many users simultaneously, this summed signal is just a hiss that is technically referred to as pseudo-random noise. But the hiss itself is really a random jumble of datastreams of numbers. This works because the coding formulas are chosen with great care to be independent of one another. (The mathematicians call them "mutually orthogonal.") The hiss is modulated over a wide bandwidth to produce a spread-spectrum signal. Figure 20.9 shows the basic concept.

This great hiss is sent through the satellite and is possibly received by many stations. A station can only pull out one of the datastreams by knowing the exact formula that was used to encode it. In fact, each station cannot discern how many other datastreams are in use. This makes the system very secure, and it is no wonder that such systems were originally developed for the military. If a station does know the formula, it multiplies the incoming "hiss" by it again, and out falls the original datastream.

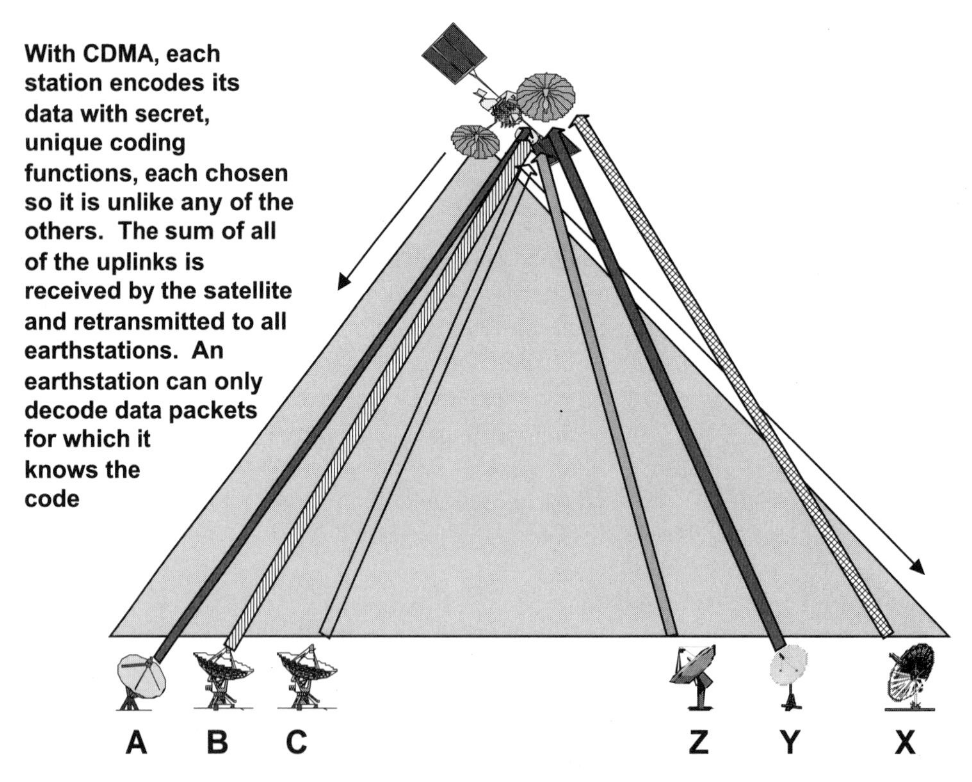

Figure 20.9 In a multiple-access system using code division, each sender uses a distinct, often secret encoding formula to manipulate its data before sending it to the transponder. This signal is sent whenever ready, for there is no assignment of frequency or time slots. The transponder relays all of the signals back to Earth as one combined signal. Only an earthstation that is the intended recipient and knows the proper encoding formula can decode just the part of the signal sent to it.

Even more amazingly, CDMA is the only multiple-access technique for which the efficiency of communication actually goes up as the number of users increases. Of course, there is a trade-off: as the number of users rises, the signal quality of each individual user, measured by the bit error rate, decreases, but decreases slowly with increasing number of users so that CDMA can have a higher capacity than other multiple-access schemes.

A way to comprehend what is going on is to go back to our cocktail party analogy in the Embassy of the Duchy of Grand Fenwick. Suppose you and another person are at first the only ones in the room, and are conversing in English. There is one "traffic flow" going on. Now another couple enters the room, but they are speaking Japanese (that is, their signal is "coded" differently). As long as they are not too loud, and because Japanese is not at all like English, they only minimally interfere with your conversation, but they have increased the "traffic" in the room. Now other groups enter, speaking German, Russian, Urdu, Swahili, and other languages. The background noise in the room increases, but so does the total amount of traffic in the room.

20.6 PCMA: paired-carrier multiple access

A newer multiple-access technique is made possible by the ability to precisely manipulate digital datastreams. In *paired-carrier multiple access*, or PCMA, earthstations communicating with one another send their signals at the same time to the same transponder (or part of a transponder). This means, of course, that both earthstations receive both the signal from the other earthstation and the signal they themselves sent out about a quarter of a second before.

You may recall in Chapter 17 that we discussed the problem of echo in telephone channels, in which your voice gets reflected back to you from the distant earthstation, and thus you hear not only the other person, but an echo of yourself. One method of solving this is the echo canceler, which stores what you say and, applying the proper time delay, inverts your signal and uses it to cancel out the echo.

In PCMA, a very similar technique is used. Each earthstation knows what it sent, and how long the round-trip is to the satellite. Thus, when its own signal comes back to it about a quarter-second after transmission, it can invert this and use it to cancel the echo of itself, while preserving the signal from the remote earthstation. Such cancellation is not perfect, but can nonetheless allow both earthstations to use the same frequency at the same time, conserving bandwidth. Of course, PCMA can be combined with other techniques to further improve throughput.

20.7 Demand assignment

Most SDMA systems, since they use spot beams, are permanently assigned to regions of the earth, although a few satellites have moveable beams. They are an example of what is called *pre-assigned multiple access*, abbreviated PAMA. FDMA and TDMA systems may assign the capacity either on a fixed (PAMA) basis, assigning each a fixed amount of bandwidth or time to use, or on a demand-assignment basis, in which they use the capacity of the system as they need it and request it.

Demand-assigned multiple access, DAMA, can work in two basic ways. The first is called *contention*, in which users vie for attention. We saw an example of such a system in Aloha, where each station simply transmitted whenever it was ready. The more users of such a system, the more likely you will have traffic collisions.

The other technique is called *polling*, in which some central controller asks each potential user if it has any traffic. If a station does, it transmits its data and relinquishes the channel so that the controller can ask the next station. This can maximize throughput, since only stations with demand use the system's resources. On the downside, it can produce delays for particular users, since it may take a while for the controller to get back to polling the station.

DAMA systems are often more economical, both in terms of use of the transmission resources, but also in the equipment needed by the earthstations. They usually improve the overall network connectivity.

Demand-assigned systems are sometimes called "bandwidth on demand" as that is just what they provide users. It could just as well be called "bitrate-on-demand."

Before selecting a method, you need (as usual) to know the users' demands for their traffic. You can combine several types of multiple-access methods to further increase capacity. For example, satellites are designed as FDMA devices by assigning a number of transponders to specific frequencies. However, within a transponder, the earthstations might use TDMA or CDMA to partition its capacity.

20.8 Multiplexing onboard satellite

All of the multiple-access methods discussed so far take place at the earthstations and apply to transmissions through bent-pipe transponders. This is the way most multiple access is done, because most satellites are bent-pipes. Some particular applications might be better done if a satellite could multiplex a number of independent uplinks onboard.

Take, for example, a DBS system. In order to multiplex together the several digitally compressed television channels that are sent together to a user's home antenna, the individual program sources must first send their raw signals to the DBS system operator. The operator then compresses them, multiplexes them together at its uplink earthstation and sends the multiplexed signal to the appropriate transponder. This requires expensive links to the DBS operator from each programming source.

An attractive alternative would be a system that allows each programmer to independently uplink its signal to a satellite where the signals could be combined. A fairly recent system that allows this is Skyplex, developed by the European Space Agency for use on the Eutelsat fleet. Skyplex is an onboard multiplexer that combines the digital uplinks from small dishes at each programmer's site. Dishes no more than 2 m in diameter are required, and the satellite may accept either a number of FDMA SCPC uplinks, or up to six TDMA uplinks using a wide range of datarates. The circuitry aboard the satellite multiplexes them together and forwards the combined signals to a 33-MHz transponder in standard DVB format. This allows the programmers to independently send their signals to the satellite, and for each to use its own choices of encryption, conditional access, and other techniques.

The trade-off here is that the uplinking to the satellite is simpler and less expensive, but the satellite's communications payload electronics are much more expensive.

20.8.1 Multiple multiplexing

It is possible to combine multiplexing methods. For example, a satellite could have several beams (SDMA, allowing frequency reuse), and each of the beams could use frequency-, time-, or code-division to apportion the capacity of the transponders. This would work fine for a simple bent-pipe satellite. However, if the satellite has the ability to connect signals from one beam to another, then the multiple-access method to be used must be time-division, since it takes time to process the individual uplinked signals on board and route them to another beam.

20.9 Multiple access summary

Let us summarize the characteristics of each multiple-access technique in terms of their utilization of power and spectrum resources, requirements, and efficiency:

Power Usage: Because of intermodulation problems, FDMA, CDMA, and PCMA must backoff amplifiers, and the sum of the signal powers of the individual signals will be less than the total rated transponder power. Because TDMA uses only one frequency at a time, no backoff is needed and the full transponder power can be used.

Bandwidth Usage: Because FDMA systems require guardbands between the individual signals, the sum of the bandwidths of the individual signals will be less than the full bandwidth of the transponder. TDMA, CDMA, and PCMA use the full bandwidth at all times.

Spectrum Efficiency and Capacity: As the number of users goes up, an FDMA system needs more guardbands and its spectrum efficiency decreases with increasing number of users. TDMA systems need guardtimes between signals, but these are easier to control, so that a TDMA system declines in spectrum efficiency less quickly than an FDMA system. With CDMA systems, the spectrum efficiency goes up with increasing numbers of users, but each user's signal quality decreases somewhat. PCMA almost doubles the useful bandwidth since two earthstations use the same frequency at the same time.

Synchronization and Control: No synchronization is needed for FDMA systems; each user is assigned a frequency slot and is independent of other users. A high-speed TDMA system requires tight timing control and coordination. No synchronization or control is needed for CDMA systems, since the properly chosen coding formulas keep the signal separate. PCMA requires that each earthstation has the equipment to cancel out its own returned signals.

Allocation to Users: Both FDMA and TDMA can allow use by either fixed assignment, or by demand assignment. The allocation to CDMA users is completely random and on demand.

The one big issue to keep constantly in mind is that there is no ideal multiple-access method (or anything else) for every application. No system is ideal, and every decision involves trade-offs. You must start with the users' needs, and from them figure out which methods, schemes, and protocols work best to fit in with the users' criteria.

And this brings us to the longest acronym in this book, from science fiction writer Robert A. Heinlein, TANSTAAFL: "There ain't no such thing as a free lunch!" There are always trade-offs in selecting techniques.

We have now covered just about all of the technical details of satellite telecommunication, with all of its acronyms, abbreviations, and jargon. It is now time to pull all of this together and explore some examples of real satellite systems.

Part 6

Satellite Communications Systems

Chapter 21

Satellite Communications Providers and Competitors

Satellites carry less than 10 percent of the world's total telecommunications traffic, despite the fact that the total capacity of all of the commercial communications satellites in orbit is less than the capacity of a single transoceanic fiber optic cable. In 2002, the satellite part of the telecommunications industry took in an estimated $100 billion.

Many satellite system operators, earthstation operators, and ancillary services have sprung up to accommodate users around the world. (As we will see in the next chapter, the trend has been away from governmentally controlled entities and toward privately owned multinational corporations.) This chapter will explore the strengths and weaknesses of telecommunications carried by satellite, the industry's major competitors, some of the systems and networks providing space segments, and an overview of some of the ancillary service providers.

21.1 Satellite competitiveness

Satellites are only one way of carrying telecommunications signals, and they are often used in conjunction with other telecommunications technologies. The signals destined for carriage through a satellite must somehow reach the originating earthstation, and be sent onward from the receiving earthstation. This is frequently done terrestrially, by fiber optic cable, coaxial cable, and microwave tower relay. Increasingly, so-called satellite companies have come to realize that they are not just satellite companies: they are telecommunications providers, and satellites just happen to be the major technology that they use for transmission. Satellite communications providers are niche players in the global telecommunications industry; their total revenue is less than a tenth of global telecommunications revenue.

You might ask, when are satellites a competitive means of carrying information? While each situation is unique, there are some general cases in which satellites are particularly useful, cost-effective, or, in some cases, the only way to go. Some of the main ones:

- *When terrestrially based systems are deficient, incompatible, or too expensive*

For example, Inmarsat began because high-frequency radiocommunication to mid-ocean ships was unreliable and often of poor quality. Even though the first ship terminals cost close to a quarter of a million dollars, if a shipping company could save a few days of travel time rerouting a supertanker that costs $100,000 per day to operate, they could soon recoup the cost. As another example, the global mobile telephony systems were designed to compensate for the lack of terrestrial cellular telephone systems in many parts of the world. Not only do satellites provide services where terrestrial services cannot, they are highly reliable. The typical continuity of service for satellites is better than 99.99%, meaning that service is unavailable for less than one hour per year. Terrestrial systems often charge by the kilometer, whereas every earthstation is 72,000 km from every other earthstation. Satellites "avoid the middleman." Terrestrial connections often involve going through several systems, with different protocols, which complicates the flow of information and slows it down.

- *When geographic obstructions make terrestrial systems difficult*

  For example, in the Andes of South America, many countries are putting in cellular telephone systems, but find it difficult to link cities or remote sites due to rugged terrain. In some cases, satellites are used to link the cell sites or cities. Maritime communication is another example of a geographical obstacle overcome by satellites. In another case, a company needed to provide telephone and data communications for a few weeks from an island close to shore—so close that you could easily see it from the beach. Point-to-point microwave towers were out of the question because the line-of-sight crossed a busy shipping channel. Instead, a satellite link was used as the most cost-effective way. It was easier and cheaper to send the signal across 72,000 km via satellite than across the 30-km channel.

- *When capacity flexibility is important*

  Satellite systems can be made very flexible, capable of providing bandwidth on demand. Many users need occasional capacity, sometimes on a random basis, and satellites work very well in such cases. One example might be VSATs, very small aperture terminals, connecting point-of-sale terminals or reservations-systems traffic to company headquarters' or a central reservations computer. Because of this on-demand ability, satellites are good for asymmetric services where the amount of information flowing in one direction may be greatly different from that flowing in the opposite direction. For example, a telephone call is an approximately symmetric information flow, whereas Internet access is highly asymmetric, with only a few bits typically going out from a workstation to request a webpage or other file and often megabits being returned in response.

- *When route flexibility, alternate routing, or route redundancy is important*

  One prime example is the way that satellites serve as a backup for transoceanic fiber optic cables when fishing trawlers or curious sharks sever the cable.

Another example is the vaulting of a signal over a country that does not allow transit of communications.

- *When quick implementation is needed, or when the requirement is for a short time only*

  The example above of communicating from island to shore applies again here. Any special event in a location without permanent (or adequate) terrestrial connections is a case as well, such as the site of a disaster, a political convention, or a local remote newsfeed. Satellite newsgathering (SNG) is an exemplar.

- *When the cost of the terrestrial link of the "last mile" is prohibitive*

  Many communications applications via satellite began as ways to bypass expensive by-the-distance terrestrial systems, such as long-distance carriers and wide-area networks. The "last mile" problem is a common case in any wired communication system—which naturally requires a continuous connection from end to end. You might be able to get your signal across the country or across the globe, but not across a river, street, a political boundary, or someone's right-of-way. Sometimes the "last 22,000 miles" is easier than the last mile!

- *When you want the cost of transmission independent of the distance the signal travels*

  Terrestrial transmission facilities usually charge by the kilometer, but all satellite stations are the same distance from a satellite. A satellite-carried signal costs no more to send from Washington to Baltimore than to send from Washington to London or Bonn.

- *When recipients are spread over a wide geographic area*

  Such a case is best suited for some broadcast-type transmission, which could be terrestrial or satellite. However, satellites cover much larger areas of the Earth, usually making them more cost-effective. Another good example is the interconnection of networks, which avoids incompatible protocols and the large numbers of routers in a terrestrial packet network. The economic case for using satellites will, of course, depend on the extent of the deployment of terrestrial infrastructure, particularly optical fiber networks. A recommendation by the International Engineering Consortium estimates that satellites will be more economical when the task is to distribute signals to more than about 20 geographically dispersed sites in North America, more than 8 sites in multiple countries in Europe, or more than 4 sites in multiple countries in the Asia-Pacific area.

- *When users are mobile*

  It is self-evident that a mobile user cannot be connected to a permanent system.

- *For thin routes and medium routes*

  Satellites are usually not competitive when two linked sites have a large amount of traffic to exchange; in such cases, permanent optical fiber connections are more cost-effective. But for locales with less traffic, especially sporadic traffic, satellites may well be the best solution. For a number of developing countries, satellites are the only international connections that they have.

Keep in mind that users' requirements change, and that a user requirement that begins as a little traffic may grow to the point where another technology becomes more appropriate.

Satellite telecommunication is basically a bunch of niche markets within the much larger context of global telecommunications. Many of the niches are large, some are fairly stable, but many come and go. There is money to be made by satellite systems if this reality is acknowledged and the changing demands are answered.

21.2 Satellite's competitors

Of course, all terrestrial services compete with satellites. One of the major reasons for the economic failures of the LEO and MEO satellite telephony systems Iridium and GlobalStar was the faster-than-expected expansion of terrestrial cellular telephone systems. Terrestrial cable television systems as well as traditional broadcast stations compete with DBS systems; terrestrial dispatch systems for utilities compete with such low-datarate satellite systems like OrbComm; traditional radio broadcasts and audio over cable competes with DARS systems.

21.2.1 Fiber optics

Certainly the greatest terrestrial competitor to satellites is optical fiber. (It is also a complementary technology in many ways, as many earthstations make their terrestrial connections to other networks or users via fiber.) Optical fiber has tremendous capacity. It is worth repeating that all of the communications in space do not have a total carrying capacity equal to that of one transoceanic fiber cable.

Even worse for satellite operators, while the cost of installation of a fiber is very high, the cost to carry a bit through a fiber has been dropping by around 50% a year. There is a lot of so-called "dark fiber" (strands not yet used) that can be brought online if demand picks up.

These economics are the reason that fiber will always be much, much less costly than satellites for heavy point-to-point routes. On the other hand, the high installation costs of running fibers will keep satellites competitive for mobile and broadcast services, as well as for short-term requirements and the other uses noted above.

21.2.2 Stratospheric platforms

Possibly on the horizon is a proposed competitive technology variously called *high-altitude, long endurance* (HALE) platforms, *high-altitude, long operation* (HALO), or *high-altitude platform systems* (HAPS, the official ITU terminology).

Some sources also call them *aerostats*, *atmospheric satellites*, or *sub-orbital satellites*. Such a platform could establish a metropolitan area network quickly and relatively cheaply.

The basic concept involves a balloon, dirigible, or aircraft about 20 km above ground supporting electronics and antennas much like those on satellites and providing the same kinds of services. The platform is maneuverable, refuelable, upgradeable, and repairable, and could derive its electrical needs from solar power. A series of such stratospheric relay stations could be a powerful competitor to GSO satellites.

A HAPS station has other advantages. Its beams could cover a region hundreds of kilometers wide, a region that could be partitioned into cells much like cellular telephony. The delay in the ground-to-station link would be negligible. A station could link to other nearby stations, and could also have links to GSO satellites. A HALE station could be positioned to be stationary over any location on Earth, so the elevation angles of the links between earthstations and platform would be high, minimizing atmospheric effects. Some preliminary economic studies claim that one could be built for a tenth of what a GSO satellite would cost. As the cost of deployment and operation would be lower than those for satellites, the owners could charge less.

So far, only one potential HALE competitor, Skystation, has been licensed, but it has yet to produce even a proof-of-concept station. Other systems using manned or unmanned aircraft have been partially tested. Several projects are in the development stages in Japan and Europe for deployment in the mid-to-late part of this decade. We can only wait and see whether such high-altitude platforms will turn out to be a real competitor to satellites, a complementary technology, or a "pie-in-the-sky" idea that proves impractical.

21.3 Satellite system economics

Obviously, the costs to deploy and operate a telecommunications system determine what the owners of the system must charge customers for services. So what does a satellite system cost? Of course, all systems are different, but some rough numbers can give an idea of the economic scale involved.

Let's suppose that you want to launch a single FSS satellite. If you go to one of the major manufacturers for a typical run-of-the-mill satellite, it might cost you \$75 to \$100 million. Let's assume the lower figure. To launch it, you would probably have to spend another \$75 million, although you might get away with less due to competition or by sharing a launch with another satellite. Your satellite and its concomitant expenses are costly, so you probably want to insure the \$150 million cost of satellite-plus-launch. Over the years, the rates have fluctuated, but let's assume a medium rate of 15%, which will cost you around \$23 million. Thus, your total cost to buy and launch and insure the satellite is \$173 million.

You will probably finance the system. If you can borrow the $173 million at an interest rate of 8%, that will cost you $95 million over the 12-year lifespan of the satellite. You also have to operate the satellite (or hire another company to do so); a cost of $5 million a year might be about right, for a total of $60 million over the 12 years.

Thus, the total cost to deploy and operate the satellite for 12 years is $328 million, or about $27 million per year. If your typical satellite has 24 transponders, and knowing that there are 8760 hours in a year, this tells you that your break-even cost is $128 per hour per transponder.

If you were trying to build a BSS service, with its more powerful satellites and more complicated earthstations, you might double or triple the costs.

21.3.1 Satellite networks and systems

In almost all telecommunications using satellites, there are at least three major components: the originating uplinking earthstation, the satellite providing the space segment bandwidth, and the terminating downlinking earthstation. If the communication is two-way, then both earthstations will be uplinking and downlinking. In broadcasting applications, there may be one or several uplinkers and millions of downlinkers. In hubbed VSAT networks, there may be thousands of small two-way earthstations, a large two-way hub earthstation, and use of multiple transponders or even multiple satellites.

To pull together all of the information we have developed in the previous chapters, this chapter will discuss the providers of space segment capacity, providers of uplinking and downlinking services and terrestrial interconnections, and how end-users can obtain telecommunications services through satellites.

21.4 Categorizing satellite systems

There are many ways of categorizing satellite systems, such as orbit, use, geographic coverage and ownership. In a previous chapter, we saw the official service designations as defined by the ITU and regulatory agencies. From a business and legal perspective, one important way of distinguishing space segment operators is by the type of ownership.

We can categorize satellite systems according to ownership under three major headings:

- Treaty-based intergovernmental organizations
- Nationalized companies or organizations
- Private companies.

A little bit of history is useful to understand the how the current status of satellite service providers has developed.

From the first commercial satellite communications in the 1960s to the 1990s, international communications via satellite was performed solely by international treaty-based monopolies. Transborder satellite traffic could only be carried by one of the intergovernmental organizations (IGOs) such as Intelsat, Intersputnik, and Inmarsat, and the links to those systems' satellites could only be performed by a single designated company in each country. Most countries' domestic satellite traffic was the monopoly of the national post, telegraph, and telephone administrations, generically referred to as PTTs.

In more market-oriented countries, led by the United States, domestic satellite communication was left to private companies, and slowly, during the 1980s and 1990s, other countries liberalized and privatized their telecommunications markets to a greater or lesser extent

In the mid-1980s, a foresighted and determined businessman, Rene Anselmo, asked the U.S. government for permission to build a private satellite system for international and domestic satellite communications. Against the resistance of the entrenched monopolies, he founded PanAmerican Satellite Company, now PanAmSat, and opened the way to wider private global satellite communications.

As demands for satellite links greatly grew, several of the IGOs found it difficult to be flexible enough to keep up with markets. Political and business pressures to privatize resulted in most of these IGOs becoming private companies. Over the same period, various satellite system operators purchased one another or merged, resulting in a considerably consolidated industry. The companies also sometimes acquired earthstations, fiber optic links, and other ancillary capabilities in order to serve their customers better.

In the overview below, we look at some of the major players in the modern satcomms marketplace.

21.4.1 Treaty-based operators

Treaty-based organizations are multinational entities established by treaties among nations. The nations themselves, as sovereign powers, are the members and owners of the organization. Legally they are known as *Intergovernmental Organizations*, IGO. As such, the constitutions, charters, or other founding documents of the organizations have the status of law in the signatory nations and are thus binding upon telecommunications users in those nations. National telecommunications laws and regulations in the member nations are implemented in conformance with the IGO's rules, often allowing for individual nations to differ, with limits, from the international regulations. Nations invest in these organizations, and hold ownership usually as a percentage of their use of the system. Such organizations are invested with many quasi-governmental privileges. (The ITU, described in some detail in Chapter 2, is a body established and organized similarly, but it is not an operator of satellite services and so is not considered in this chapter.)

Intelsat, originally a shortened name for the International Satellite Telecommunications Organization, was the first such international body, founded in 1962. Inmarsat came along about a decade later. Like several such organizations, they are now a private company, divested of their intergovernmental status and privileges. This has become both necessary and possible because of vast changes in the economics of telecommunications, the spread of democracy, and the liberalization of regulatory strictures.

While Intelsat, Inmarsat, and Eutelsat (to be described more fully below) have now become fully private commercial entities, there are still two important intergovernmental treaty-based satellite systems.

21.4.1.1 Arabsat

Officially known as the Arab Satellite Communications Organization, Arabsat was founded in 1976 by the members of the Arab League to provide the space segment for communication among its members. It is headquartered in Riyadh, Saudi Arabia. The 21 member countries (as of 2002) are located mostly in the Middle East (Palestine is also a member). The individual member states provide the ground segment facilities as needed. Although membership is still composed of Middle Eastern and North African countries, the system is also used by others with significant Muslim populations who are not members.

The first Arabsat satellites were launched in 1985, and the satellites themselves are also named Arabsat. The satellites provide both regional telecommunications services and domestic services for individual member nations. Traffic on the satellites is a mix of FSS services: video, radio, telephony, and data. As you would expect, the satellites' footprints primarily cover the Middle East and North Africa to about the southern edge of the Sahara. Lower power coverage on some satellites extends eastward to India and northward into much of central Europe.

Current information about Arabsat is available on the web at www.arabsat.com.

21.4.1.2 Intersputnik

The International Organization of Space Communications, Intersputnik, was formed by and in the Soviet Union in 1971 during the Cold War as an iron curtain equivalent to Intelsat, but smaller. It began with 9 members, and has grown to 24 (as of 2002). It began operations in 1979, serving Communist nations, with some joint agreements with other nations. It has its headquarters in Moscow, Russia.

To date, it has used satellite capacity leased from the Russian federation. These included the Gorizont satellites, the newer Express satellites that replace the Gorizonts, and the DBS satellites named Gals. Having only a few satellites, Intersputnik serves the once-Communist countries in Europe and Asia, as well as a few countries in the Americas. It has no Pacific-region satellites. However, use of the system is open to anyone, not just members. Cooperative ventures with firms in other countries have allowed Intersputnik to offer a variety of services, including VSAT, to a wide area.

Further, current information may be found at www.intersputnik.com.

21.4.2 National domestic and regional satellite systems

Two dozen or so nations now have satellite systems for internal use. Some are strictly or primarily for telecomms traffic, some are for direct-to-home video services, and some are for mobile communications. (Two nations, the U.S. and Russia, also operate global position-determination systems, which will be discussed separately below.) Some nations have only a single satellite, some have dozens. Because of the inherently international nature of a satellite-delivered signal, however, the operational boundaries are not as rigid as they once were. Some of these systems are used outside the country of ownership as a regional system or for transborder services with neighboring nations by bilateral agreement.

Furthermore, the country of origin, ownership, and/or operation is not always the area where the services are provided. Many systems have extensive international ownership and financing. A prime example of such a system (detailed below in the section on the U.S. because that is where it is registered) is Worldspace Inc., which provides DARS services in several regions of the globe, not including the U.S. For business reasons, some satellites may be legally registered in such places as Bermuda or Gibraltar, which have no satellite systems of their own.

Several of the domestic and regional systems are members of "alliances" or joint ventures with other satellite systems and/or with terrestrial providers so to be able to offer customers a wider range of geographic coverage and services.

As systems come, go, and merge, it is difficult to list them all. You are referred to the several satellite industry directories listed in Appendix C.

21.4.3 Private satellite operators

The six satellite system operators described here are the world's major providers of telecommunications via satellites. The vast majority of commercial satellite telecommunications traffic passes through their transponders. Some of these are companies that own and operate their own satellites, and a couple are organizations that combine the capacities of other firms or of subsidiaries. Again, there are often changes in ownership and details, so the best up-to-date information is in the industry directories listed in the bibliography.

21.4.3.1 Eutelsat

The European Telecommunications Satellite Organization, or Eutelsat, was founded as a treaty-based organization in 1977 by European PTTs, but it was not formally established until 1985, with 17 members. Its headquarters is in Paris, France.

Eutelsat is a regional system of 18 satellites (as of 2002) providing a wide mix of fixed, broadcast, and mobile services, such as video, radio, data, satellite newsgathering, land mobile, and other services to its members for domestic, regional, and international telecommunications. Its satellites send more than 1000 television channels and 550 radio stations to European homes. The satellites have footprints that cover all of Europe, the Mediterranean and North Africa, the Middle East, southwest Asia, and the Americas. Eutelsat is also the operator of two other satellite services. EutelTracs is an RDSS service to land and marine fleets. Emsat is a satellite-delivered mobile telephony and low-speed data service for private voice networks, with connections to the PSTN. The organization also sells capacity on several nonEutelsat satellites to enlarge coverage. Revenues in 2000 were €686 million.

An overview of the Eutelsat system is online at www.eutelsat.com.

21.4.3.2 Inmarsat

Founded in 1979 as the International Maritime Satellite Organization, Inmarsat was a treaty-based organization. It was to be similar to Intelsat, but serve a maritime market for voice and low-speed data services and serve as part of the Global Maritime Distress and Safety System, GMDSS. It began service in 1982, catering almost entirely to shipping, at which time terminals cost a quarter of a million dollars. Inmarsat later changed the word "maritime" to "mobile" to recognize the growing importance of nonmaritime satellite-delivered communication (but did not change its moniker to "Inmobsat."). Usage grew, prices dropped, and new land-based and aircraft-based terminals expanded the market. By 2002, there were more than 210,000 terminals in use, the cheapest costing only a few thousand dollars. Its headquarters is in London, UK, and just before its privatization, there were 86 member countries.

When it went into operation, it was unique in the satcomms industry because the end users—at the mobile end of the links—communicated directly with the space segment through ship-based terminals, rather than through an intermediary organization. Connections to the PSTN are through land earthstations (LES) typically owned and operated by the telecommunication authorities in several nations. There are about 40 LESs in 30 nations as of 2002. Links to mobile users take place in the L-band; links from satellite to land-based stations are in the C-band.

Before privatization, Inmarsat spun off a private commercial company, ICO Global Communications, to finance, build, and operate satellites for mobile communications. Unfortunately, ICO went bankrupt and its remnants were bought by others.

On April 15, 1999, Inmarsat converted itself into a public limited corporation, the first intergovernmental organization ever to do so. While it made the organization more responsive, it also cast aside its intergovernmental protections and privileges.

A small residual intergovernmental body called the International Mobile Satellite Organization, IMSO, continues to provide Inmarsat's public service obligations, including those of the Global Maritime Distress and Safety System (GMDSS). The new company's headquarters remains in London, and IMSO will be governed by a 15-member board of directors, one of whom will be the CEO and three of whom will be drawn from developing countries.

Inmarsat services include telephony, telex, facsimile, electronic mail, and slow-speed data (up to about 64 kbps) for marine, aeronautical, and land mobile users. The data services include position reporting, status monitoring, and fleet management functions. It is worth noting again that the Inmarsat satellites and services do not include the capacity for broadcast-quality video, but use of larger stations that can handle datarates up to 64 kbps allows the use of slow-scan video and highly compressed teleconferencing video.

The Inmarsat space segment consists of four primary satellites, covering the Pacific Ocean region, POR, Indian Ocean region, IOR, and two Atlantic Ocean regions, AOR-East and AOR-West. The other satellites serve as backup for the primary ones.

Inmarsat's services are categorized into service types that roughly determine the kinds of traffic that may be carried and the technical specifications of the terminals used. Inmarsat-A, Inmarsat-B, Inmarsat-C, and Inmarsat-M are the best known. A new Inmarsat-E service is designed for emergency position and communication to maritime users as part of GMDSS. Inmarsat establishes the standards for these services for both user and network terminals, and then "type certifies" equipment meeting these standards manufactured by a wide range of companies.

Current information is available at http://217.204.152.210/about_inm.cfm.

21.4.3.3 Intelsat

The International Telecommunications Satellite Organization, Intelsat, was established in 1964 as the world's first intergovernmental satellite organization, and launched its first satellite, Intelsat I (also called Early Bird), in 1965. It built a fleet of two dozen satellites and acquired almost 200 member countries over three decades. For some undeveloped nations, Intelsat was and is the only external telecommunications link.

As the need to privatize grew, Intelsat split off six of its satellites and services into a new private company called New Skies Satellites, N.V. On July 18, 2001, Intelsat Limited became a Bermuda-based private company with more than 200 stockholders, most of them its previous members. Intelsat is headquartered in Washington, D.C., USA. Its 2001 revenue was $1.084 billion. As of 2002, it operates a fleet of 20 satellites.

Its voice and data services provide both international and domestic connectivity between and within countries large and small. There are connections to the PSTN, as well as private networks for applications such as VSATs, corporate voice and data networks, and others. The basic service is called IDR, intermediate datarate, which can be used for a wide range of applications from 64 kbps to 155 Mbps. A thin-route-on-demand service provides as-needed communications to low-usage PSTNs. A higher-speed TDMA service provides almost any kind of digital connection.

Video services carry contribution feeds, backhaul, and distribution feeds of broadcast quality globally. Customers include networks, producers, news organizations, cable systems, and others. Some capacity is also used for DTH services in some parts of the world.

Intelsat operates its satellites in four service regions: Atlantic Ocean region, AOR, which includes the Americas and Caribbean, Europe, Africa, Middle East, and India; Indian Ocean region, IOR, which covers Europe, Africa, Asia, Middle East, India, and Australia; Asia Pacific region, APR, serving Europe, Africa, Asia, Middle East, India, and Australia; and the Pacific Ocean region, POR, covering Asia, Australia, the Pacific, and the western part of North America.

Intelsat's website is at www.intelsat.com.

21.4.3.4 Loral

Loral Space & Communications is both a major operator and a major manufacturer of communications satellites. Loral got into the operator business in 1997 with the purchase of the Skynet system, and later with Orion Network Systems. As of 2002, Loral has 7 Telstar satellites operating in the FSS. It has enhanced its customer offering by joining with satellite operators in other countries to form the Loral Global Alliance. Loral fully or partially owns several of these allied systems.

It is a part-owner of Satelites Mexicanos, of a joint venture with Alcatel Espace called Europe*Star, and through that company is part owner of a joint venture called Stellat. It wholly owns a subsidiary called XTAR, whose markets are defense departments, and Loral Skynet do Brasil. Loral Cyberstar is a satellite/terrestrial system for high-speed data. It was also the principal in GlobalStar, the mobile satellite telephony system.

Details are available at www.loral.com.

21.4.3.5 New Skies Satellite

New Skies, located in the Netherlands, was established as a company to acquire half a dozen of Intelsat's satellites before that organization privatized to reduce monopoly concerns. New Skies has five satellites operational (as of 2002) and plans more. Being a new company, it is totally independent of any corporate legacy, of

PTTs, and of manufacturers. It provides the same range of voice, data, and video services as its competitors. Its 2001 revenues were $209 million.

Its website is at www.newskies.com.

21.4.3.6 PanAmSat

PanAmSat began in the mid-1980s as Rene Anselmo's effort to have a private firm compete with the satcomms monopolies of Intelsat and Eutelsat. It succeeded at this, and was even one of the goads toward making those rivals privatize. Over the years it has built a large fleet of satellites, 21 as of 2002. In 1996, it was acquired by Hughes Electronics Corporation, but retained its name. Its revenue in 2001 was $1.024 billion.

PanAmSat is a growing global system with coverage of almost all of the world. Like most satcomms companies, it provides customers with a variety of video, voice and data services. To support the satellite fleet, it owns teleports and terrestrial fiber optic networks, including a transAtlantic link

Further information is available at www.panamsat.com.

21.4.3.7 SES

Originally Société Européene des Satellites, SES was the first European satellite television provider through its Astra satellites, and was founded in 1985 in Luxembourg. By 1995, it had 60 million television viewers.

Through acquisitions and partnerships, it has grown to one of the world's largest satcomms systems. SES Global, the parent firm, is a group management organization that provides global satellite telecommunications services through its own subsidiaries and partner firms. SES Global totally owns SES Astra, which provides European services through 13 satellites (as of 2002); SES Americom which it bought from GE, with 16 satellites; and Columbia Communications, another U.S.-based firm. SES Americom, in turn, owns half of Americom Asia-Pacific, serving that region of the world. In addition, SES Global has a 28.75% share of Nahuelsat, the Argentinian system; 50% of Nordic Satellite AB, which operates the Sirius satellite covering Europe; 34.1% of AsiaSat, Asia's largest satellite operator; and 19.99% of Star One, a Brazilian satellite fleet covering Latin America. SES Global's revenues for 2000 were €1.487 million.

Further information is at www.ses-global.com.

21.5 Using communications via satellite

Users of satellite communications often have a choice about how they get services. In the simplest cases, users simply purchase a turnkey service from a service provider, and need give little thought to the details of the technology, or indeed whether they are using satellites at all. In other cases, a user may wish to acquire and interconnect the technical resources necessary in order to provide a customized service perhaps not available from an existing service provider. This is very similar to other business service decisions common to all businesses.

As we have seen, providing service through a satellite often includes terrestrial connectivity from the source of the information to the uplink site—sometimes called the "terrestrial tail," the uplink, space segment, downlink, and possible destination-side terrestrial links. A company owning its own satellites must also provide TT&C for them.

Some services offer turn-key convenience. For example, if you want to make telephone calls over the Inmarsat system, you just buy an Inmarsat type-approved terminal, establish an account with Inmarsat, use the terminal as directed, and pay your bills. For another example, if you want DTH television to your home, you sign up with one of the DTH providers, buy compatible equipment, and connect it to your television.

A further instance of a prepackaged service might be VSATs, in which a service station or retail store need not be concerned with the details of the fact that each sale is relayed to some central office for processing and credit card authorization. Nor would a truck driver need to be constantly aware that RDSS equipment on the rig is providing the company fleet manager with updates on the truck's position and other details.

But consider another, more complex, example: suppose you want to conduct a secure two-way audio-visual teleconference between two of your firm's offices. This involves arranging for meeting space and equipment at both ends, establishing terrestrial links from the venues to uplink and downlink earthstations, and obtaining space segment and any other special requirements. You must be concerned that all of the equipment and communications protocols are compatible.

The alternative to doing all of this yourself is to hire a teleconferencing service to provide the end-to-end service. Then, your only responsibility is to tell them when and where you want to connect. At the designated time, you would simply show up and start talking when the little red light comes on, and receive a single bill for everything later. You will, of course, be paying more than the sum of the component costs, but you will have saved yourself a lot of work.

If you do decide to "roll your own," then you need to arrange for the various components that connect the users.

21.5.1 Obtaining space segment

How exactly does one go about obtaining space segment time? The answer to this is simple: one either buys it and owns it, or buys it from someone who has some.

To own some space segment itself, the most expensive way is to buy, launch, and operate a satellite of your own. Since even a modest commsat can cost several hundred million dollars to bring into operation, this is usually done only by those with a lot of traffic such as a system operator or someone who hopes to make money selling off transponder time. Once on orbit, a typical FSS satellite has an asset value of about $400–500 million.

For those with fairly continual major requirements for transponder time, one way to go is to take out a long-term lease on a transponder, or part of a transponder, aboard a satellite positioned within range of your expected users. You can get leases for year or two or three, or for the expected life of the satellite. The satellite owner then controls the satellite—providing TT&C—and you have full control of your transponder to use as and when you wish. You will typically pay a set amount per year for the transponder. The amount varies from satellite to satellite and region to region. If you don't use all of the transponder's capacity—such as not using it all of the time or not using it at its full bandwidth—you are free to sell some of it to others as "occasional time."

And that is the third way you can obtain space segment: by buying occasional time from a transponder owner. Here you will pay by the watt, the bandwidth, and the time. It can be on a totally random basis, or scheduled, such as half an hour every afternoon. Again, costs vary with satellite and region.

Where do you find this occasional time? One place is to inquire of satellite system operators themselves. In other cases, you can approach transponder owners who may have spare capacity. Transponder brokers are firms that keep track of what is available and act as an intermediary between the buyer and seller, similar to commodities brokers.

21.5.1.1 Transponder brokers

The service business that supports much of the use of FSS services is that of transponder brokers, sometimes known as bandwidth exchanges or satellite capacity exchanges. Like other brokerages, they bring together buyers and sellers. In the telecommunications industry, there are brokerages for satellite capacity (also called space segment), for fiber optic capacity, and for wireless capacity. Some of the space segment brokers work in cooperation with fiber brokers to provide one-stop-shopping for users needing capacity.

Sometimes these are the satellite operators themselves, offering long-term leases, short-term leases, or "occasional time" to would-be users. Thus, they market excess capacity on their own satellites, or on other systems when they themselves do not have the capacity required.

Some transponder brokers are parts of firms that have taken long-term leases on transponders themselves, and have some excess capacity to sell off. This excess capacity may be small, such as unused bandwidth alongside a video signal on a transponder, or even a few minutes or hours of transponder use during times when the owner does not need it. There are even firms that simply lease transponder time wholesale from a system operator to broker it at a retail level to anyone who needs some space segment. Thus, many of the cable channel providers may broker time on their transponders, or those of others.

There are also many smaller firms, especially including teleport operators, that can also act as brokers to provide space segment capacity. A more recent innovation is the online satellite capacity exchange that lists offerings of available capacity and matches them with requests for capacity. The first such was the London Satellite Exchange (www.e-sax.com). Others now include SatCap International (www.satcap.com), iacto (www.iacto.com), and Global Satellite Exchange (www.gsatx.com).

Such transponder time is very much like a perishable commodity market: you can lease it, buy it, option it, barter it, use it, or even give it away. The transponder brokers are also much like other commodity brokers, in that they know each other and often work together. If you call one broker for a half-hour slot tomorrow afternoon, and she hasn't got any available, she can probably get it from another broker for you. Like other commodities, transponder capacity can be bought, leased, rented, reserved, optioned, traded, and even given away.

However, unlike some tangible commodities, transponder time is volatile or perishable: if an hour's time is not used this afternoon, it is lost. Thus, the economics for a satellite operator is rather like filling airline seats or hotel rooms: you know how much it costs you to supply the service and how many transponder hours are available; if you fill a certain percentage of them you break even; fill more and you make a profit; but an hour not filled on a transponder is lost forever.

There are three main parameters that determine what it will cost to use a transponder: time, bandwidth, and power. In other words, you pay by the hour, the megahertz, and the watt. Secondary parameters include time of day (late afternoon and prime time evening hours cost more than wee hours of the morning), specific dates (days when the World Cup finals are on will be more expensive than on Christmas morning), and frequency (Ku-band may cost a bit more than C-band in some parts of the world, less in others). All of these items, however, are subject to change because of other factors, such as the number of satellites in each band that are visible to the would-be users, or local factors such as national regulatory rules and tariffs. Charges for the space segment or transponder time usually run in the few hundreds to few thousands of dollars per hour range.

During periods in which there is a dearth of transponder capacity, obviously prices rise. This happened in the early 1990s because several satellites were ending their planned operational lifetimes, a couple failed, and replacements did not reach orbit in time. Over about six months, C-band occasional-time rates quintupled, and many users found it difficult or sometimes impossible to get occasional time. Conversely, in the 1980s, many satellite operators would launch a satellite with less than 50% of its capacity prebought, leaving a lot of transponder time available for occasional users. Today, most operators want more than 75% of a satellite's capacity to be reserved before launch. Some of this capacity is bought up by brokers, leaving even less occasional time available. During the crunch of the early 1990s, many would-be occasional users found it impossible to get transponder time unless they had reserved it months or years in advance.

In countries that have market-based pricing, the price trend over the past decade has been variable. Many new transponders have become available, but others have left the market due to planned end-of-life or satellite failure. In countries that had strong, overpriced telecommunications monopolies, or in regions that until recently had few satellites serving them, the price trend has been somewhat downward.

A good way to follow the trends in pricing space segment time is to use the online transponder brokerage exchanges, such as www.esax.com.

The number of transponders on orbit has continued to grow quickly during the last decade, especially over areas of the world that were deficient in capacity, such as the Asia-Pacific region. Total capacity is still growing, but slowed due to the telecomm "crash" of the early years of this decade.

FSS services will, within the next decade, also incorporate the proposed high-speed data systems, sometimes called mega LEOs, although some of them will use GSO as well. These will push satellite operations to higher frequencies in the Ka-band, V-band, and W-band.

Depending on needs and location, space segment capacity may be like a commodity (much as bandwidth on fiber optics systems is), or it may not be because of specialized requirement and services provided by the space segment owner. These value-added services might include uplinking and downlinking, signal processing, or some other activity.

21.5.2 Obtaining ground segment services

The ground segment services needed will, of course, depend on the details of what you are doing. You may need to arrange to get a signal to a teleport, or to the teleport's communications hub. You may need to rent a portable uplink and/or downlink to connect sites without permanent facilities, or you may need to rent a satellite newsgathering truck to attend some event.

There are many operators of terrestrial telecommunications networks. You may pay for a fixed service or by duration. You may need some so-called value-added services such as encryption, turnaround, format conversion, or post-production services. The link providers, particularly teleports, can often provide these for you. Again, you can find lists of providers in standard industry directories.

21.5.3 Obtaining TT&C

An end-user who is merely using transponder capacity need not even be aware of the details of the satellite's operation, much less the need to control it. About the only major operations concern of such a user is to know if and when the satellite will be out of service due to eclipses or sun outages.

A satellite owner, on the other hand, must monitor and keep the satellite under control at all times. Some satellite owners only have one or two satellites. In such cases, it may be more economical for them to hire-out the TT&C functions for their satellites. A small number of companies will provide TT&C services for a yearly fee. Of course, such firms must have control stations in location(s) that can see your satellite. You may have your own satellite control center and just use the linking capacities of the TT&C stations, or may have them do the whole job.

If you own more than a few satellites, it may be more economical for you to build and operate your own TT&C stations. Again, the stations must be visible by your satellites. Most TT&C operations have geographically dispersed redundant control stations for reliability.

For some satellite networks, there are also network control stations. These deal with providing the communications capacities, assigning channels, cross-strapping transponders, and otherwise configuring the system for the end users. These functions are often carried out at network operations centers (NOC) separate from the TT&C stations and from the satellite operations centers that control the satellites.

In this chapter, we have looked at the various satellite services and applications and how they can be used. The satellite telecommunications industry, like all parts of telecommunication in general, is looking to the future to increase its offerings and revenues. To do this, it faces issues and problems that must be confronted. The following chapter tries to summarize some of these concerns.

Chapter 22

Issues, Trends, and the Future

"It is hard to predict, especially the future" a wry commentator once said, and that is certainly true of the arena of telecommunications in general, and of satellite communications in particular. Specific predictions (especially those expensive market reports), should always be taken with several kilograms, not grains, of salt. Furthermore, market predictions and projections should always be considered, at best, valid only when made. They are continually subject to change. In a field as dynamic and even turbulent as telecommunications, many caveats are in order.

Consider the previous decade. In the 1990s, huge demands were predicted for mobile satellite telephony, as was a shortage of launch vehicles and consequently of insurance underwriting capacity. Contrary to that the mobile systems went bust, telecommunications crashed, and today we are oversupplied with both transponder capacity and launcher capacity. On the other hand, no one at the time was predicting demands for Internet via satellite, which is happening now.

Nevertheless, we can try to discern some of the present trends, issues, and concerns that will influence the industry over the first decade of the new century. These can simplistically be categorized into technical issues, business issues, and political (or regulatory) issues. Of course, there is much overlap and interaction among them.

22.1 Rapid changes in the telecommunications industry

Topping the list of concerns is the rapidity with which the telecommunications industry as a whole is changing, both quantitatively and qualitatively: people want to communicate more and in more different ways. Satellite communications are only a few percent of the total telecomms traffic and must respond to changes in markets, regulatory environments, and technology like all other parts of the industry.

One of the most insidious changes is the quick change of technology. The most famous illustration of this is "Moore's Law," which has correctly predicted the huge drop in cost for computer processing power and memory for decades. Much of the telecommunications industry had become accustomed to buying equipment and amortizing it over decades, and took a hard hit when equipment still carried on their long-term depreciation schedules became obsolete. Others overbought or overbuilt during the go-go days of the dot-com bubble. Some of them went bankrupt, and others bought them up cheaply, incurring no large debt but gaining huge amounts of installed capacity.

This has been particularly true in the fiber optics industry, causing the price-per-bit to fall by half or more every year. The same trends have not been active in the satellite side of telecommunications, making fibers even more of a threat to the satellite industry. More so than ever before, satellite operators have to concentrate on niche markets where they are competitive.

22.2 Some major telecommunications and satellite issues

Many of the major satellite industry issues have been around for decades. Most are just getting more troublesome. Ironically, the successes of the industry have spurred increasing demands for more services, exacerbating the problems. While we have discussed most of these issues in previous chapters, it is worth summarizing some of the major items of concern to the viability of commercial communications satellites.

22.2.1 Spectrum availability

All radiated transmission technologies—terrestrial and satellite—are affected by the increasing demands for more bandwidth to support more services. The range of radio frequencies to supply this capacity is limited by incumbent users of spectrum, by limitations caused by the atmosphere, and by limitations in technology development.

The radio spectrum remains a limited natural resource and the future expansion of its use will be almost a zero-sum game, in the sense that bandwidth given to a new or expanded service can mostly happen by taking away bandwidth from some other service. Improved technology can partially alleviate the problem, by making it possible to send more information through each hertz of spectrum, such as the paired carrier multiple-access technique explained in Chapter 20, but there is a limit to this. The push to higher and higher frequency bands can also help expand spectrum availability, but we run into further atmospheric problems and the immaturity of electronics at these frequencies above the Ka-band. Still, we will see a push to use some of these where possible. We will also see "re-farming" of parts of the spectrum, as some older services go out of use and others are forced to use bandwidth more efficiently.

These trends pit satellite systems against terrestrial systems and GSO satellites against NGSO satellites. The situation is complicated by legacy services that are difficult to displace, and by the disparate spectrum allocations that exist among the 200+ nations and regions of the world.

22.2.2 Orbit availability

There are only about 264,000 km along the Clarke orbit. The challenge is to get more communications capacity in this finite space. More efficient use of the spectrum is part of the answer, as is going to higher frequencies to allow closer spacing of satellites. Larger satellites with more directional beams will allow more frequency reuse.

Even NGSO satellites run into space problems. While technically there is an infinite number of nongeostationary orbits, not all are equally useful, and both orbital inaccuracies and gravitational and atmospheric perturbations limit our ability to vastly increase the number of NGSO satellites. The coordination of their continually moving beams, both with each other and with GSO systems, is complicated and limiting.

Furthermore, many of these NGSO satellites will be using the same frequencies as each other, or as some GSO satellites. Frequency coordination is an even tougher issue than simply avoiding one satellite running into another. Not only that, NGSO systems greatly add to the orbital debris problem. We do not yet have a track record to know how bad this will be.

A continuing problem is that of "paper satellites," filings with the ITU for satellite systems that never get built. Even so, once filed for, any later prospective satellite systems must coordinate with those already on file. Worse yet, some companies have filed for systems even though they know the satellites will never come into existence, either to "keep their oar in the water" or to deter or preclude rivals from establishing other systems. Revised ITU regulations have made a first step in cutting down on these highly speculative systems, but more needs to be done.

22.2.3 Industrial issues

One trend we have seen over the past decade is the consolidation of the industry players. Several space firms have merged in order to compete with each other. Some manufacturers have become vertically integrated to provide not just satellite hardware, but launch services, financing services, and communications services, sometimes in competition with their customers. Other firms have gotten out of the hardware business to concentrate on providing communications.

The growth in demand for satellite communications has driven manufacturers to try to produce satellites in a shorter length of time and at lower costs. This has also led to requirements for satellites to do more and last longer; that has in turn pressured launch providers to speed up both the production of launch vehicles (to produce new, more capable vehicles), and to reduce the interval between launches.

All of these pressures have led to reductions in quality control. Speeding up the production of satellites and launchers has resulted in corners cut in manufacturers' quality management, resulting in launch failures and an unusual spate of on-orbit partial or complete failures. These have led to higher insurance rates, increasing the cost of bringing a satellite system into operation.

The demands for increased capabilities of the satellites have led to more haste in incorporating new, but less well tested, technologies into the satellites, with some resultant failures.

As the industry has consolidated, not only has the range of choices for manufacturers grown more limited, but some acquisitive firms have found it difficult to assimilate their acquisitions and maintain quality. National and regional governments have grown wary of possible monopolistic trends, and related concerns about intellectual property exchanges, national security, and commercial offerings by nonmarket economies have complicated the situations.

22.2.4 Launcher and launchpad availability

There are only a handful of launchers capable of carrying satellites to GSO. Another handful can be used for NGSO missions. To maintain quality, there is a limit to how quickly they can be manufactured.

Further, every launcher requires a unique launch pad. Each launch inevitably slightly damages the launch pad, which must be refurbished. Each launch vehicle takes time to assemble and to integrate its payload. Thus, each launch pad has a "turn-around time" that limits the launch frequency. Having multiple pads is a partial solution, but there must be a concomitant increase in launch personnel as well, and pads cost money to maintain even when they are not in use.

A further problem is that most launch pads are owned by governments, in many cases controlled by military services. For both major launch sites in the United States this is the case, as it is for sites in Russia and China. In some cases, private firms have leased or made other arrangements to use government facilities for their launchers, but they are often still subject to governmental strictures. A few firms are working to provide nongovernmental launch sites.

Both existing and new firms are endeavoring to produce new launch vehicles with the goal of greatly reducing the cost of access to space. A rough estimate is that it costs around $25,000 a kilogram to launch a satellite today. The industry would like to see the figure fall to a tenth of that.

In this decade, we are likely to see an oversupply of launchers and launch capability, but new ideas and financing could cause a resurgence of demand such as we saw in the 1990s. One thing is certain: it is reliability, more than just cost, that will be the most important marketing tool of launch providers.

22.2.5 Financial capability availability and risk

After the resounding bankruptcies of Iridium and ICO, the financial markets are more cautious of such high-risk projects. Certainly not all proposed systems will get built. The telecomm crash of the early years of this decade has greatly tightened the availability of both financing and insurance.

The insurance underwriting capacity for satellite launches at any instant has varied over the years from a few hundred million dollars to a couple of billion dollars. This capacity limits the availability of insurance, and its rate, for satellite systems wishing to launch, especially if several such systems come to readiness closely in time. The changing success records of the various launch vehicles contributes to the volatility of the insurance market as well.

The financial community is also becoming more aware of the rapidity of change in telecommunications. Among the reasons for the bankruptcies of Iridium and ICO was that fact that during the long time from conception and funding to launch and deployment, the industry had changed without a concomitant alteration in the business strategies of the planned satellite systems. What are the risks of obsolescence before a system is even operational?

Financiers are also asking how important it is for a new type of system to be first to market. While such pioneers may have some market advantage, they also take much higher risks exploring unknown commercial territory. There is an old quip that you can always identify the pioneers by the arrows in their chest.

22.2.6 Multiple standards

One legacy item slowing deployment of new technologies is the continuing multiplicity of standards. Several different standards exist in various nations for wired telephony, multiplexing, wireless telephony (cellular systems), and television broadcast. Satellite television may be slowly standardizing on two digital standards, DVB and DSS, but the consumers still have televisions that receive only NTSC, PAL, or SECAM, and local phone systems that work on standards established in some cases half a century ago. There is also an ongoing debate between advocates of the DOCSIS transmission protocol and the DVB-RCS protocol.

Among the things impeding change to more global standards are such simple things as the sheer number of installed terminals of one standard or another, the "not invented here" mentalities of some companies, countries, and regulators, or the difficulties of accommodating non-Roman character sets into an increasingly globalized exchange of information. Even such seemingly simple issues as addressing limits deployment of systems: one can't call a telephone unless it has a number, and can't send data between devices that aren't identifiable to the communications systems. Some systems, designed in the days when the industry had a much more limited idea of who and what will be communicating, are limited in the amount of addressing information that they can handle.

22.2.7 Multiple regulatory environments

Not only are companies subjected to a multiplicity of technical standards, they must operate in a large number of often conflicting regulatory environments. Many regulatory bodies remain hidebound, bureaucratic, and slow. Debates continue on just what constitutes "fair" or "equitable" division of limited natural resources such as spectrum and slots.

The regulations promulgated by nations and groups of nations range from such major global concerns as national defense, security, and politics; to personal concerns such as privacy, freedom of expression, and intellectual property; to industrial issues such as the protection of industries.

There is and will continue to be a three-way tug among the desires and requirements of the developed nations, the developing nations, and the truly underdeveloped countries. There are more than 200 nations on the planet, each with a history, economic and political status, desires, and dominant culture(s). There are conflicting ideas on how much one can criticize a government or religion, whether children should appear in television commercials, and what degree of nudity can be shown, for example. There have been instances in which one progressive group in a nation set up a satellite-based service, only to have it totally abolished by reactionary forces.

The debate between developed and developing nations on use of natural resources will continue. The 1976 "Declaration of Bogota," in which nations along the equator claimed ownership rights to portions of the GSO lying over their territories is still on the books, though ignored by the heavily telecommunicating nations. However, there is some justification for the worry on the part of less-developed nations that orbital and spectrum resources will be largely taken over by more-developed states.

A further complicating factor that began emerging in the late 1900s, and which will continue to occupy debates well into this century, is the notion of national sovereignty and how it comports with the growth of multinational corporations and supranational entities like the European Community. Whose laws govern, who taxes, who owns what, where? What role do bodies such as the ITU have in the future?

The international politics of regulatory activities certainly reinforces two old adages: "Politics makes strange bedfellows" and "All politics is local." In the area of satellite communications, we have seen such effects. For example, in the early 1990s, when plans and technologies started to gel for mobile satellite services, much of Europe was against the idea of making global allocations of frequencies for such services, largely because Europe had in place a single widespread cellular telephone standard, didn't want competition, and further because European industrial interests recognized that U.S. firms could be quicker to get into such markets. Satellite firms in the U.S. were all in favor of these new services, seeing new lines of business, but in the ITU's one-nation/one-vote system of operation, the U.S. alone could hardly carry the issue. It found unlikely allies in the lesser-developed nations, whose telephone systems were small, limited, and often obsolescent. Together, the U.S. and the LDCs got the new Big LEO and Little LEO frequency allocations passed. Such alliances are likely to occur again.

Still further, we have seen a decline in what is called the "international comity of nations," the willingness to act courteously and respect standards of conduct and international law. For example, despite ITU prohibitions against trafficking in orbital slots, some nations have done so without penalty because there is no enforcement process available.

22.3 Satellite industry trends

Among the general trends we can see in telecommunications are the following:

- Increasing dominance of digital technologies for transmission
- Rapid growth in global connectivity via the Internet as a distance-insensitive, time-insensitive medium
- Increasing demands for bandwidth requiring more sophisticated and expensive spectrum sharing
- A shift in the paradigm of charging by time and distance to one that charges by the time or bit
- A trend toward privatizing government-run monopolies and toward somewhat open, more competitive markets
- Growth in use of the Internet and Internet protocol
- Increasing use of DVB protocols
- Increasing demands for "multimedia," a variety of types of signals
- Rapid deployment and increasing capacity of terrestrial fiber networks with consequent huge annual drops in cost
- A shift in the old paradigm of rigidly hierarchical networks to one of a mesh of interconnections between networks reinforcing a trend from switched to routed connectivity
- Increasing demand for mobile connectivity "anywhere, anytime"
- Possible deployment of competitive high-altitude relay platforms
- Increasing intelligence and memory built into lower-cost user terminals
- Globalization of the telecommunications industry
- Continuing problems of adapting legacy hardware and networks to provide new types of services
- Trends toward the interconnectivity of everything
- The slow "convergence" of computing and communication on a personal level

The satellite industry is responding to these by deploying both larger, more powerful, longer-lived satellites for GSO, and smaller, niche-targeted satellites in LEO and MEO. Increasingly, planned satellites are being optimized for data transmissions and versatility. With the exception of some satellites intended specifically for highly focused applications, such as feeding cable headends with video programming, more satellites will incorporate onboard intelligence, signal processing, signal routing, and other capabilities to keep them competitive with terrestrial services. Some larger satellites will have a total power consumption over 20 kW.

22.4 The future of communications via satellite

Neither I nor anyone else can claim to have a clear crystal ball for peering into the future of the industry. The hazards of making predictions are legend. Any predictions are fraught with uncertainties. The best we can hope for is to discern trends as we tried to do earlier in this chapter.

To see how difficult it is to predict the future of communications systems, witness the troubles of some of the mobile telephony systems within the past decade. And who, 10 years ago, could have predicted that a hot topic at the satellite conferences 10 years later would be broadband multimedia Internet connectivity? Some of the advances in commercial communications have come from largely unforeseen developments, ranging from secret military research becoming public, to unexpected political upheavals like the fall of European communism.

Perhaps one approach is to first consider some of the guiding principles for making predictions.

Science fiction writer (but engineer by training) Robert A. Heinlein looked at the history of technological predictions, and noted that long-term predictions tend to be too pessimistic, and short-term predictions tend to be too optimistic. That is, if we say some technology will be available in 25 years, it often takes only 10 to 20 years, but if we say something will be available in 3 years, it may take 5 to10.

For strategic planning purposes, predictions must be made, but they should be constantly reviewed, updated, and checked for the validity of the assumptions upon which they are based. Predictions should not be considered static.

Further, when consulting those expensive market reports and analyses, always check on the sources and age of the data from which conclusions are drawn. Remember the old adage that data does not mean information, information does not equate to knowledge, and knowledge is not the same thing as wisdom and experience.

Another admonition which bears repeating, but which is too often ignored: when planning a communications system, first look at the customers' requirements, and tailor the technology to suit; do not start with a technology and try to find a way to coerce customers into using it. We have seen the failure of this approach several times in our industry just in the past decade. Remember, to an end user, the service is important, and the technological details are largely irrelevant.

Finally, in designing a system, whether technical or commercial, there are no ideal solutions; there are always trade-offs, and sometimes going after cutting-edge technology (otherwise known as bleeding-edge technology) may actually result in the creation of a less useful service. Sometimes "good enough" is indeed good enough, while the pursuit of technical perfection can delay service implementation beyond a period of opportunity. As remarked in the section summarizing multiple-access techniques, remember Robert Heinlein's TANSTAAFL: "There ain't no such thing as a free lunch."

The concepts, issues, and applications covered in this book have evolved over more than a half century. We could only touch on the basics in this book, but it is hoped that you now have a general overview of the industry and its interconnected parts as a basis for learning the more detailed information necessary for your role in the industry.

The foresight of Clarke, Rosen, Pierce, and the many other unsung scientists, engineers, and business leaders who have contributed inventions, techniques, and business structures have built an industry dynamic, progressive, and exciting to work in.

Appendix A

Glossary of Common Satellite Telecommunication Terms

ACCESS: The general term for the ability of a telecommunications user to make use of a network.

ACCESS LINE: A local loop (q.v.).

ACKNOWLEDGMENT: A signal sent by a receiver to a transmitter to indicate that a message was received correctly.

ACTUATOR: A motorized device that causes a satellite dish to move so that it can point to different satellites.

ADAPTIVE DIFFERENTIAL PULSE CODE MODULATION (ADPCM): A variant of PCM which encodes only the difference between the value of the current sample and the previous one.

ADDRESS: The unique identifier of a terminal on a network.

ADDRESSABILITY: The ability of a network, especially a satellite or cable system, to individually address and thus control (usually to enable decryption) users' receivers.

ADVANCED DATA COMMUNICATION CONTROL PROCEDURE (ADCCP): A bit-oriented datalink protocol similar to HDLC.

ALLOCATION: Entry in the table of frequency allocations of a given frequency band for the purpose of its use by one or more terrestrial or space radiocommunications services or the radio astronomy service under specified conditions. This term shall also be applied to the band concerned.

ALLOTMENT: Entry of a designated frequency channel in an agreed plan, adopted by a competent conference, for use by one or more administrations for a terrestrial or space radiocommunication service in one or more identified countries or geographical areas and under specified conditions.

ALPHANUMERIC: Any character that is either a letter of the alphabet or a numeral.

ALTERNATE ROUTING: A communication path used if the primary path is not available.

AMERICAN NATIONAL STANDARDS INSTITUTE: A U.S. standards organization and the U.S. representative to the ISO.

AMERICAN STANDARD CODE FOR INFORMATION INTERCHANGE (ASCII): An American dialect of International Alphabet No. 5 which uses a 7-bit code to represent alphanumeric and control information.

AMPLIFIER: An electronic device that increases the power of a received signal.

AMPLITUDE MODULATION (AM): The technique of varying the strength of a carrier wave in proportion to the strength of a signal.

ANALEMMA: The figure-8 path followed during one day by an inclined-orbit satellite as seen from an earthstation.

ANALOG: An signal that can take on any of a continuous range of values between some minimum and some maximum. Compare Digital.

ANALOG CHANNEL: A channel on which the carrier of the information can take on any of a range of values.

ANALOG DATA: Information conveyed by continuously varying some property.

ANALOG TRANSMISSION: Transmission of a carrier wave that varies continuously and can take on any of a range of values of one or more of the properties of the wave.

APERTURE: The area of an antenna's dish receiving a signal.

APOGEE: The point in a closed orbit farthest from Earth.

ARCHITECTURE: In a network, the overall structure of the network and its components, intended to fulfill some user requirements.

AREA CODE: See Numbering plan area.

ASSIGNMENT: Authorization given by an administration for a radio station to use a radio frequency or radio frequency channel under specified conditions.

ASYNCHRONOUS: A transmission technique in which each group of bits to be sent is preceded by a "start" character(s) and followed by "stop" character(s).

ASYNCHRONOUS TRANSFER MODE (ATM): A switched digital transmission protocol used for a variety of traffic; it sends information in 53-byte cells.

ATLANTIC OCEAN REGION: One of the operating regions of communications satellites systems.

ATTENDED OPERATION: A system in which human intervention is required to establish a modem connection and shift over from voice to data mode.

ATTENUATION: The reduction in strength of a signal.

AUDIO FREQUENCY (AF): The frequencies of sound waves that can be perceived by the human ear, roughly 30 to 20,000 Hz. Also sometimes called voice frequencies (VF), but actually broader in bandwidth.

AUTHENTICATION: The verification of the identity of a telecommunications entity.

AUTHORIZED POWER: The maximum signal strength allowed for a particular class of licensed transmitting station.

AUTOMATIC REQUEST FOR REPETITION (ARQ): An error-correction system in which the receiver asks the sender to repeat information received incorrectly.

B CHANNEL: A 64-kbps channel, part of the ISDN system, usable for voice or data.

BACKHAUL: The traffic within a television network, before being broadcast. In the optical fiber industry, connection from undersea landing points to metropolitan networks.

BACKOFF: The amount by which an amplifier must be reduced in gain to avoid intermodulation distortion caused by simultaneous frequencies through the amplifier.

BANDWIDTH: The gamut of frequencies used. Also loosely used to refer to the speed (bitrate) of a digital signal.

BASE STATION: A stationary station in the land mobile service that can communicate with other mobile stations.

BASEBAND: The range of frequencies inherent in a source of information.

BASEBAND SIGNALING: Transmission of information at its original range of frequencies.

BAUD: A measure of the signaling speed of a circuit, the baud rate is the number of discrete signal conditions per second.

BAUDOT CODE: A 5-bit code for expressing alphanumeric and control information, used in telegraphy.

BEACON: A low-power signal sent from a satellite to enable ground controllers to track the satellite and often to receive telemetry data from it, or a similar signal transmitted to a satellite from an earthstation to enable it to maintain its orientation.

BEAM: The pattern of distribution of radiation sent out from a transmitting antenna or the pattern of sensitivity of a receiving antenna.

BEL: A unit of strength of a signal, equal to 10 dB (q.v.).

BENT PIPE: Slang for a transponder or satellite that does no onboard processing of the received signals other than amplifying and relaying them back to Earth.

BINARY-CODED DECIMAL: An encoding scheme in which each decimal digit is encoded as a 4-bit binary number.

BINARY SYNCHRONOUS (BISYNC): A character-oriented synchronous datalink protocol that allows frames of data of arbitrary length. ASCII or EBCDIC is used for encoding the characters.

BIRD: Slang term for a satellite.

BIT: Binary digit, the smallest piece of information possible, usually expressed as either a "1" or a "0."

BIT ERROR RATE: The fraction of bits transmitted which are received in error. Sometimes expressed as the number of bits received correctly per bit received in error, or its inverse, usually as a power of 10.

BIT-ORIENTED PROTOCOL: A datalink protocol in which the bit, rather than the character, is the fundamental unit.

BIT STUFFING: A technique of inserting a 0-bit in a transmitted datastream used in bit-oriented protocols to ensure that six 1s never appear consecutively.

BITRATE: The rate at which bits are transmitted, expressed as bits per second.

BLOCK CHECK CHARACTER: A set of bits sent at the end of a binary synchronous frame used to check the frame for accuracy.

BLOCK CONVERTER: An electronic device that takes a wide range (block) of frequencies and converts them to some other band, preserving the bandwidth.

BORESIGHT: The direction toward which a dish is pointed.

BRIDGE: A telecommunications device that connects two similar networks.

BROADBAND: Referring to a bandwidth greater than the baseband bandwidth, or greater than a voice frequency bandwidth.

BROADCAST SATELLITE SERVICE (BSS): The satellite service designed to bring primarily video entertainment from satellites directly to consumers via high-power satellites and small user antennas. More commonly referred to as DBS, Direct Broadcast Service.

BUFFER: A computer storage device used to receive, hold, and then release a datastream to allow other devices of different transmission speeds to communicate.

BUNDLING: The practice of offering several services as a package.

BURSTY: A characteristic of datastreams in which the bitrate changes.

BYPASS: The use of transmission facilities other than those of the local telephone company network.

BYTE: Eight bits.

C BAND: Frequencies in the 4 to 6 GHz range used both by terrestrial microwave links and for satellite links.

CABLE: One or more conductors contained within a protective insulating cover.

CABLE TELEVISION: Any system that receives transmissions from program sources and distributes them to users (usually homes) via coaxial cable, usually for a fee.

CARRIER: A continuous (usually high-frequency) electromagnetic wave that can be modulated by a signal to carry information.

CARRIER-SENSE MULTIPLE ACCESS WITH COLLISION DETECTION (CSMA/CD): A multiple-access technique in which all stations monitor the network for activity; if there is none, any station may send data; if two stations send at the same time, they detect the "data collision" and retry.

CARRIER SYSTEM: A technique of frequency multiplexing in which individual signals are carried on separate carrier waves of different frequencies.

CARRIER-TO-NOISE RATIO (C/N): The ratio of the strength of the carrier signal to that of the noise in a channel, usually expressed in decibels.

CASSEGRAIN FEED: An optical system used in telescopes and satellite antennas in which the feed is placed behind the dish, and the radio waves are brought to a focus there after being reflected first from the main dish and then from a subreflector near the dish's prime focus.

CELLULAR MOBILE RADIO: A land-mobile telephone service provided by a grid of low-power computer-connected and controlled cells, rather than by one high-power station. The user may travel from cell to cell, as the system maintains track and switches the links accordingly.

CENTRAL OFFICE: The office in a telecommunications network that is directly connected to the end user.

CENTUM CALL SECONDS: A measure of traffic on a telephone network, equal to 100 seconds of usage.

CHANNEL: An signal path between two devices. Also called link, path, or line.

CHANNEL GROUP: A block of 12 telephone channels, transmitted together.

CHARACTER: Any of the letters, numbers, and other symbols used as part of a written message.

CHARACTER STUFFING: In the binary synchronous protocol, a technique for allowing any character to be a part of the datastream.

CIRCUIT SWITCHING: A network technique in which a physical connection is made between the sender and receiver for the duration of the transmission.

CIRCUIT: See Channel.

CIRCULAR POLARIZATION: See Polarization.

CLIENT: In a client-server network, the entity that makes a request (of the server) for some service.

CLOSED-CIRCUIT TELEVISION: A television signal carried, usually via microwave or coaxial cable, between two or more locations, but not broadcast for general reception.

CLUSTER CONTROLLER: A device used to control a number of terminals.

COASTAL EARTH STATION (CES): One of the land-based nodes in the maritime satellite service which is connected to the public telephone network. Also called Land Earth Station.

COAXIAL CABLE: A cable in which one electrical carrier is a wire running down the center of the cable, and the other conductor forms a cylindrical shell coaxial with the central conductor.

CODE-DIVISION MULTIPLE ACCESS: See Spread-spectrum multiple access.

CODEC: Coder/decoder, a device that converts an analog signal into and/or from a digital signal.

COMMON CARRIER: A regulated telecommunications company that will carry messages for anyone for a fee.

COMMON CHANNEL INTEROFFICE SIGNALING: The network that switching systems within a telephone network use to exchange control information.

COMMON CHANNEL INTEROFFICE SWITCHING: A means of telephone network control in which control signals are carried independently of the voice channels, enabling greater security and enhanced features.

COMMON CONTROL SWITCHING ARRANGEMENT: A service that provides users with intercity private switched lines.

COMMUNICATIONS SATELLITE: An artificial satellite, usually placed in geostationary orbit, used to relay radio transmissions.

COMMUNITY ANTENNA TELEVISION (CATV): An earlier name for what are now simply called cable TV companies.

COMPANDOR: Compressor/expander, a device to improve the signal-to-noise ratio of a transmitted signal by compressing the dynamic range of the signal at the sender and expanding it to its original value at the receiver.

COMPONENT VIDEO: A video signal in which the brightness and color information is sent as individual signals.

COMPOSITE VIDEO: A video signal in which all of the brightness and color information is combined as one integrated signal.

COMPRESSED VIDEO: A video signal that has had much of the redundancy removed from it, thus greatly decreasing its required bitrate.

COMPRESSOR: A device to compress the dynamic range of a transmitted signal.

CONSULTATIVE COMMITTEE ON INTERNATIONAL RADIO (CCIR): The old name for a working body of the ITU that provides recommendations for telecommunications standards and practices in the fields of radiocommunication. Now the ITU-R.

CONSULTATIVE COMMITTEE ON INTERNATIONAL TELEGRAPHY AND TELEPHONY (CCITT): An old name for a working body of the ITU that provides recommendations for telecommunications standards and practices in the fields of telegraphy and telephony, including data transmission. Now the ITU-T.

CONTENTION: A method by which terminals on a network transmit when they are ready to. If two such transmissions collide, they try again.

CONTRIBUTION FEED: A transmission of material (audio or video) that will be used in a program, either live or delayed.

CONTROL CHARACTER: A character that is interpreted as a signal to control the transmission, rather than as an item of the message being sent.

CONUS: The contiguous United States; i.e., all of the states (including the District of Columbia) except Alaska and Hawaii.

COORDINATION: The process by which multiple users of a frequency band analyze their usage and potential mutual interference, and revise their systems to minimize or eliminate it.

CROSS-POLARIZATION: Arranging the sense of two polarized signals so that they are orthogonal (for linear polarization) or of opposite handedness (for circular polarization).

CROSS-POLARIZATION DISCRIMINATION: The amount by which two cross-polarized signals are independent of each other, usually measured in decibels.

CROSS-SUBSIDIZATION: The practice of using revenues from (usually unregulated) services to pay for other (usually regulated) services.

CROSSTALK: The unwanted leakage of signal between supposedly independent channels.

CUSTOMER PREMISES EQUIPMENT (CPE): Telecommunication equipment located at the user's site rather than at a central office.

CYCLICAL REDUNDANCY CHECK: A mathematical algorithm to produce an error-checking code within a datastream.

D CHANNEL: A 16-kbps channel within the ISDN system, usable for either packet data or network signaling.

DATA CIRCUIT-TERMINATING EQUIPMENT: A device that is the interface between a data transmission facility and data terminal equipment.

DATA COMMUNICATION: The transfer of digital data by electronic or electrical means.

DATA ENCRYPTION STANDARD (DES): A symmetrical encryption method approved by the U.S. government.

DATA OVER CABLE SERVICE INTERFACE SPECIFICATION (DOCSIS): An electronic specification of a method of carrying multimedia services over cable television networks.

DATA SERVICE UNIT (DSU): A device that accepts input data from a user's terminal (a DTE) and formats it for proper transmission over a network.

DATA SET: A device that provides the modem and control functions enabling data communication between terminals over some communications facilities.

DATA TERMINAL EQUIPMENT (DTE): A device capable of transmitting digital data over a communications circuit.

DATA UNDER VOICE: A method of transmitting data over a telephone circuit being used by a voice, in which the data is carried by frequencies below the voice frequencies.

DATAGRAM: A single packet of information transmitted over a packet network.

DATALINK: The circuit and equipment that allows data to be transmitted between terminals.

DATALINK PROTOCOL: A standard that governs how two terminals exchange data at the datalink level.

DATAPHONE DIGITAL SERVICE (DDS): A digital private-line service offered by AT&T.

DECIBEL (dB): One-tenth of a bel, a unit of measure of signal strength. Bels and decibels are relative units using a logarithmic scale.

DELAY DISTORTION: Distortion of a transmitted signal due to different frequencies traveling through a transmission path at different speeds.

DELAY EQUALIZER: A device to eliminate the effects of delay distortion in a transmission path.

DEMAND ASSIGNMENT MULTIPLE ACCESS (DAMA): Any system whereby many users may access a network on demand, rather than be permanently connected.

DATAGRAM: The unit of data that is routed by the Internet Protocol.

DEMODULATION: The process of extracting the original signal from a modulated carrier.

DEMUX: Demodulation.

DENSE WAVELENGTH-DIVISION MULTIPLEXING (DWDM): A technique for increasing the capacity of existing fiber optic lines by sending and receiving simultaneously.

DIAL PULSE: The pulse of DC current produced by a dial telephone.

DIAL-UP: The use of a telephone to place a call between two users.

DIGITAL SIGNAL: A signal that can take on only certain, discrete values.

DIGITAL SPEECH INTERPOLATION: A statistical multiplexing system that time-assigns channels to telephone calls only when a user is actually speaking.

DIGITAL TERMINATION SYSTEM: A high-speed microwave transmission service for local or intercity traffic.

DIGITAL VIDEO BROADCAST (DVB): A family of digital transmission protocols for video and video-like data.

DISCONNECT SIGNAL: A signal within the telephone network indicating that the circuit should be disconnected.

DISTORTION: Any undesirable change in the form of a wave as it is transmitted.

DOMAIN NAME SYSTEM (DNS): The system of databases that allow network names to be interconverted to IP addresses.

DOMSAT: Domestic communications satellite.

DOWNCONVERTER: A piece of equipment that reduces the frequency of an input signal to some lower frequency.

DOWNLINK: The signal sent from a communications satellite to the earthstation.

DUPLEX: A mode in which there exists two-way simultaneous transmission between the users.

DYNAMIC ALLOCATION: The allocation of transmission capacity depending upon demand.

E-MAIL: See Electronic mail.

EARTHSTATION: The Earth-based link in a communications satellite transmission system.

ECHO: The unwanted reflection of a signal to its originator.

ECHO CANCELER: A device that eliminates the echo in a channel.

ECHO CHECK: An error-checking technique in which data sent to a receiver is sent back to the sender for verification.

ECHO SUPPRESSOR: A device that reduces the signal coming back to a transmitter to reduce echo.

EFFECTIVE ISOTROPICALLY RADIATED POWER (EIRP): A measure of the effective power emitted by a transmitter, or a measure of the signal strength received on Earth from a satellite.

ELECTRONIC MAIL (E-MAIL): The sending and receiving of typewritten messages via electronic means.

ELEVATION-AZIMUTH MOUNTING: Also called an el-az mount. A method of mounting a satellite dish such that the dish can change its angle above the horizon (elevation) and the compass direction to which it points (azimuth).

ENCODER: A device that converts an analog input into a digital output.

ENCRYPTION: A technique of altering a signal so that it is unusable to unauthorized receivers.

EQUALIZATION: The technique of compensating for differences in attenuation of a signal at different frequencies.

EQUALIZER: A device to perform equalization.

EQUATORIAL MOUNTING: A method of mounting a satellite dish on an axis aligned parallel to the earth's axis. Rotation around this polar axis causes the dish to sweep along the geostationary arc. Also called a polar mounting.

ERLANG: A measure of intensity of traffic on a telephone circuit, equal to the average number of circuits in use.

ERROR-DETECTING CODE: A technique by which a message is sent along with additional information derived from the message in order that the receiver can determine if any errors occurred during transmission.

ESCAPE MECHANISM: A technique in which a control character signifies to the receiver that one or more following codes are to be given some alternative meaning.

EUTELSAT: European Telecommunications Satellite Organization.

EXCHANGE: A local hub in a telecommunications network connected to end users.

EXPANDER: A device that expands the dynamic range of the strength of an input signal.

EXTENDED BINARY-CODED DECIMAL INTERCHANGE CODE (EBCDIC): An IBM-originated 8-bit code used to represent characters internally in IBM equipment and occasionally for external communications.

FACSIMILE (FAX): Any system intended for the scanning, transmission, and reproduction of images of documents.

FAST-CONNECT CIRCUIT SWITCH: A circuit switch that establishes its connection typically in milliseconds.

FEDERAL COMMUNICATIONS COMMISSION: A U.S. executive branch administrative body, headed by a panel of seven commissioners appointed by the President, which has authority over all of the interstate and foreign electromagnetic communications originating in the U.S.

FEED: For a transmitter, the part of the antenna from which the signal originates; for a receiver, the place where the reflector brings the received signal to a focus. In the broadcast industry, the sending of programming.

FIBER OPTICS: The technology in which a modulated beam of light carries information through a thin glass or plastic fiber.

FIGURE OF MERIT: Also called G/T, it is a measure of receiving sensitivity, the ratio of the gain of a receiving satellite dish to the noise entering the antenna. It is usually measured in decibels. Higher G/T means a more sensitive dish.

FILTER: A device designed to allow only a predetermined range of frequencies to pass through it.

FIXED SATELLITE SERVICE (FSS): The telecommunications service between nonmoving earthstations (although the antennas may be moveable, just not in motion at the time of use).

FLOW CONTROL: Any technique for governing the flow of information through a network.

FLUX DENSITY: The strength of a received signal, usually measured in dB/m^2.

FOOTPRINT: The geographical range of reception from (or transmission to) a satellite, usually given with signal strength.

FOREIGN EXCHANGE SERVICE: A service that connects a user's telephone to a telephone exchange that normally does not serve the user's location. Sometimes called a tie-line.

FORWARD ERROR CORRECTION: A technique for transmitting enough additional information with an item of information to enable the receiver to reconstruct the original information if it is received with errors.

FOUR-WIRE CIRCUIT: A circuit in which a pair of wires or coax is used to carry a signal in each direction between two terminals.

FOUR-WIRE EQUIVALENT CIRCUIT: A two-wire circuit that allows simultaneous two-way communication by utilizing different carrier frequencies in opposite directions.

FRAME: A unit block of data.

FREQUENCY MODULATION (FM): A technique whereby a carrier wave is made to carry information by changing its frequency in proportion to variations in strength of a lower-frequency signal.

FREQUENCY-DERIVED CHANNEL: A channel within a frequency-divided multiplexed group of channels.

FREQUENCY-DIVISION MULTIPLE ACCESS: A technique for allowing many users to share a transmission bandwidth by assigning each of them a small bandwidth such that the sum of all such user bandwidths is equal to the allowed bandwidth.

FREQUENCY-DIVISION MULTIPLEX: A system in which each baseband signal occupies a small part of a wideband channel, each at a different frequency.

FREQUENCY-SHIFT KEYING (FSK): An FM technique in which different states of the original signal are transmitted as different frequencies of the carrier wave.

FULL DUPLEX: A circuit which carries information in both directions at the same time.

GATEWAY: Any device that is an interface between different networks.

GEOSTATIONARY ORBIT (GEO): A special geosynchronous orbit that is circular and lying over the equator such that the satellite seems to remain stationary in the sky as seen from a location on the surface of Earth.

GEOSYNCHRONOUS ORBIT (GSO): An orbit around the earth with an average distance from the center of Earth of about 26,000 miles, in which a satellite would have a period equal to the rotation period of the earth.

GEOSYNCHRONOUS TRANSFER ORBIT: An elliptical orbit into which a launch vehicle initially injects a satellite. Typically, it has a perigee of a several hundred kilometers and an apogee at the geostationary distance.

GLOBAL BEAM: A satellite beam that provides service to the entire region of the earth visible from the satellite.

GRADE OF SERVICE: A measure of the probability of a user being unable to use a network because it is busy.

GUARD BAND: A narrow band of frequencies separating frequency channels in an FDM system, used to avoid crosstalk between them.

GUARD TIME: The brief pauses between data packets in a time-division multiple access system.

HALF-DUPLEX (HDX): A circuit that can carry information in both directions, but only in one direction at a time.

HALF-POWER BEAMWIDTH (HPBW): The angle, centered on the boresight of a satellite dish, within which the transmitted signal strength or the sensitivity to received signal is at least half of the value measured along the boresight.

HAND-OFF: In a satellite system, the transfer of a communication session from one satellite to another as they move out of and into the area being served.

HANDSHAKING: Those predetermined signals which allow the establishment of a connection between two devices or networks.

HARMONIC DISTORTION: Overtones (harmonics) produced when a simple sine wave is sent through a circuit with nonlinear electrical characteristics.

HERTZ: One cycle per second of a wave.

HIERARCHICAL NETWORK: A telecommunications network in which a message is passed through nodes of different classes.

HUB: An earthstation that is the central locus and often the controlling entity of a satellite network.

HUB POLLING: A polling technique in which a polling signal is sent sequentially from one terminal to the next.

HYBRID: Referring to a satellite, one that can operate in more than one of the major frequency bands. Referring to satellite Internet access, a system in which uplinks from the customer travel by telephone lines, but downlinks to the customer are by satellite. In the telephone industry, it is an interface device between parts of the network.

ILLUMINATION LEVEL: The received signal power from a satellite, usually expressed in decibels per square meter.

INCLINATION: The angle the plane of a satellite's orbit makes with the plane of the earth's equator.

INCLINED ORBIT: A satellite that has been allowed to move slightly north and south from the Clarke orbit. During one day, as seen from an earthstation, the satellite moves several degrees above and below the equator and a few tenths of a degree east and west of its assigned slot.

INFORMATION UTILITY: A service that provides its customers access to sources of information usually by means of public data networks.

INTEGRATED DIGITAL NETWORK (IDN): A telecommunications network employing both digital switching and digital transmission of information.

INTEGRATED RECEIVER DECODER (IRD): A satellite receiver that contains a decoding circuit for scrambled signals.

INTEGRATED SERVICES DIGITAL NETWORK (ISDN): A telecommunications network capable of carrying any form of traffic (voice, data, etc.) in a digital form.

INTEREXCHANGE CARRIER: A common carrier that carries traffic between LATAs.

INTEREXCHANGE CHANNEL: A link between two exchanges.

INTERMEDIATE FREQUENCY (IF): A frequency used within electronic equipment, or within an earthstation, which is higher than the baseband frequency of the signals being processed, but lower in frequency than the radio frequencies used for transmission.

INTERMODULATION DISTORTION: The self-induced distortion (noise) created in a nonlinear electronic component, usually an amplifier, when multiple frequencies are present and interact with one another.

INTERNATIONAL ALPHABET NO. 2 (IA2): A CCITT-standard 5-bit character code.

INTERNATIONAL ALPHABET NO. 5 (IA5): A CCITT-standard 7-bit character code widely used for data transmission. ASCII is a dialect of IA5.

INTERNATIONAL STANDARDS ORGANIZATION (ISO): An international organization that sets standards in many fields, including telecommunications.

INTERNATIONAL TELECOMMUNICATION UNION (ITU): The world's oldest telecommunications standards organization, now a part of the United Nations. Its working bodies include the ITU-T and ITU-R.

INTERNET PROTOCOL (IP): The data transmission protocol of the Internet that ensures the transmission of packets (datagrams) across the network(s).

IP ADDRESS: A 32-bit number, usually written in the dotted-decimal form 000.000.000.000, which uniquely identifies an entity on the Internet.

IN-PLANT: Those components of a telecommunication system which are located in a user's facilities and do not utilize a common carrier.

INDIAN OCEAN REGION: One of the operating regions of communications satellites.

INTEROFFICE TRUNK: A trunk line between telephone exchanges.

ISOTROPIC: Equally in all directions.

K (or Ku) BAND: Frequencies in roughly the 10 to 20 GHz range.

LAST MILE: The colloquial term for the final stretch of the connection to a customer. The local loop of a telephone system or a local drop from a cable network are examples.

LEASED FACILITY: a telecommunications facility leased for the sole use of one customer. Also called a private line.

LINE CONDITIONING: See Conditioning.

LINE SWITCHING: Circuit switching.

LINE TERMINATION DEVICE: Any device, such as a modem, that connects a data machine to a transmission facility.

LINEAR POLARIZATION: See Polarization.

LINK: See Channel.

LINK ACCESS PROTOCOL (LAP): A CCITT-standard bit-oriented protocol. Link access protocol balanced (LAPB) is a variant. Both are subsets of HDLC.

LOADING: The adding of loading, or inductance, coils to a transmission circuit to minimize distortion.

LOCAL ACCESS AND TRANSPORT AREA (LATA): A geographical area within which a local telephone company provides service. Service between LATAs is carried by IECs.

LOCAL AREA NETWORK (LAN): A private data communication network connecting terminals usually within a limited geographical range, and operating at high speed.

LOCAL EXCHANGE: Also called an end office, the exchange in a telephone network where the user's line terminates.

LOCAL LOOP: A (usually two-wire) connection between a telephone exchange office and a subscriber.

LONGITUDINAL REDUNDANCY CHECK: A technique for detecting errors in transmitted blocks of information.

LOW NOISE AMPLIFIER (LNA): Equipment located at the focus of the earth-station antenna which amplifies the very weak signal from the satellite sufficiently enough to send on to the receiving equipment. If the LNA also downconverts (lowers) the received frequency to some lower frequency, it may be known as a low noise converter (LNC) or low noise blockconverter (LNB).

MARGIN: The amount by which a signal is, or is desired to be, above some specified measurement.

MARITIME SATELLITE SERVICE (MSS): The satellite telecommunications service designed for communication between ships on the high seas and coastal earthstations that are connected to the switched telephone network.

MEAN TIME TO FAILURE (MTTF): The average time for which a system operates without a fault.

MEAN TIME TO REPAIR (MTTR): The average time needed to correct a malfunction of a system.

MEASURED-USE: A telecommunications service that charges by the actual time used by the customer.

MESH NETWORK: A network configuration in which each node is connected with every other one. Also called multipoint-to-multipoint.

MESSAGE SWITCHING: The communications technique that receives, temporarily stores, and then transmits a message as the network becomes available. It does not establish a connection between sender and receiver as in circuit switching.

METROPOLITAN AREA NETWORK (MAN): A telecommunications network serving a metropolitan area.

MICROTERMINAL: A very small earthstation, often called a VSAT, installed on customer premises as part of a satellite network for exchange of data.

MICROWAVE: The portion of the electromagnetic spectrum above about 1 GHz in frequency.

MODEM: A modulator combined with a demodulator.

MODULATION: A process of having a signal change the characteristic(s) of another (usually higher-frequency) wave.

MOST ECONOMICAL ROUTING SCHEME: A system in which the network attempts to send a message first via the least expensive route, and if that is not available, through the next most expensive, and so on, until a connection is made.

MOTION PICTURE EXPERTS GROUP (MPEG): A standards-setting body, and the standards set for compressed video. MPEG-2 is the most common video compression scheme.

MULTICAST: A distribution of a signal to many receivers, but not to all (compare broadcast).

MULTIDROP: See Multipoint.

MULTIPLE ACCESS: Any technique that allows many users to access a given transmission facility.

MULTIPLE CHANNEL PER CARRIER (MCPC): A multiplexed signal that carries several programs on a single radio carrier.

MULTIPLEXER: A device that allows several signals to use the same communication channel.

MULTIPLEXING (MUX): The combining of separate signals in a way that allows one communication channel to carry them all. Multiplexing may be done by dividing up space, frequency, time, or encoding technique.

MULTIPOINT: A network that has many sources or destinations.

MULTIPOINT DISTRIBUTION SERVICE: A one-way microwave radio service from a central hub to several receiving stations.

MULTIPOINT-TO-MULTIPOINT: A network configuration in which each node is connected with another. Also called a mesh network.

MUX: Multiplex.

NAK: Negative acknowledgment: A signal sent by a receiver to a transmitter to indicate that a message was received incorrectly.

NATIONAL TELECOMMUNICATIONS AND INFORMATION ADMINISTRATION (NTIA): An agency of the U.S. Department of Commerce that manages government use of the electromagnetic spectrum and advises the government on telecommunications.

NATIONAL TELEVISION STANDARD COMMITTEE (NTSC): The television broadcast standard in the United States and many other nations.

NOISE: Any unwanted contribution to a signal, either natural or created by interference from other signals.

NOISE FIGURE: The level of noise expressed in decibels.

NOISE TEMPERATURE: The level of noise expressed as the equivalent temperature of an ideal radiator, measured in degrees Kelvin.

NOTICE OF PROPOSED RULEMAKING: An FCC request for comments on proposed regulations or on existing regulations proposed for change.

NUMBERING PLAN AREA: A scheme for numbering geographical areas within the U.S. and Canada to allow direct dialing.

OCCASIONAL TIME: Transponder capacity sold by the minute with little or no preplanning; contrasted with scheduled time or leased capacity.

OFF-HOOK: Operational, such as a telephone set whose receiver has been lifted from its cradle.

ON-HOOK: Not in operation, such as a telephone sitting on its cradle.

ONBOARD PROCESSING (OBP): Any of several kinds of signal processing techniques performed onboard a satellite, as contrasted with those performed at the earthstation. Most satellites do not have OBP ability.

ONLINE: The mode of a system in which input to and output from a processing facility are communicated directly to and from the input and output devices.

OPEN SKIES POLICY: The 1971 FCC decision to allow private domestic services by communications satellites.

OPEN SYSTEMS INTERCONNECT REFERENCE MODEL (OSI): A layered network architecture proposed by the ISO to coordinate telecommunications standards.

OPTICAL FIBER: See Fiber optics.

ORBITAL SLOT: The longitude assigned to a geostationary satellite.

ORIENTATION: The process of keeping a satellite's antennas pointed toward the intended service region on Earth.

ORTHOGONAL FREQUENCY-DIVISION MULTIPLEXING: A modulating/multiplexing scheme to increase data capacity by splitting the signal into smaller signals of different frequencies sent at the same time.

OSI REFERENCE MODEL: See Open systems interconnect reference model.

PACING: A technique for regulating the flow of information in a network.

PACKET: A unit of data sent over a network.

PACKET ASSEMBLY AND DISASSEMBLY (PAD): The device or process of splitting a data message into packets for transmission over a packet network.

PACKET SWITCH: A switch in a packet network that receives an incoming packet, examines the address, and forwards it to the next node in the network.

PACKET SWITCHING: The system of sending packets of data through a network with mesh topology. No physical path is established between sender and receiver, and various packets of one message may travel different paths to be reassembled at the receiving end.

PAL: Phase alternation by line; the color television broadcast standard for the U.K., West Germany, and many other nations.

PAPER SATELLITE: Colloquial term for a satellite that has been notified to the ITU but will probably never be made or enter into service.

PARALLEL TRANSMISSION: A system in which bits making up a character are sent simultaneously along several channels.

PARITY: An error-detection technique in which an extra bit is added to the bits of an encoded character so that the character sent over a transmission line contains either an even number (even parity) or an odd number (odd parity) of 1s.

PAY-PER-VIEW (PPV): A system whereby a cable or DBS customer can order individual programs, such as films or sporting events, either immediately or for some specific later time.

PERIGEE: The point in an orbit closest to Earth.

PERSONAL EARTHSTATION (PES): M/A-Com term for a microterminal.

PHASE MODULATION: The technique of changing the phase of a carrier wave in response to changes in the state of a signal.

POINT-TO-MULTIPOINT: A network configuration in which a major hub is connected to many subsidiary nodes. Also called a star network.

POINT-TO-POINT: A network in which two nodes are connected by a single dedicated line not used by any other nodes.

POLAR MOUNT: See Equatorial mounting.

POLARIZATION: Constraining a vibrating wave either to vibrate in only one plane (linear polarization), or to vibrate in a place which rotates uniformly (circular polarization).

POLLING: A means of regulating communication on a channel by sequentially asking each node if it has any traffic to send.

PACIFIC OCEAN REGION (POR): One of the operating regions of communications satellites.

POWER FLUX DENSITY (PFD): A measurement of received signal strength, usually measured in decibels per square meter per hertz.

PREEMPTIBLE: A transponder (or other communications channel) that can be shut off to allow its resources (usually the amplifier) to be used to restore service to another transponder.

PRIMARY CENTER: A Class 3 office in the telephone network, connecting toll centers.

PRIMARY STATION: The station of a datalink that controls the transmission of data.

PRIME FOCUS: The point where a lens or mirror brings collected waves to a focus. A type of satellite dish with the feed placed in front of the dish at the prime focus.

PRIVATE LINE: A communications line used exclusively by one customer.

PROPAGATION DELAY: The time it takes for a signal to go from a sender to a receiver.

PROTECTED TRANSPONDER: A transponder whose operation can be restored following amplifier failure by co-opting the amplifier of another (preemptible) transponder.

PROTOCOL: The standards and rules that govern the exchange of information between parts of a network.

PROTOCOL CONVERTER: A device that converts the protocol used by one device in a network to the protocol used by another device or another network.

PTT: Post, Telephone, and Telegraph Authority; the governmental body which, in some countries, controls the operation of all postal and telecommunication services.

PUBLIC DATA NETWORK: A public network designed for the transmission of data.

PUBLIC-SWITCHED NETWORK: A circuit-switched network available to the public.

PULSE-AMPLITUDE MODULATION (PAM): A modulation scheme whereby the strength of the modulating signal controls the amplitudes of a series of pulses.

PULSE-CODE MODULATION (PCM): A modulation scheme whereby the strength of the modulating signal is expressed by a digital code.

PULSE MODULATION: Any modulation scheme in which the modulating signal is carried by a series of pulses. Includes pulse amplitude, pulse duration (pulse width), pulse position, and pulse-code techniques.

PUSHBUTTON DIALING: The use of keys rather than a rotary dial to enter a called number into a telephone network.

QUALITY OF SERVICE (QOS): The specification of the performance of a network, such as BER, outages, committed datarates, etc.

RADIODETERMINATION SATELLITE SERVICE (RDSS): A satellite telecommunications service intended to provide navigation and message services to mobile users on land.

RAIN FADE: A drop in received signal strength due to the effects of rain between the satellite and the earthstation.

REALTIME: A term used to describe any system that operates such that input, processing, and output take place over a short period of time and without any long delays or storage of input, or of intermediate, or final results.

REGIONAL ADMINISTRATIVE RADIO CONFERENCE (RARC): Policy-making meetings of the ITU that consider issues for one or two regions.

REPEATER: Any device which receives a signal and repeats it to parts of a network farther down the line, usually amplified or restored to proper shape and intensity. On a communications satellite, a transponder.

RESPONSE TIME: The time it takes for a device to react to some input to it.

RESTORABLE: A transponder whose operation can be restored following amplifier failure by co-opting the amplifier of another (preemptible) transponder.

RING NETWORK: A network configuration in which the nodes are connected sequentially on a loop.

ROLL CALL POLLING: A polling technique in which the primary station polls each terminal in the network in order to determine if the terminal has traffic for transmission.

ROUTER: A node device in a packet-switched network that receives packets and sends them on to the next router.

ROUTING: The path, or the assignment of a path, for traffic through a network.

RS232C: A very popular physical level data communication standard, useful for distances up to about 50 feet and bitrates up to about 20 kbps.

SATELLITE MASTER ANTENNA TELEVISION (SMATV): A cable television system, sometimes serving a large area, but more commonly referring to service to an apartment building, housing complex, etc.

SATURATION: The signal level above which no better or more powerful signal is achieved. The maximum useful input level of an electronic component.

SCRAMBLING: Any system intended to render a received message unintelligible or unviewable without authorization. Also called encryption and encoding.

SECAM: *Sequence couleur avec memoire*; the color television broadcast standard of France and several other nations.

SELECTION: Accessing a particular terminal in a network.

SELECTIVE ADDRESSING: The ability of a restricted distribution system to allow only authorized receivers to decode the signal.

SELECTIVE CALLING: A feature whereby a sending station can specify which of several receiving stations on a network is to actually receive the message.

SERIAL: A transmission technique in which the bits of a character are sent one after another.

SHANNON'S LAW: A "law" that relates the maximum theoretically possible datarate in a circuit to the signal-to-noise ratio of that circuit.

SIDEBAND: The band of frequencies on either side of a carrier frequency produced by the process of modulating it by other frequencies.

SIGNAL-TO-NOISE RATIO (SNR): The ratio of the strength of the signal to the strength of noise in a channel, usually expressed in decibels.

SIMPLEX: Transmission in one direction only.

SINGLE CHANNEL PER CARRIER (SCPC): A version of frequency-division multiplexing and multiple access in which the transmission channel is electronically divided into many identical channels, each one assigned to carry a single signal.

SLOT: See Orbital slot.

SLOW-SCAN TELEVISION: A television technique in which a still image is transmitted over a low-bandwidth channel. A new picture is usually transmitted periodically.

SOLID STATE POWER AMPLIFIER (SSPA): A radio-frequency amplifier used to increase the power of a signal at a transmitter.

SPACE: A condition of a circuit indicated by a lack of current, usually indicating a digital 0.

SPECIALIZED COMMON CARRIER (SCC): A carrier which provides point-to-point common carrier services other than telephone service.

SPECTRUM: The entire range of frequencies of waves, such as sound waves or electromagnetic waves, or the graphical depiction of the strength of such waves as a function of frequency.

SPREAD SPECTRUM: A modulation technique in which the signal is spread over a wider-than-minimum bandwidth in order to provide better signal-to-noise ratio and/or communications security.

SPREAD-SPECTRUM MULTIPLE ACCESS: Also called CDMA, code-division multiple access, a digital technique that allows multiple users access to a transmission path by assigning each a unique code, which is combined with the data to produce the transmitted signal.

STABILIZATION: Maintaining the correct orientation of a satellite.

STAR NETWORK: A network configuration in which one major hub is connected to many nodes, which are not connected with one another. Also called point-to-multipoint.

STATIONKEEPING: Maintaining a geostationary satellite near its assigned orbital slot.

STATISTICAL MULTIPLEXING: A technique of time-division multiplexing that allows the combining of many low-speed signals onto fewer high-speed channels in order to use them more efficiently. Examples are TASI and DSI.

STORE AND FORWARD: The technique of temporarily stopping the flow of a message from sender to receiver by storing it *en route*, and then forwarding it to the receiver some time later.

SUBCARRIER: A minor signal modulated alongside a larger one, used to carry low-bandwidth signals that may or may not be associated with the larger signal. Most common with video channels.

SUBVOICE-GRADE CHANNEL: A communications channel with a bandwidth insufficient to carry a telephone-quality voice signal.

SUN OUTAGE: Those periods around the times of the equinoxes during which the sun appears to move directly behind geostationary satellites, thereby blinding the earthstations.

SUN TRANSIT: See Sun outage.

SUPERVISORY SIGNAL: Signals used within a network to indicate its operating state.

SUPPRESSED CARRIER: A transmission technique in which one or both sidebands of a modulated carrier are transmitted, but the carrier is partially or totally removed.

SWITCHING CENTER: Any node in a network which has many incoming and outgoing lines and is able to switch traffic among them.

SWITCHOVER: The transfer of traffic from one part of the network to another in case of failure of one link, for example.

SYNCHRONOUS: Referring to any system in which the bits or characters are of fixed duration and are sent at instants determined by some master clock with which all stations are synchronized.

SYNCHRONOUS DIGITAL HIERARCHY: The ITU version of the SONET standards for transmitting data on high-speed optical networks.

SYNCHRONOUS OPTICAL NETWORK (SONET): A set of standards for transmission of data traffic over optical fibers. The ITU version is known as synchronous digital hierarchy.

SYNCHRONOUS TRANSMISSION: A system for data transmission in which all stations operate at the same frequency determined by a master clock.

SYSTEMS NETWORK ARCHITECTURE (SNA): A network architecture from IBM comprising protocols and standards intended for data communications.

T1: A carrier system in which 24 digitized telephone channels (12 two-way calls) are combined and transmitted at 1.544 Mbps.

TARIFF: The established rate for and specifications of a telecommunication service.

TELECOMMUNICATIONS: The conveying of information by electrical or electronic means.

TELEPORT: A group of co-located earthstations providing a variety of links to satellites, as well as a variety of value-added services (e.g., turn-around, format conversion).

TELEPROCESSING: A data-processing facility that uses telecommunications.

TELETYPE: Any of a series of teleprinters from Teletype Corporation.

TERMINAL: Any device that can send and/or receive information.

THRESHOLD: The minimum signal that will get a response from an electronic component.

TIME-ASSIGNED SPEECH INTERPOLATION (TASI): A technique of dynamic channel assignment in which a channel is actually assigned only when someone is actually speaking at any given instant.

TIME-DERIVED CHANNEL: A communications channel obtained by time-division multiplexing.

TIME-DIVISION MULTIPLE ACCESS: A technique in which many users are assigned time slots during which the may make use of a transmission facility.

TIME-DIVISION MULTIPLEXING: A technique of combining many slower speed signals into one higher speed signal to maximize efficient use of a transmission facility.

TOKEN PASSING NETWORK: A network, common in LANs, in which an electronic signal, the token, is passed from station to station. When received by a station, it may either pass the token to the next station, or transmit a message followed by the token.

TOLL CENTER: A telephone central office at which many channels terminate.

TOLL CIRCUIT: The American term for a circuit connecting two exchanges. In the U.K., it is called a "trunk circuit" if it is over 15 miles in length, but called a "junction circuit" if shorter.

TONE DIALING: See Pushbutton dialing.

TRACKING, TELEMETRY, AND CONTROL: The "housekeeping functions" for a satellite, intended to keep it on-station and to monitor and control its operation.

TRANSCEIVER: A device that can both transmit and receive.

TRANSLATOR: A device that converts information expressed in one system to the proper expression in another system.

TRANSMISSION CONTROL PROTOCOL: The protocol governing data transmissions on the Internet, used together with Internet Protocol, and referred to collectively as TCP/IP.

TRANSPARENT: A property of a device or system whereby it appears not to exist, so that end users need not be aware of its operation.

TRANSPARENT TEXT: A mode of operation of a datalink in which the message may contain any sequence of bits.

TRANSPONDER: The circuit aboard a communications satellite that receives the uplink signal sent from the ground, shifts its frequency to the downlink frequency, amplifies it, and transmits it to the ground.

TREE NETWORK: A hierarchical network configuration in which primary nodes are connected to secondary nodes that are in turn connected to tertiary nodes, and so on.

TSAT: A VSAT station capable of handling T1-rate traffic.

TVRO: Television receive-only; the "backyard-dish" satellite reception market.

TWO-WIRE CIRCUIT: A circuit in which a pair of wires or coax is used as the transmission link.

UNATTENDED OPERATIONS: A station capable of operating without human intervention.

UPLINK: The signal sent by an earthstation to a communications satellite.

VALUE-ADDED CARRIER: A carrier that provides special features in addition to carriage of telecommunications traffic. Also called value-added network.

VIDEOCIPHER II: A widely used encryption technique, and the equipment that performs the encryption and decryption, invented by M/A-Com and now sold by General Instrument.

VIRTUAL: A property of a circuit or device whereby it appears to exist even though it does not.

VOICE FREQUENCY (VF): The range of frequencies carried in the public telephone network without special line conditioning, from about 300 Hz to 3000 Hz. Sometimes called audio frequencies (AF), but actually smaller in bandwidth.

VOICE GRADE CHANNEL: A channel capable of carrying a telephone-quality voice signal.

VOICE OVER IP: A technique of carrying voice signals using packets and the Internet Protocol.

VSAT: Very small aperture terminal; a one- or two-way earthstation 1–2 meters in size. Also called personal earthstation, or microterminal.

WIDE-AREA NETWORK: A network which reaches beyond a local area network or sometimes beyond a metropolitan area network.

WORLD (ADMINISTRATIVE) RADIO CONFERENCE (WRC or WARC): The policy-making meetings of the ITU which consider global issues.

WIDEBAND: See Broadband.

X-Y MOUNT: A method of mounting an earthstation dish such that the dish can move up and down, right and left, rather like a camera tripod.

Appendix B

List of Common Acronyms and Abbreviations

ACI	Adjacent channel interference
AceS	Asia cellular satellite system
ACK	Acknowledge
ACTS	Advanced communications technology satellite
ACU	Automatic calling unit Antenna control unit
A/D	Analog-to-digital
ADACS	Attitude determination and control system
ADM	Adaptive delta modulation
ADPCM	Adaptive differential pulse code modulation
ADSL	Asymmetric digital subscriber line
AF	Audio frequency
AGC	Automatic gain control
AKM	Apogee kick motor
AM	Amplitude modulation
A-MAC	Type-A multiplexed analog component television standard
AMF	Apogee motor firing
AMS	Apogee and maneuvering stage
AMSS	Aeronautical Mobile Satellite Service
ANSI	American National Standards Institute

AOR	Atlantic Ocean Region
AOS	Attitude and orbit control subsystem Acquisition of signal
APC	Adaptive predictive coding
API	Applications programming interface
APK	Amplitude phase keyed
APS	Auxiliary power supply
ARQ	Automatic repeat request
ASAT	Antisatellite (weapon)
ASCII	American standard code for information interchange
ASIU	ATM satellite interworking unit
ASU	Acquisition and synchronization unit
ATC	Adaptive transform coding
ATM	Asynchronous transfer mode Automatic teller machine
AZ	Azimuth
B	Byte (8 bits)
BAPTA	Bearing and power transfer (or takeoff) assembly
BBS	Baseband switch
BCC	Block check character
BCD	Binary-coded decimal
BDLC	Bisync data link control
BECO	Booster engine cut off
BER	Bit error rate
BEX	Broadband exchange
B-MAC	Type-B multiplex analog component
BO	Backoff

BOC	Bell Operating Company
BOL	Beginning of life
BPF	Bandpass filter
BPS	Bits per second
BPSK	Binary phase-shift keying
BSBD	Baseband
BSC	Binary synchronous protocol
BSS	Broadcast Satellite Service
BW	Bandwidth
CATV	Community antenna television = cable TV
C-BAND	Frequencies in the 4–6 GHz range
CBR	Constant bitrate
CC	Common carrier
CCAFS	Cape Canaveral Air Force Station
CCAM	Collision and contamination avoidance maneuver
CCI	Co-channel interference
CCIR	Consultative Committee for International Radio (now ITU-R)
CCITT	Consultative Committee for International Telegraph and Telephone (now ITU-T)
CCS	Centum call seconds
CDMA	Code-division multiple access
CDR	Critical design review
CEPT	Council of European Post and Telecommunications (agencies)
CES	Coastal earthstations
C/I	Carrier-to-interference ratio
C-MAC	Type C multiplexed analog component

CMR	Cellular mobile radio
C/N	Carrier-to-noise ratio
C/N_o	Carrier-to-noise-density ratio
CMR	Cellular mobile radio
CNES	*Centre National d'Études Spatiales* (French NASA)
CNR	Carrier-to-noise ratio
CO	Central office
CODEC	Coder-decoder
CONUS	Contiguous United States (i.e, not Alaska and Hawaii)
COTS	Commercial off-the-shelf software
CPE	Customer premises equipment
CPES	Customer premises earthstation
CPFSK	Continuous phase frequency shift keying
CPSK	Coherent phase shift keying
CPS	Characters per second Customer premises services
CPU	Central processing unit
CR&T	Command, ranging, and telemetry
CRTC	Canadian Radio, Television, and Telecommunications Commission
CRTS	Cellular Radio Telephone Service
CSC	Common signaling channel
CSG	*Centre Spatial Guyanais* (Guiana Space Center)
CSMA	Carrier-sense multiple access
CSSB	Companded single-sideband
CW	Continuous wave
D/A	Digital-to-analog
DAMA	Demand-assigned multiple access

DARS	Digital Audio Radio Service
DAVIC	Digital Audio Visual Council
dB	Decibel (also DB or db)
DBS	Direct Broadcast Satellite
DC or D/C	Downconverter Direct current
DCE	Data communications equipment
DCPSK	Differentially coherent phase shift keying
DDI	Direct digital interface
DEMOD	Demodulator or demodulation
DEMUX	De-multiplex
DES	Digital encryption standard
DL or D/L	Downlink
DM	Delta modulation Digital multiplex
D-MAC	Type-D multiplex analog component
D2-MAC	Type-D2 multiplexed analog component
DNI	Digital noninterpolated
DNS	Domain name system
DOC	(Canadian) Department of Communications
DOCSIS	Data over cable service interface specification
DOMSAT	A commsat used for domestic communications
DPCM	Differential pulse code modulation
DPSK	Differential phase shift keying
DQPSK	Differential quadrature phase shift keying
DS1, DS2, etc.	Digital services hierarchy transmission levels
DSB	Double sideband
DSCS	Defense satellite communications system
DSI	Digital speech interpolation

DSL	Digital subscriber line
DTE	Data-terminating equipment
DTH	Direct-to-home broadcast
DTS	Digital termination system
DUV	Data under voice
DVB	Digital video broadcast
DVB-RCS	DVB with return channel by satellite
DVB-S	Digital video broadcast by satellite
DVB-RCS	Digital video broadcast return channel by satellite
DWDM	Dense wavelength-division multiplexing
E	East
E1, E2, etc.	European digitized telephone hierarchy
EBCDIC	Extended binary coded decimal interchange code
E_b/N_o or E_bN_o	Energy-per-bit to noise-density ratio
EBU	European Broadcast Union
EDTV	Enhanced-definition television
EHF	Extremely high frequency
EHT	Electrothermal hydrazine thrusters
EIRP	Effective isotropically radiated power
EL	Elevation (angle above horizon)
ELV	Expendable launch vehicle
EMS	Electronic Message Service
EOL	End of (satellite) life
EPIRB	Emergency position-indicating radio beacon
ESA	European Space Agency
ETSI	European Telecommunications Standards Organization

EuroDOCSIS	European version of DOCSIS
E/W	East/West
FCC	Federal Communications Commission
F/D	Focal length to diameter ratio
FDM	Frequency-division multiplex
FDMA	Frequency-division multiple-access
FDX	Full duplex
FEC	Forward error correction
FM	Frequency modulation
FSK	Frequency shift keying
FSS	Fixed Satellite Service
G	Giga (1,000,000,000)
G&N	Guidance and navigation
GAN	Global area network
G/T	Gain-to-noise temperature ratio (also called "figure of merit")
GATT	General Agreement on Tariffs and Trade
GATS	General Agreement on Trade in Services
GCE	Ground control equipment
GEO	Geostationary (or geosynchronous) Earth orbit
GHz	Gigahertz
GII	Global information infrastucture
GMDSS	Global maritime distress and safety system
GMPCS	Global Mobile Personal Communication Service
GPS	Global Positioning System
GSE	Ground support equipment

GSM	Global system for mobile
GSO	Geostationary (or geosynchronous) Earth orbit
GTO	Geosynchronous transfer orbit
HALE	High altitude long endurance
HALO	High altitude long operation
HDLC	High-level data link control
HDTV	High-definition television
HDSL	High-bitrate DSL
HDX	Half duplex
HEO	Highly-elliptical orbit
HF	High frequency
HLLV	Heavy lift launch vehicle
HPA	High power amplifier
HVCS	Half voice circuits
Hz	Hertz = 1 cycle per second
IA2, IA5, etc.	International alphabet number 2, number 5, etc.
IBS	Intelsat business services
ICO	Intermediate circular orbit
IDR	Intermediate datarate
IDTV	Improved-definition television
IEEE	Institute of Electrical and Electronic Engineers
IETF	Internet engineering task force
IF	Intermediate frequency
IFRB	International Frequency Registration Board
IGSO	Inclined geosynchronous orbit
IM	Intermodulation

IMO	International Maritime Organization
INMARSAT	International Mobile Satellite Organization
INTELSAT	International Telecommunications Satellite Organization
IOC	Intelsat operations center Initial operational capability
IOR	Indian Ocean Region
IOT	In-orbit test
IP	Internet protocol
IPoS	Internet protocol over satellite
IPSEC	Internet protocol security
IRC	International record carrier
IRD	Integrated receiver-decoder
IRU	Indefeasible right of use
ISDN	Integrated services digital network
ISL	Inter-satellite link
ISO	International Standards Organization
ISS	Inter-Satellite Service
ITU	International Telecommunication Union
ITU-R	ITU Radiocommunications Sector
ITU-T	ITU Telecommunications Sector
IUS	Inertial upper stage
IXC	Interexchange channel
k	Kilo (1000)
K	Kelvin, a scale of (noise) temperature
kbps	Kilobits per second (also kb/s)
kHz	Kilohertz
KSC	Kennedy Space Center
Ku-Band	Frequencies in the 10–18 GHz range

LAN	Local area network
LATA	Local access and transport area
LEO	Low Earth orbit
LH or LH_2	Liquid hydrogen
LHCP	Left-handed circular polarization
LMDS	Local multipoint distribution service
LMSS	Land mobile satellite service
LNA	Low-noise amplifier
LNB	Low-noise block converter
LNC	Low-noise converter
LO or LO_2	Liquid oxygen
LPC	Linear predictive coding
LV	Launch vehicle
M	Mega (1,000,000)
MAC	Multiplexed analog component
MAN	Metropolitan area network
MATV	Master antenna television
MCC	Mission control center
MCS	Maritime communications subsystem
MCU	Monitor and control unit
MDS	Multipoint Distribution Service
Mbps	Megabits per second (also Mb/s)
MHz	Megahertz
MEO	Medium Earth orbit
MMDS	Microwave multipoint distribution system
MECO	Main engine cut-off
MES	Main engine start

MERS	Most economical routing scheme
MFTDMA	Multiple-frequency time-division multiple access
MMSS	Maritime Mobile Satellite Service
MOD	Modulator or modulation
MODEM	Modulator-demodulator
MPEG	Motion Picture Experts Group
MSS	Mobile Satellite Service
MTBF	Mean time between failures
MTTR	Mean time to repair
MUX	Multiplex
N	North Newton
N/S	North/South
NACK	Negative acknowledgment
NASA	National Aeronautics and Space Administration
NCC	Network control center
NF	Noise figure
NGSO	Non-geostationary orbit
NiCd	Nickel cadmium (battery)
NiH or NiH_2	Nickel hydrogen (battery)
NOC	Network operation center
NOI	(FCC) Notice of inquiry
NPRM	(FCC) Notice of proposed rulemaking
NTIA	National Telecommunication and Information Administration
NTSC	National Television Standards Committee (U.S. color television standard)

OBE	Out-of-band emission
OFDM	Orthogonal frequency-division multiplex
OIS	Orbit insertion stage
OW	Order wire
PAD	Packet assembly and disassembly
PAL	Phase alternation by line (a television standard)
PAM	Pulse-amplitude modulation Payload assist module
PAMA	Pre-assigned multiple access
PBX	Private branch exchange
PCM	Pulse code modulation
PCMA	Paired carrier multiple access
PDM	Pulse duration modulation
PFD	Power flux density
PKM	Perigee kick motor
PM	Phase modulation
POR	Pacific Ocean Region
POTS	Plain old telephone service
PPM	Pulse position modulation
PPV	Pay-per-view
PSK	Phase-shift keying
PSN	Packet switching network Private satellite network
PTT	Postal, Telephone and Telegraph (Authority)
PWM	Pulse-width modulation
QAM	Quadrature Amplitude Modulation
QC	Quality control

QoS	Quality of service
QPSK	Quarternary phase-shift keying
QZS	Quasi-zenith satellite
RARC	Regional Administrative Radio Conference
RCC	Radio common carrier
RCVR	Receiver
RCS	Reaction control system
RDSS	Radiodetermination Satellite System
RDT&E	Research, development, testing and evaluation
RF	Radio frequency
RFC	Request for comments
RFI	Radio frequency interference
RHCP	Right-handed circular polarization
RNSS	Radionavigation Satellite Service
RR	Radio Regulations (of the ITU)
RS232	A standard for a serial interface cable to connect equipment
RTS	Request to send
RX	Receiver or receive
S	South
SARSAT	Search and rescue satellite
SC or S/C	Spacecraft Suppressed carrier
SCC	Spacecraft control center
SCPC	Single carrier per channel
SDARS	Satellite Digital Audio Radio Service
SDCU	Satellite delay compensation unit

SDH	Synchronous digital hierarchy
SDLC	Synchronous data link control
SDMA	Space-division multiple access
SECAM	Sequential color with memory
SECO	Sustainer engine cut off
SHF	Super high frequency
SIT	Satellite interactive terminal
SMATV	Satellite master antenna television
S/N	Signal-to-noise ratio
SNA	System network architecture
SNG	Satellite news gathering
SNR	Signal-to-noise ratio
SOC	Space operations center
SOCC	Satellite operations control center
SONET	Synchronous optical network
SPADE	Single-channel-per-carrier PCM multiple-access demand assignment equipment
SPELDA	*Structure porteuse externe pour lancement double Ariane*
SRB	Solid rocket booster
SRM	Solid rocket motor
SS	Spread spectrum
SSB	Single sideband
SSMA	Spread spectrum multiple access
SS-TDMA	Satellite-switched TDMA
SSCP	System services control point
SSPA	Solid state power amplifier
SS-TDMA	Satellite-switched time-division multiple access
SSUS	Solid spinning upper stage

STDN	Spaceflight tracking and data network
STS	Space transportation system (space shuttle)
SYLDA	*System de lancement double ariane* (allows 2 payloads)
T	Tera (1,000,000,000,000)
TANSTAAFL	"There ain't no such thing as a free lunch"
TBPS	Terabits per second T carrier A hierarchy of Bell digital systems, T1, T2, etc.
T1	Transmission system 1, 1.544 Mbps
TASI	Time-assigned speech interpolation
TAT	Trans-Atlantic telephone (cable)
TBD	To be determined
TC	Telecommand
TCP/IP	Transmission control protocol/Internet protocol
TDM	Time-division multiplexing
TDMA	Time-division multiple access
TDRSS	Tracking and data relay satellite system
TM	Telemetry
TT&C	Tracking, telemetry and control
TTY	Teletype
TV	Television
TVRO	Television receive-only earthstation
TWT or TWTA	Traveling wave tube amplifier
TWX	Teletype
TX	Transmit(ter)
UC or U/C	Upconverter
UHF	Ultra high frequency

UL or U/L	Uplink
USAT	Ultra small aperture terminal
VAB	Vehicle assembly building
VAN	Value-added network
VC II	VideoCipher II
VF	Voice frequencies
VHF	Very high frequency
VOIP	Voice over Internet protocol
VOW	Voice order wire
VPN	Virtual private network
VSAT	Very small aperture terminal
W	West
WAN	Wide area network
WARC	World Administrative Radio Conference
WDM	Wavelength-division multiplexing
WRC	(ITU) World Radio Conference
WSMC	Western Space and Missile Center (Vandenberg Air Force Base, CA)
WWW	World Wide Web
XPNDR	Transponder
XPD	Cross-polarization discrimination
XPOL	Cross-polarization

Appendix C

Selected Bibliography for Additional Reading

Bhargava, Vijay H. *Digital Communications by Satellite: Modulation, Multiple Access, and Coding*. Melbourne, FL: Krieger Publishing, 1991.

Cochetti, Roger. *Mobile Satellite Communications Handbook*. New York: John Wiley & Sons, 1998.

Elbert, Bruce R. *Introduction to Satellite Communication*, 2nd ed. Norwood, MA: Artech House, 1987.

Elbert, Bruce R. *The Satellite Communication Applications Handbook*. Dedham, MA: Artech House, 1997.

Elbert, Bruce R. *The Satellite Communication Ground Segment and Earth Station Handbook*. Dedham, MA: Artech House, 2000.

Feher, Kamilo. *Digital Communications: Satellite/Earth Station Engineering*. Englewood Cliffs: Prentice-Hall, 1981.

Freeman, Roger L. *Telecommunication Transmission Handbook*, 3rd ed., New York: Wiley Interscience, 1991.

Gomez, Jorge M. *Satellite Broadcast Systems Engineering*. Dedham, MA: Artech House, 2002.

Goodman, Robert. L. *Digital Satellite Service: Installation and Maintenance*. New York: McGraw-Hill, 1996.

Gordon, Gary D. and Walter L. Morgan. *Principles of Communications Satellites*. New York: Wiley-Interscience, 1993.

International Telecommunication Union. *Handbook on Satellite Communications*, 3rd ed. New York: John Wiley & Sons, 2002.

Ippolito, Louis J. Jr. *Radiowave Propagation in Satellite Communication*. New York: Van Nostand Reinhold Company, 1986.

Kadish, Jules, E. and Thomas W. R. East. *Satellite Communications Fundamentals*. Dedham, MA: Artech House, 2000.

Kennedy, Charles H. and M. Veronica Pastor. *An Introduction to International Telecommunications Law.* Boston: Artech House, 1996.

Logsdon, Tom. *Mobile Communication Satellites.* New York: McGraw-Hill, 1995.

Luther, Arch, and Andrew Inglis. *Satellite Technology: An Introduction*, 2nd ed. Burlington, MA: Focal Press, 1997.

Maral, Gerard, and Michel Bousquet. *Satellite Communication Systems: Systems Techniques and Technology*, 4th ed. New York: John Wiley & Sons, 2002.

Mead, Donald C. *Direct Broadcast Satellite Commmunications: MPEG Enabled Service.* New York: John Wiley & Sons, 1999.

Miller, Michael J., Branka Vucetic, and Les Berry, eds. *Satellite Communications: Mobile and Fixed Services.* Norwich, MA: Kluwer Academic Publications, 1993.

Morgan, Walter L., and Gary D. Gordon. *Communications Satellite Handbook.* New York: John Wiley & Sons, 1989.

National Telecommunications and Information Administration. *Telecommunications: Glossary of Telecommunications.* Rockville, MD: ABS Group, Inc.,1997.

Ohmori, Shingo, Hiromitsu Wakana, and Seiicchiro Kawase. *Mobile Satellite Communications.* Dedham, MA: Artech House, 1997.

Page, David and William Crawley. *Satellites Over South Asia: Broadcasting Culture, and the Public Interest.* Thousand Oaks, CA: Sage Publications, 2001.

Parsons, Patrick R., and Robert M. Frieden. *The Cable and Satellite Television Industries.* Boston, MA: Allyn & Bacon, 1997.

Pascall, Stephan C. *Commercial Satellite Communications.* Burlington, MA: Butterworth-Heinemann, 1997.

Pelton, Joseph N. *Global Talk.* Netherlands: Sijthoff & Noordhoff, 1981.

Pelton, Joseph N. *Wireless and Satellite Telecommunications: The Technology, The Market & The Regulations.* Upper Saddle River, NJ: Prentice Hall, 1995.

Pratt, Timothy, Charles W. Bostian, and Jeremy E. Allnutt. *Satellite Communications.* New York: John Wiley & Sons, 2002.

Richharia, M. *Satellite Communication Systems: Design Principles*, 2nd ed. New York: McGraw-Hill, 1999.

Richharia, M., and Manu Richmaria. *Mobile Satellite Communications: Principles and Trends.* New York: Addison Wesley, 2001.

Roddy, Dennis. *Satellite Communications*, 3rd ed. New York: McGraw-Hill, 2001.

Schwartz, Rachael, and Rachel E. Schwartz. *Wireless Communications in Developing Countries: Cellular and Satellite Systems*. Dedham, MA: Artech House, 1996.

Sullivan, Thomas F. P., ed. *Official Telecommunications Dictionary: Legal and Regulator Definitions*. Rockville, MD: ABS Group, Inc., 1997.

Tedeschi, Anthony M. *Live Via Satellite, The Story of COMSAT and the Technology that Changed World Communication*. Washington: Acropolis Books Ltd., 1989.

Werner, M., Erich Lutz, and Axel Jahn. *Satellite Systems for Personal and Broadband Communication*. New York: Springer Verlag, 2000.

Whalen, David J. *The Origins of Satellite Communications: 1945-1965*. Washington, DC: Smithsonian Institution Press, 2002.

Appendix D

Periodicals and Newsletters

Asia-Pacific Satellite
ICOM Publications Ltd.
Chancery House
St. Nicholas Way
Sutton, Surrey SM1 1JB
UK
www.asiapacificsatellite.com

Aviation Week & Space Technology
McGraw-Hill, Inc.
1221 Avenue of the Americas
New York, NY 10020
USA
www.aviationnow.com

Cable & Satellite International
Perspective Publishing Limited
402 The Fruit & Wool Exchange
Brushfield Street
London E1 6EP
UK
www.cable-satellite.com

Interspace
Phillips Business Information, Ltd.
The Forum
Stevenage
Herts SG1 1EL
UK
www.telecomweb.com/satellite

Satellite News
PBI Media, LLC
1201 Seven Locks Road
Potomac, MD 20854 USA
www.satellitetoday.com

Space News
Space Holdings, Inc.
6883 Commercial Drive
Springfield, VA 22159-0500
USA
www.space.com/spacenews

TELE-satellite International
P.O. Box 1124
Ascot, Berks SL5 0XH
England
www.tele-satellite.com

Via Satellite
PBI Media, LLC
1201 Seven Locks Road Potomac, MD 20854
USA
www.viasatellite.com

Appendix E

Mathematical Background and Details

This information is intended as background for readers with a bit more mathematical preparation who want to see more of the details of formulas and calculations and some examples. These are not necessary for the nontechnical reader.

E.1 Units and measurements

Measurements are the heart of any technical endeavor, and the units used are chosen to be the ones most appropriate for the application.

So-called common units, such as statute miles, pounds, and degrees Fahrenheit, are used primarily in the United States. Elsewhere, and in technical fields, the preferred system is the International System (SI) of meters, kilometers, grams and kilograms, and degrees Celsius. Both systems use seconds, minutes, hours, days, and years for time measurements.

In the SI, the preferred usage is to adopt unit measurements, such as grams and meters. For larger and smaller units, most measurements go up and down in factors of 1000, or 10^3; thus we have kilograms and kilometers. Prefixes (usually of Greek origin), such as kilo- and milli-, are used with the units to denote these larger and smaller quantities. Here is a list of some of these prefixes:

Power of 10	Prefix	Symbol	Common name
18	exa	E	
15	peta	P	quadrillion
12	tera	T	trillion
9	giga	G	billion
6	mega	M	million
3	kilo	k	thousand
0			(units)

Power of 10	Prefix	Symbol	Common name
−3	milli	m	thousandth
−6	micro	μ	millionth
−9	nano	n	billionth
−12	pico	p	trillionth
−15	femto	f	quadrillionth
−18	atto	a	

A few very special measurements, because of historical longevity, use other multiples; thus we have decibels (tenths of a bel) and a few others.

Care must be taken because some words relating to measurements and quantities, such as "billion" and "ton" have different meanings in different countries or in different systems of units.

Temperature scales can be confusing also. The United States uses the Fahrenheit scale, on which the freezing point of water is 32 degrees; most other nations use the Celsius scale, on which the freezing point of water is 0 degrees. Both are relative scales, based on the properties of water. As noted in Chapter 8, for physical calculations, an absolute temperature scale is needed. The most common one is the Kelvin scale: its origin is at absolute zero, the temperature at which all molecular motion ceases and there is no heat. Note that the notation of the Kelvin scale does not use the degree symbol (°), so that instead of writing 273°K, one writes just 273 K.

E.2 Logarithms and decibels

Logarithms were invented to simplify calculations and to enable very large and small values to be expressed conveniently.

Logarithms are powers—exponents—of numbers. They are based on some chosen number, called the *base*, for which the logarithm is defined to be exactly one. The logarithms we use in this book are called "base-10." (Another base number, useful in science and engineering, is the irrational number $e = 2.71828...$, but we will not be using it here.)

With 10 as our base, we see that $10 \times 10 = 100$, also denoted as 10^2 (ten squared). Note that the exponent tells you the number of times the base is multiplied by itself to get the number in question. Similarly, $10^3 = 10 \times 10 \times 10 = 1000$. By definition, any number raised to the zeroth-power equals exactly 1; in mathematical notation, $10^0 = 1$. Look back at the table in section E.1 of this appendix and you will see that the number in the left-hand column is the logarithm of the number named in the right-hand column.

Another way of saying this is that the logarithm of 100, written $\log_{10}(100)$ is 2; similarly, $\log_{10}(1000) = 3$. Note that the subscript 10 denotes that we are using "base-10" logarithms. This subscript is usually dropped for convenience, and base 10 is assumed. There can also be fractional exponents; these are also logarithms. While these are somewhat less intuitive than whole-number exponents, they are useful. For example, $10^{0.3} \approx 2$, therefore $0.3 \approx \log 2$. For another example, $10^{½}$ is the same as the square root of 10, or 3.1622, (and therefore $1/2 = \log 3.1622$).

To multiply numbers, you simply add their logarithms. For a simple example, $10 \times 1000 = 10^1 \times 10^3 = 10^{1+3} = 10^4 = 10{,}000$. The same rule applies to noninteger exponents, or logarithms. Thus $10^{1.5} = 10^1 \times 10^{½} = 10 \times 3.1622$ or $= 10^{1.5} =$ 31.622.

You can look up the values of logarithms in tables in books, but because of the ubiquity of calculators today, that is the easier way.

E.2.1 Decibels

Here, for the record (engineers need to memorize this, but you probably don't), is the official definition of decibels. Suppose you have two values denoted by P_1 and P_2 (they must be measured in the same units). Then the two values differ by a number of decibels calculated from the formula

$$dB = 10 \cdot \log(P_1/P_2).$$

(The formula is spoken as "dB difference equals ten times log-to-the-base-ten of the ratio of P_1 to P_2.")

In telecommunications, the *P*s are usually power levels in watts or milliwatts, but they could also be temperatures, bandwidths, or other measurements.

Remember that the logarithm of 1 is 0, so if two signals are the same strength, they differ by 0 dB. The important concept to keep in mind is that differences in decibels correspond to factors in the numbers they represent.

Again just for the record, here are the decibel differences from 0 to 10, with their corresponding "real number" factors, correct to 2 decimal places:

0 dB	a factor of 1.00
1 dB	a factor of 0.79 or 1.26
2 dB	a factor of 0.63 or 1.58
3 dB	a factor of 0.50 or 2.00
4 dB	a factor of 0.40 or 2.51
5 dB	a factor of 0.32 or 3.16

6 dB	a factor of 0.25 or 3.98
7 dB	a factor of 0.20 or 5.01
8 dB	a factor of 0.16 or 6.31
9 dB	a factor of 0.13 or 7.94
10 dB	a factor of 0.10 or 10.00
20 dB	a factor of 0.01 or 100
30 dB	a factor of 0.001 or 1000
etc.	

In other words, if you have two signals that differ by 1 dB, the stronger of the signals is 1.26 times as strong as the weaker (equivalently, the weaker one is only 0.79 times the strength of the stronger).

If you have more than 10 decibels, for each 10 dB, you have another factor of 10. For example, if you have a signal amplification of 100 times, that corresponds to 10×10, which is two 10s multiplied together, or 20 dB. Likewise, if two signals differ by 14 decibels, they differ by a factor of $10^{1.4}$. Since $14 = 10 + 4$, the 10 dB is a factor of 10, the 4 dB is a factor of 2.51, so one of the two signals is $10 \times 2.51 = 25.1$ ($= 10^{1.4}$) times the strength of the other.

E.3 Bandwidth expressed in decibels

To convert bandwidth (*BW*) in hertz into bandwidth in decibels, we use the formula

$$BW_{\text{db}} = 10 \log(BW_{\text{Hz}}).$$

The units are called dB-Hz. To give some examples, a 4-kHz telephone channel would have a bandwidth of 36.02 dB-Hz; a 27-MHz transponder would have a bandwidth of 74.31 dB-Hz; and a 36-MHz transponder would have a bandwidth of 75.56 dB-Hz.

E.4 The binary number system

Essentially, bits are numbers in a counting scheme. In our everyday decimal number system, we have 10 symbols to represent quantities: 0, 1, ... 9. Once we get to a number greater than 9, we have to use a second "decimal place" and go on to count to 10, 11, 12, ..., etc. Each decimal place represents a higher power of 10, so a number like 137 really means $(1 \times 10^2)+(3 \times 10^1)+(7 \times 10^0) = 100 + 30 + 7 = 137$.

Similarly, in the binary system we have only two symbols, or numbers, 0 and 1, which allows us to count all the way up to 1. When we want to use bigger numbers, we add another "binary place," or bit. Thus, a binary number such as 1011 means $(1 \times 2^3) + (0 \times 2^2) + (1 \times 2^1) + (1 \times 2^0)$, which equals 8+0+2+1 and is 11 in the decimal numbers. Here is a table that gives some common powers of 2:

$2^0 = 1$	$2^6 = 64$
$2^1 = 2$	$2^7 = 128$
$2^2 = 4$	$2^8 = 256$
$2^3 = 8$	$2^9 = 512$
$2^4 = 16$	$2^{10} = 1024$
$2^5 = 32$	$2^{11} = 2048$
etc.	

This table also shows us how many bits it takes to characterize an analog signal. Suppose, for instance, that you want to measure the volume of a voice signal and characterize the volume by electrical signals of 256 different levels (which is common in the telephone industry). You would then need a group of bits that is 8 bits long to provide for that many discrete symbols. Such an 8-bit group is often called a *byte*. Other fixed, defined groups of bits are sometimes called *words*.

When you were in school, did you perhaps make up a secret code in which you let a =1, b=2, c=3, etc.? Did you write notes to your friends in this code? If you had known then about the binary numbers, you could just as easily have made a = **00000**, b =**00001**, c = **00010**, etc. That sort of thing is just what is done to make binary numbers—digits—carry information.

E.5 Noise temperature and noise figure

Again, if we want to perform a calculation using decibels, all values must be in decibels. You can convert back and forth between *NT* and *NF* using the formula

$$NF_{\text{dB}} = 10 \cdot \log(1 + NT/290).$$

Noise temperature is measured in degrees on the Kelvin temperature scale, which has its zero-point at absolute, where there is no molecular motion and thus no noise.

To take some numerical examples, if an LNB had a noise temperature of 75K, its noise figure would be 1.0 dB; an LNB with *NT* = 120K has a *NF* = 1.5 dB; a noise source with *NT* = 2000K is producing 8.9 dB of noise.

E.6 Shannon's Law of channel capacity

This law states the maximum speed of a digital channel. Mathematically, the formula is

$$R = \mathrm{W} \times \log_2(C/N + 1).$$

R is the maximum bitrate possible (bits per second), W is the bandwidth of the channel (in Hz), and C/N is the carrier-to-noise ratio (expressed as a real number, not in decibels). Note that the logarithm is to the base two, not 10.

In the real world, communication channels never come close to achieving this maximum throughput. Real speeds of around a third of the maximum are common.

E.7 Kepler's Laws of orbits

Between 1609 and 1619, Johannes Kepler used the earlier accurate observations of Mars made by Tycho Brahe to derive the laws by which one small object moves about another larger one. Later in that century, Isaac Newton showed that Kepler's Laws were a natural result of the law of gravity.

Kepler's first law states that orbits take the geometrical form of conic sections. Closed orbits take the form of an ellipse; a special case is that of a circular orbit. Open orbits, of no interest to us for communications satellites but used by space agencies to send probes to other planets, take the form of a parabola or hyperbola.

Kepler's second law is basically the law of conservation of angular momentum. A line drawn from a focus of the ellipse (for instance, the center of the earth) to the satellite sweeps across equal areas in equal times: this means that the satellite moves faster when close to the focus and slower when farther away.

Kepler's third law states that the period of a closed orbit depends only on the average distance between the two objects. Mathematically, if P is the period and a is half of the average distance (called the *semimajor axis* in orbital mechanics), then the relationship between them is

$$P = Ka^{3/2}.$$

The value of K depends only on the mass of the larger object, and will have different values for objects orbiting Earth, planets orbiting the Sun, etc. The most important item to note is that the period does not depend on the mass of the satellite, its size, or its orbital orientation.

E.8 Orbital parameters

Several orbital elements are necessary to fully describe an orbit. Two describe the size and shape of the orbit:

a, the semi-major axis, is half of the maximum dimension of the ellipse; from Kepler's third law, specifying a is equivalent to specifying the period of the orbit.

e, the eccentricity, measures how elliptical the orbit is; it ranges from exactly 0 for a circular orbit to just less than 1.0 for a highly elliptical orbit.

All orbits are measured with respect to some coordinate system which has a defined fundamental plane and a fundamental zero-point of direction. (For analogy, on Earth the fundamental plane is that of the earth's equator, and the zero-point direction is the longitude meridian of Greenwich.) For Earth satellites, the fundamental plane is again that of the earth's equator. The fundamental direction is toward a defined point in the sky called the Vernal Equinox (or sometimes called the "First Point in Aries.")

One more orbital element called the "argument of perigee" specifies the orientation of the ellipse in the orbital plane.

Two more orbital elements specify the orientation of the orbit with respect to the fundamental plane: ***i*** is the inclination of the satellite's orbit to the plane of the equator; the angle around the equatorial plane to the point at which the satellite crosses the plane in its orbit (going in a northerly direction); this is called the "longitude of the ascending node."

One final element is needed to specify the exact position of the satellite along the orbit. This is done be noting one of the times that the satellite is at its closest point to Earth, called the time, or epoch, of perigee.

E.9 Antenna properties

The antenna properties of directionality and gain depend on both the diameter of the dish and the frequency in use. Diameter, denoted commonly by the symbol d, is measured in meters. Frequency f is measured in GHz.

E.9.1 Beamwidth

The formula for the size (in degrees) of the main beam, or lobe, of a parabolic dish antenna is given by

$$HPBW = \frac{21}{f \cdot d} \cdot$$

Note that this is the full width of the main beam, so it extends to half that value on either side of the boresight direction. This is the total angle through which the transmitted power, or receiving sensitivity, is at least half that of the value on-axis. The half-power beamwidth is sometimes also called the 3-dB beamwidth.

To take some numerical examples, you can see that it would be unwise to try to use a 1-m dish at C-band (4 GHz) for satellites spaced along the Clarke orbit only 2° apart: the HPBW is 5.25°, and half of that value (the angle either side of the boresight direction) is less than the orbital spacing. Such an attempted link would have much adjacent-satellite interference. However, you could use a 2-m dish, for which the HPBW is 2.63°, because half of that is less than the orbital spacing. But, if you try the same thing at Ku-band (12 GHz), the HPBW is 1.75, so this might well work. At Ka-band frequencies, say 30 GHz, the HPBW is only 0.7°. Thus, higher frequencies allow closer orbital spacing, meaning more satellites in orbit.

E.9.2 Dish gain

The gain of a dish is the amount of signal strengthening provided by the focusing power of the geometry of the dish, compared to an isotropic (i.e., omnidirectional) antenna, and is called dbI. The formula for the gain of a dish, in decibels (dbI) is

$$G = 10 \log (60 f^2 D^2),$$

assuming an average value for dish efficiency of 55%.

Dish gain is a major factor in the link budget. To take some numerical examples, we can calculate that the gain of a 1-m dish at 4 GHz is about 30 dBI, an amplification factor of a thousand. The same size dish at the Ku-band frequency of 12 GHz is 39 dBI. At Ka-band, 30 GHz, the same dish has a gain of 47 dBI. Thus, higher frequencies allow even small dishes to have quite large gains. This higher gain partially offsets the greatly increased losses in signal due to atmospheric attenuation at the higher frequencies (see Chapter 18).

E.9.3 Antenna gain rule (sidelobe gain rule)

Regulators limit the amount of power that a transmitting antenna may radiate to directions off the boresight direction of the antenna. While there are several such formulas, the main one recommended by the ITU is (in decibels)

$$G \leq 29 - 25 \log A,$$

where A is the angle in degrees from the boresight direction. The formula applies to angles greater than 1° from the boresight.

In Fig. 16.10, this formula is plotted as the smooth line that serves as an envelope limiting the maximum emission of an antenna, depicted by the squiggly line.

E.10 Space loss

Space loss is the amount of signal power lost (measured in decibels) due to the inverse-square law spreading of the radiation from a point source.

By the way that space loss is defined and used by radio engineers, the amount of space loss depends on both the distance from the sender to the receiver, and on the frequency of the signal. If we let the letter S denote the distance, and the frequency in use, measured in gigahertz, be f, then there is a general formula for the amount of space loss, L, in decibels:

$$L = K + 20 \cdot \log S + 20 \cdot \log f.$$

The value of K depends on which distance units are used for S:

If S is in kilometers, use $K = 92.45$;
If S is in statute miles, use $K = 96.58$;
If S is in nautical miles, use $K = 97.80$;
If S is in Earth radii, use $K = 168.53$.

E.11 Digital link budget

For digital links, it is more appropriate to use not C/N, but the energy-per-bit-to-noise-density ratio, denoted by E_b/N_0. As you would expect, the faster the datarate, the less energy that can be put into each bit. Thus, the formula for converting between C/N and E_b/N_0 is

$$E_b/N_0 = C/N_0 - 10 \cdot \log BR$$

where BR is the datarate in bits per second.

This value, combined with the modulation method and the error-correction technique in use will determine the average bit error rate, BER. The details are highly complicated. If you want to know more, consult the more technical books listed in Appendix C.

E.12 Combining uplink and downlink *C/N*

It is the overall combined uplink+downlink *C/N* that determines the signal quality at the receiving end of the channel.

To combine the quality of the links, you first calculate separately the *C/N* of the uplink and of the downlink. To combine them is a bit messy and complicated. You must do the calculation in "real numbers," not in decibels, so you must turn the decibel values of each link into their equivalent ratios, average them by the formula below, and turn them back into decibels again. You probably will not have to do this yourself, but just to show you how, this is what you do.

For simplicity, let's represent the *C/N* of a link by the symbol *Q* (for "quality"), still in decibels. We first convert the *Q* of the uplink and downlink to ratios, symbolized by *R*, using the equation

$$R = 10^{(Q/10)}.$$

Then we combine the values of *R* for the uplink and downlink using this special averaging formula

$$1/R_{overall} = 1/R_{up} + 1/R_{down}.$$

Finally, we convert the ratio $R_{overrall}$ back into decibels using the usual formula

$$C/N_{overall} = Q_{overall} = 10 \cdot \log(R_{overall}).$$

Let's take an example. Suppose you have calculated that the *Q* (=*C/N*) of the uplinked signal will be 14 dB, and the *Q* of the downlinked signal will be 16 dB. Then $R_{up} = 10^{14/10} = 25.1$, and $R_{down} = 10^{16/10} = 39.8$. So $1/R_{overall} = 1/25.1 + 1/39.8 = 0.039 + 0.025 = 0.064$. Thus, $R_{overall} = 1/0.064 = 15.6$. Converting that ratio to decibels, we get $Q_{overall} = 10 \cdot \log(15.6) = 11.9$. Thus, the overall *C/N* is only 11.9 dB, much less that the quality of either link.

It is always true that your signal always gets worse, and that the combined *C/N* is always lower than the lower of the uplink and downlink *C/N*s. If you have identical *C/N* values for both uplink and downlink, the overall C/N will be 3 dB lower.

It should be emphasized here that this applies only to bent-pipe satellites. If a satellite performs onboard processing, the averaging procedure given here is not applicable.

Index

B

C

D

E

M

N

O

Q

R

S

U

V

W

X

Y

Z

Mark R. Chartrand holds a Ph.D. in Astronomy from Case Western Reserve University. He was at the American Museum-Hayden Planetarium (in its previous incarnation) for ten years, seven of those as Chairman. During that time, he was also Adjunct Associate Professor at Fordham University, Lincoln Center Campus, and taught at the New School. Following that, he spent four years as Executive Director and Vice President of the National Space Institute (now the National Space Society) in Washington, D.C. Since 1984 he has run his own space science consulting firm, the main activity of which has been the conducting of introductory satellite industry seminars in the United States and many other countries. He typically holds 6 to 15 two-day seminars each year, both to public groups (marketed by Applied Technology Institute) and as on-site training within firms such as Intelsat, AsiaSat, and ING Bank.

Dr. Chartrand is the founding editor of a series of publications, *Strategic Directions in Satellite Communications*, and of an industry reference *Satellite* on CD. He is also a contributor to several satellite industry publications and chairman of and/or speaker at several satellite industry conferences. He has been a consultant to such firms as Intelsat, Andrew Corporation, Delmarva Power, Lockheed Martin Telecommunications, the U.S. Navy, and Boeing Capital, among many others. He has also authored several popular science books and hundreds of magazine articles on astronomy and space, and was the monthly space science columnist for *Omni* for two years. His *Skyguide* and *Audubon Field Guide to the Night Sky* continue to be among the best-selling books on the subject.

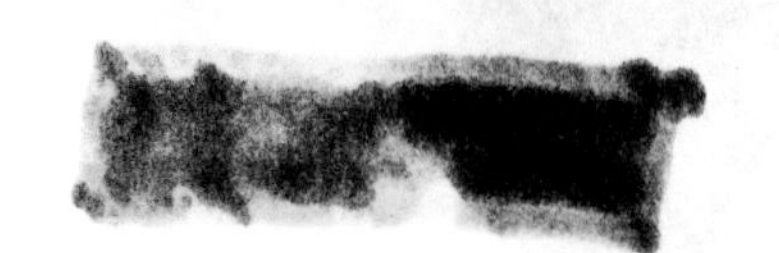